DESIGN AND MANUFACTURING OF ADVANCED COMPOSITES

*Proceedings of the Fifth Annual ASM® /ESD
Advanced Composites Conference*
25-28 September 1989
Dearborn, Michigan

Sponsors:

ASM INTERNATIONAL®

Engineering Society of Detroit

Published by

Metals Park, Ohio 44073

Copyright © 1989
by
ASM INTERNATIONAL®
All Rights Reserved

No part of this book may be reproduced, stored in a retrieval system, or transmitted, in any form or by any means, electronic, mechanical, photocopying, recording, or otherwise, without the prior written permission of the publisher. No warranties, express or implied, are given in connection with the accuracy or completeness of this publication and no responsibility can be taken for any claims that may arise.

Nothing contained in this book is to be construed as a grant of any right or manufacture, sale, or use in connection with any method, process, apparatus, product, or composition, whether or not covered by letters patent or registered trademark, nor as a defense against liability for the infringement of letters patent or registered trademark.

Library of Congress Catalog Card Number: 89-85108
ISBN: 0-87170-366-1
SAN: 204-7586

Printed in the United States of America

ACCE '89 ORGANIZING COMMITTEE

Irvin Poston
General Motors Corporation
Warren, Michigan
Executive Chairman

Claude DiNatale
General Motors Corporation
Warren, Michigan
General Chairman

Kaneyoshi Ashida
University of Detroit
Detroit, Michigan

Norman Chavka
Ford Motor Company
Dearborn, Michigan

Harry Couch
Inland Division-GMC
Dayton, Ohio

Douglas Denton
Chrysler Corporation
Detroit, Michigan

Lawrence T. Drzal
Michigan State University
E. Lansing, Michigan

Joseph Gaynor
Cincinnati Products
Cincinnati, Ohio

Nabil Grace
Lawrence Institute of Technology
Southfield, Michigan

Paul Graves
Dow Chemical
Southfield, Michigan

J. Michael Grubb
DeComa International
Southfield, Michigan

William E. Haskell
Army Materials Tech Lab
Watertown, Massachusetts

James E. Hill
FMC Corporation
San Jose, California

Josh Kelman
Davidson Technology Center-Textron
Dover, New Hampshire

Thomas Kosakowski
Mobay Corporation
Troy, Michigan

Edward J. Lesniak, Jr
Chrysler Corporation
Southfield, Michigan

Steve Loud
Composites Market Report
San Diego, California

Carl Luther
General Dynamics
Birmingham, Michigan

Donald J. Melotik
Ford Motor Company
Dearborn, Michigan

Hiroshi Nagahara
Nissan
Ann Arbor, Michigan

Harold Visser
CertainTeed Corporation
Toledo, Ohio

Gullmar Nelson
Davidson Technology Center-Textron
Dover, New Hampshire

Ralph Salansky
Chrysler Corporation
Auburn Hills, Michigan

Jerry Scrivo
Ardyne, Inc.
Grand Haven, Michigan

Sharad B. Shah
University of Toledo
Toledo, Ohio

Allan B. Isham
Owens Corning
Granville, Ohio

Richard Jeryan
Ford Motor Company
Dearborn, Michigan

Carl F. Johnson
Ford Motor Company
Dearborn, Michigan

Roy H. Sjoberg
Chrysler Corporation
Detroit, Michigan

Richard Streeper
3M Corporation
Southfield, Michigan

PREFACE

These conference proceedings are the culmination of a year long effort by the 1989 ASM/ESD Advanced Composites Conference and Exposition (ACCE) Planning Committee to bring together an advanced composites conference of international scope, focussing primarily on automotive applications. The committee and staff members at ESD and ASM have worked diligently in this endeavor, resulting in a comprehensive program for the current year.

The major theme for ACCE '89 is **Composites, A Key to Global Competitiveness** and competitiveness starts with continued education. We have expanded the tutorial segment of our program this year and it starts off the conference on Monday. Dr. Stephen W. Tsai is with us again with an updated seminar on **Designing with Composite Materials.** In addition, Mr. Stephen B. Driscoll from Lowell Technological Institute/University brings a new subject matter to the tutorial program. It is on the **Dynamic Mechanical Properties of Reinforced Plastics/Composites.** Following Mr. Driscoll, the afternoon is devoted to a comprehensive overview of the many processes used to manufacture advanced composites. This is an excellent opportunity for a novice in plastics to become familiar with the processing side of the composites business. The added tutorial segments run concurrently with Dr. Tsai's seminar on Composite Design.

An Executive Plenary Session is scheduled to kick off the technical program on Tuesday morning and features the leaders from three newly created support organizations. The highlighted organizations are the Michigan Materials and Processing Institute (MMPI), the Automotive Composites Consortium (ACC), and the National Institute of Technology Standards (NITS). These three speakers will describe how they assist the Automotive Industry in the competitive world market. In addition, Mr. Jack Simon, a Senior Policy Analyst to the Office of Science and Technology Policy will describe how to leverage government resources toward composite research and development. These talks will be followed by an exciting presentation on the technology that resulted from the B-2 Bomber program and a luncheon keynote address on the European Car Market through the eyes of Mr. Colin R. Spooner, Special Projects Director of Lotus Engineering, England.

The technical program for the conference will cover the traditional subjects of **Composite Design, Manufacturing, Material Science, NDT** and **Material Characterization**. It will also include special sessions to address more specific technologies such as **Composite Preforming, Liquid Molding, Recycling,** and the **Ceramics/Metal Matrix** disciplines. This interchannge of technical information has been very instrumental to the continued expansion of advanced composites into a growing number of applications throughout the world. The **ACCE** event has certainly gained an excellent reputation since its inception over five years ago and it continues to be **"The Conference"** that engineers and scientists in the Automotive Industry wait for to disclose their case studies and revolutionary advanced composite applications.

I would like to personally thank the members of the planning committee and staff members of ASM and ESD for their enthusiastic support and the many hours they have committed to the preparation and execution of this 5th Annual Conference. I would also like to acknowledge the management of the Advanced Engineering Staff of the General Motors Corporation for their support of my involvement in this conference.

Irvin E. Poston
Executive Chairman

TABLE OF CONTENTS

DESIGN

Using Variations in Stacking Sequence to Improve Delamination Toughness 1
W. M. Jordan, Louisiana Tech University, Ruston, LA

**Strain Concentrations Around Embedded Optical Fibers by FEM
and Moiré Interferometry** 11
A. Salehi, A. Tay, D. A. Wilson, D. G. Smith, Tennessee Technological University,
Cookeville, TN

High Performance Composites for the Automotive Industry 21
L. Dodyk, GenCorp Automotive, Marion, IN

Process Driven Design of a Plastic Bumper Beam 29
R. G. Dubensky, Chrysler Motors, Highland Park, MI; D. E. Jay, Chrysler Motors,
Detroit, MI; R. K. Salansky, Creative Industries Group, Auburn Hills, MI

**Design Synthesis and Assessment of Energy Management in a
Composite Front End Vehicle Structure** 33
R. L. Frutiger, S. Baskar, CPC Group/General Motors, Pontiac, MI;
K. H. Lo, R. Farris, Shell Development Company, Houston, TX

A CAE Methodology for Plastic Component Design 45
R. G. Dubensky, Chrysler Motors, Highland Park, MI

Need for a CAE Based Plastics Technology Program 53
R. G. Dubensky, Chrysler Motors, Highland Park, MI; D. E. Jay, Chrysler Motors,
Detroit, MI; R. K. Salansky, Creative Industries Group, Auburn Hills, MI

**Case Histories of an Adhesive Interleaf to Reduce Stress Concentrations
Between Plys of Structural Composites** 61
R. B. Kreiger, Jr., American Cyanamid Company, Havre De Grace, MD

LIQUID MOLDING

Improving the Processing Characteristics of Structural RIM Systems 69
T. B. Howell, R. E. Camargo, D. A. Bityk, ICI Polyurethanes, Sterling Heights, MI

**Mold Filling Analysis of Structural Reaction Injection Molding (SRIM)
and Resin Transfer Molding (RTM)** 77
M. J. Liou, W. B. Young, K. Rupel, K. Han, L. J. Lee, The Ohio
State University, Columbus, OH

MATERIALS SCIENCE

The Probabilistic Nature of Fracture in Carbon-Carbon Composites 83
    H. Aglan, A. Moet, Case Western Reserve University, Cleveland, OH

Mechanical Responses from Scale Models of Carbon Fiber Reinforced Composites 89
    Y. Chen, A. K. Srivastava, M. S. Madhukar, Michigan State University, East Lansing, MI

Durability in Aluminum-SMC Adhesively Bonded Systems 97
    J. G. Dillard, J. W. Grant, I. Spinu, Virginia Polytechnic Institute and State University, Blacksburg, VA

Non-Destructive Determination of Fiber Volume and Resin Content of Fiber Reinforced Composites 111
    C. Salvado, Applied Sciences Corporation, Carlsbad, CA

MANUFACTURING

Heat Transfer and Cure Analysis for Pultrusion 121
    G. L. Batch, C. W. Macosko, University of Minnesota, Minneapolis, MN

The Composite Intensive Vehicle—The 3rd Generation Automobile!— A Bio-Cybernetical Engineering Approach 129
    J. Köster, MOBIK GmbH, Gerlingen, FRG

Ultrasonic Molding of Plastic Powders 139
    S. K. Nayar, A. Benatar, The Ohio State University, Columbus, OH

Correlation of Dielectric Cure Index to Degree of Cure for 3501-6 Graphite Epoxy 147
    D. R. Day, D. D. Shepard, Micromet Instruments, Inc., Cambridge, MA

Microwave Processing of Polymer Composite Materials 153
    M. C. Hawley, J. D. DeLong, Michigan State University, East Lansing, MI

Thermal Analysis of Composite Tooling Materials 161
    N. M. Ham, M. S. Molitor, Ciba-Geigy Corporation, East Lansing, MI

Moen Heating Systems for High Volume Production of Thermoplastics 167
    R. W. Aukerman, Heat Transfer Technologies, Inc., Sun Valley, CA

Surface Waviness and Observed Reflections—Mathematical Modeling for Various Inspection Methods 175
    C-C. Lee, GenCorp Research, Akron, OH

CERAMICS/METAL MATRIX

Aluminum Alloys Matrix Composites Using Particle Dispersion 187
   H. Ohtsu, Nippondenso Co., Ltd., Aichi, Japan

**Effects of Reaction Products on Mechanical Properties of
Alumina Short Fiber Reinforced Magnesium Alloy** 201
   H. Hino, M. Komatsu, Nissan Motor Co., Ltd., Yokosuka, Japan;
   Y. Hirasawa, Ube Industries Ltd., Ube, Japan; M. Sasaki, Atsugi Motor Parts Co., Ltd.,
   Atsugi, Japan

Continuous SiC Fiber Reinforced Metals (Abstract) 209
   M. A. Mittnick, Textron Specialty Materials, Lowell, MA

Commercialization of *DURALCAN* Aluminum Composites 211
   W. R. Hoover, Dural Aluminum Composites Corporation, San Diego, CA

Numerical Study of Thermal Conductivity of Fiber-Matrix Composite Materials 219
   W. Cha, J. V. Beck, Michigan State University, East Lansing, MI

Metallic Glass Reinforcement of Glass-Ceramics 227
   R. Vaidya, K. N. Subramanian, Michigan State University, East Lansing, MI

Internal Friction Characteristics of a Kevlar/Epoxy System 233
   K. S. Burson, Aerojet Solid Propulsion Company, Sacramento, CA;
   W. N. Weins, University of Nebraska-Lincoln, Lincoln, NE

RECYCLING

Analytical Chemistry as Applied to Recycled Plastics 245
   W. H. Greive, Monarch Analytical Laboratories, Inc., Toledo, OH

Plastics Recycling–Markets and Applications 249
   R. A. Bennett, The University of Toledo, Toledo, OH

**Recycled Post-Consumer HDPE: Properties and Use as a Matrix
for Wood-Fiber Composites** 255
   S. E. Selke, Michigan State University, East Lansing, MI

PREFORM

Preforming for Liquid Composite Molding 259
    E. P. Carley, J. F. Dockum, Jr., P. L. Schell, PPG Industries, Inc., Pittsburgh, PA

1990 Corvette Rear Underbody—The Case for Preform 275
    T. P. Schroeter, R. Keith Leavitt, Molded Fiber Glass Company, Ashtabula, OH

Production of an Automotive Bumper Bar Using LCM 283
    D. A. Kleymeer, Ardyne, Inc., Grand Haven, MI; J. R. Stimpson,
    CPC Group-GMC, Warren, MI

**Evaluation of Fabric Preforms for High Volume Manufacture
of Automotive Composites** 291
    J. J. Kutz, F. K. Ko, Drexel University, Philadelphia, PA

**Development of an Automated Chopped Fiber Glass Preform
Manufacturing System** 301
    D. M. Perelli, General Motors Corporation, Warren, MI

Composite Preform Fabrication by 2-D Braiding 307
    H. B. Soebroto, F. K. Ko, Drexel University, Philadelphia, PA

NDT AND MATERIALS CHARACTERIZATION

Characterization of Intralaminar Hybrid Laminates 317
    C. H. Luther, General Dynamics Land Systems Division, Troy, MI

New Techniques in Ultrasonic Imaging for Evaluation of Composite Materials 325
    Bong Ho, R. Zapp, Michigan State University, East Lansing, MI

High Strength Sheet Molding Compound—Property/Processing Interaction 331
    J. Collister, Premix, Inc., North Kingsville, OH

**Using Laser Doppler Velocimetry for the Dynamic Evaluation
of Damage in Composite Materials** 337
    J. P. Nokes, G. Cloud, Michigan State University, East Lansing, MI

*The following oral presentations were unavailable
in written format for publication in these proceedings*

DESIGN

Development of a 2.4 Meter Satellite Dish in Structural RIM
R. Lonardo, PLASTEK Corp., Newburyport, MA

Evaluation of Sandwich Composite Bedplates for Marine Application
G. Leon, General Dynamics, Electric Boat Division, Groton, CN

Starship Concepts
E. Hooper, Beechcraft, Wichita, KS

Development and Testing of Liteflex Truck Trailer Springs
J. Mutzner, D. Richard, Inland Division, GMC, Dayton, OH

Case History of GMC Heavy Duty Truck Door
E. Gray, Premix, E.M.S. Inc, Lancaster, OH

LIQUID MOLDING

Rheological Properties and Instrumented Impact Behavior
S. Driscoll, D. Gallagher, D. Harrington, University of Lowell,
Lowell, MA; M. Rao, National Aeronautical Laboratory, Bangalore, India

MANUFACTURING

Advanced Composite Tooling Processes
R. Bogart, Advanced Polymer Industries, Inc., Plymouth, MI

Fluidized Bed Opposed Jet Milling for Making Ultra-Fine Thermoplastic Powders
D. Eddington, Alpine American Corporation, Natick, MA

CERAMICS/METAL MATRIX

Computer-Aided Structural Characterization of Composite Ceramic Materials
D. Rourk, P. Glance, Concept Analysis Corporation, Plymouth, MI; W. Bryzik,
U.S. Army, Warren, MI; B. Katz, U.S. Army, Watertown, MA

RECYCLING

Creation of a New Product Equals Successful Recycling
F. J. Stark, Jr., Rubber Research Elastomerics, Inc., Minneapolis, MN

NDT AND MATERIALS CHARACTERIZATION

Damage Accumulation in a Discontinuous Glass Reinforced Polypropylene Composite
R. G. Kander, E. I. duPont de Nemours & Company, Wilmington, DE

Thermoplastic Sheet for Automotive Applications
D. W. Adkins, E. I. duPont de Nemours & Company, Wilmington, DE

USING VARIATIONS IN STACKING SEQUENCE TO IMPROVE DELAMINATION TOUGHNESS

William Mark Jordan
Mechanical and Industrial Engineering Department
Louisiana Tech University
Ruston, LA 71272 USA

ABSTRACT

In many composite structural applications a higher delamination toughness would be desirable if it did not decrease the stiffness of the structure. One option is to use a more ductile resin, but this will typically lower the composite stiffness. Another option is to vary the stacking sequence without changing the engineering elastic constants of the laminate. This study examines whether toughness is affected by variations in stacking sequence. An example of this is to vary how many plies of like angles are grouped together, as shown below.

[-45(4)/0(4)/45(4)/90(4)] S
[-45(2)/0(2)/45(2)/90(2)] 2 S
[-45/0/45/90] 4 S

The differences between these layups are the amount of splicing that has occurred (the last layup being the most heavily spliced one).

In this study the effect of splicing and fiber orientation on mode I and mode II delamination fracture toughness was examined on four composite material systems. Mode I toughness was determined from double cantilevered beam (D.C.B.) tests and mode II toughness was determined from end notch flexure (E.N.F.) tests. The materials chosen were AS4/3501-6, AS4/3502, T6T145/F155, and AS4/APC2 (PEEK). The more heavily spliced the layup was, the greater was the mode I and mode II delamination toughness.

For the AS4/3501-6 both the mode I and mode II toughness increased by about 15%. For the AS4/APC2 the mode I toughness increased about 70% and the mode II toughness increased about 23%.

These results are significant to the designer, for the less highly spliced laminate is probably easier to manufacture and would otherwise be chosen unless toughness is an important parameter.

AVOIDING DELAMINATION FAILURE of composite materials is one of the major goals of the designer. This has been approached from several directions. One possibility is that a change in stacking sequence may improve the toughness of the laminate.

There are at least three different aspects of this issue. The first aspect is whether changing the interface where delamination occurs changes the delamination toughness. Chai (1) has done a careful study where he initiated cracks at different interfaces within a multi-axial laminate layup. He found that the different interfaces of one laminate all had approximately the same fracture toughness.

Two other aspects of this issue will be examined in this paper. The first aspect is whether changing the stacking sequence affects the delamination fracture toughness. In changing the stacking sequence, the modulus of the laminate is also allowed to significantly change. The second aspect is whether changing the stacking sequence, but staying within a quasi-isotropic layup (therefore not changing the modulus of the laminate) will affect the delamination toughness of the system.

A quasi-isotropic layup is a common composite stacking sequence in structural applications. A frequent method of accomplishing this is for the composite to be symmetric with respect to its midplane and have equal numbers of 0, +45, -45 and 90 degree plies. The question dealt with in this phase of the project is to determine if either of the following options will improve toughness: grouping plies of the same direction together or splicing them (separating layers of the same orientation).

OBJECTIVES OF THE RESEARCH PROJECT

The main objective of this project was to determine if the stacking sequence of orthotropic composite laminates affected their mode I and mode II delamination fracture toughness. Four different systems were examined: (1) AS4/3502, a relatively brittle graphite/epoxy, (2) AS4/3501-6, a relatively brittle graphite/epoxy very similar to number (1), (3) T6T145/F155, a rubber toughened, moderate toughness graphite/epoxy and (4) AS4/APC2 (peek), a very ductile graphite/thermoplastic system.

Much delamination fracture toughness work has been done on unidirectional composite systems. Unidirectional systems are easier to fabricate. They are stiffer than quasi-isotropic systems, frequently making data analysis simpler as well. Quasi-isotropic systems, while more common in structural applications, have less frequently been examined experimentally.

Six laminates of T6T145/F155 and four laminates of AS4/3502 were tested (for specific layups see Table 1). The first three F155 based laminates were 24 plies thick with only the center two plies changed to examine the effect on toughness of different fiber orientations at the crack interface. The last 3 F155 laminates had 67%, 50% and 0% zero degree plies with the rest being at plus or minus 45 degrees. The four AS4/3502 laminates were arranged so that there were either 100%, 67%, 50% or 0% zero degree plies with the rest being at plus or minus 45 degrees.

Four laminates of AS4/3501-6 and AS4/APC2 were tested. All of them are quasi-isotropic (for specific layups see Table 2). The difference between the layups was how many plies of each orientation were stacked together. For the AS4/3501-6 system groups of 4, 3, 2, or 1 ply in each orientation were stacked together. That resulted in three laminates of 32 plies and one of 24 plies. For the AS4/APC2 system groups of 6, 3, 2, or 1 ply of each orientation wer grouped together (all four laminates had 48 plies).

EXPERIMENTAL PROCEDURE

PREPARING AS4/3502 AND T6T145/F155 PANELS--The T6T145/F155 and AS4/3502 laminates were laid up in 30 centimeter square panels. A strip of teflon was placed at the midplane about 4 cm into one edge of the laminate to provide a starter crack. Crack growth was in the zero degree direction. They were cured in a microprocessor controlled press according to the manufacturer's suggested cure cycle (of time, temperature, and vacuum).

PREPARING AS4/3501-6 AND AS4/APC2 PANELS--The AS4/3501-6 panels were layed up according to the specified orientations in 25 centimeter square panels (See Table 2 for details). A strip of teflon was placed at the midplane about 4 cm into one edge of the laminate to provide a starter crack that would grow in the 90 degree direction (parallel to the fibers at the center plane). The AS4/APC2 was layed up in 10 by 15 cm laminates with an aluminum insert to provide a starter crack at the midplane. The starter cracks went into the laminate a distance of approximately 4 cm.

The specific AS4/3501-6 and AS4/APC2 layups chosen were used to more easily examine the effect of splicing on delamination fracture. One difficulty in choosing the stacking sequence relates to the problem of twisting occurring during testing. During Mode I testing, only one half of each laminate is actually being bent. To eliminate coupling between twisting and bending each half laminate has to be symmetric. However, during mode II testing, the entire laminate is bending and to eliminate coupling between bending and twisting an overall symmetric layup is required. The choice made during this project was to use a symmetric stacking sequence, knowing there will be some slight twisting during mode I testing. (A calculation of the laminate sifffness showed that for the systems used in this study, the values of the [B] matrix are small with respect to that of the [A] matrix, indicating twisting should not be a large problem).

The laminates were cured in an autoclave according to standard practices for such materials. The cured laminates were C-scanned and X-rayed to establish that there were no internal defects other than the cracks caused by the edge inserts.

MODE I TESTING--The Mode I double cantilevered beam (DCB) test specimens were cut to be approximately 2 to 2.5 cm wide and up to 20 cm long. Mode I tests on the AS4/3501-6 and AS4/APC2 laminates were performed on a screw driven Instron tensile test machine operating in displacement control. Mode I tests on the AS4/3502 and T6T145/F155 laminates were performed on an MTS Model 810 tensile machine operating in displacement control. The crosshead rate used was .125 cm/min.

The load and displacement were continually recorded on a strip chart. The side of the specimen was coated with a brittle white coating to aid in determining the growth of the crack. The crack length was monitored at discrete intervals (every 1.25 cm of crack growth) and this length was marked on the load-displacement chart. The crack was grown from its original length to a length of 11.5 cm on each specimen. At least once during the crack growth period, the specimen was unloaded so that the shape of the unload/load curve could be determined.

MODE II TESTING--The mode II end notch flexure specimens were cut to be approximately 2 to 2.5 cm wide and about 18 cm long. Mode II tests on the AS4/3501-6 and AS4/APC2 laminates were performed on a screw driven Instron tensile test machine operating in displacement control. A crosshead rate of .125 cm/minute was used. The test fixture used for this is a three point bend geometry. The span of the two bottom supports was 10.2 cm, with the load being applied at the center of the specimen. For most of the tests, the crack tip was placed one-half way between one of the lower supports and the applied center load (this provided a starter crack of 2.5 cm).

The load was slowly applied until the crack began to grow. Crack growth was usually unstable, with the crack growing in one step to the center loaded region. One of the advantages of this test geometry is that once the crack has grown, the specimen can be slid over so that a new crack length of 2.5 cm is obtained, and another test can be performed on the same specimen. This is why the mode II specimens were cut considerably longer than the span of the three point test fixture.

Mode II tests on the AS4/3502 and T6T145/F155 laminates were performed on an MTS Model 810 tensile machine operating in displacement control. An end load split laminate test fixture was used for these mode II tests. This test geometry is in effect one half of a three point bend geometry. For more details of this method see references (2,3).

EXPERIMENTAL ANALYSIS AND RESULTS

MODE I ANALYSIS--Mode I tests using double cantilevered beam test specimens have been done by a number of investigators, including this one (1,2,4,5,6,7). The most common analysis has been one using linear beam theory (1,2). Devitt, et. al, (4) allowed for non-linear elastic behavior but not for permanent damage in the composite caused by crack growth.

The mode I results in this study were all in the linear elastic region as defined by Devitt except as noted below. For the last 3 laminates of both T6T145/F155 and AS4/3502 Devitt's non-linear elastic analysis was used. These were the layups with 33%, 50% or 100% plus/minus 45 degree plies. This method allows for non-linear effects, but not for far field damage.

The layups so marked in Table 2 had significant permanent deformation as a result of crack growth. (This deformation is in the form of a permanent opening displacement after all loads had been removed.) This permanent deformation occurred even though the specimens stayed within the linear elastic region according to Devitt's analysis. As a result of this analysis, all of the results in Table 2 were calculated using linear elastic beam theory. The ones with significant permanent deformation were labelled as upper bound limits to G_{Ic}. The remaining layups did not have significant permanent deformation and the fracture toughness results reported can be considered an accurate representation of mode I toughness.

MODE I RESULTS--Mode I test results are reported in Tables 1 and 2. The fracture toughness values reported in this study are for steady state crack growth.

The effect of fiber orientation across the plane of delamination on delamination fracture toughness G_{Ic} is seen in Table 1. For a specimen with plus or minus 45 degree plies across the interface that is debonding, but with a stiffness similar to the unidirectional composite, the delamination critical energy release rate was found to be very similar to the results for unidirectional laminates. Chai (1) has previously noted a similar result; namely, that ply orientation across the plane of delamination does not significantly affect the delamination fracture toughness for laminates with similar stiffness.

The much larger value of G_{Ic} indicated for the third multiaxial F155 layup may be a result of far field damage as a result of nonlinear viscoelastic behavior by the resin. The resin carries a siginificant load in the axial direction for a composite laminate with all +-45 plies. This significant resin loading not only in the crack tip region but at other locations removed from the crack tip will cause the resin to undergo nonlinear viscoelastic deformation. This energy dissipation in the far field should not be counted in the crack tip energy dissipation per unit area of crack extension. However, this G_{Ic} calculation using the area method does not distinguish between crack tip and far field energy dissipation. Thus, the G_{Ic} value of 1333 J/m^2 indicated in Table 1 is not really a G_{Ic}, but includes G_{Ic} plus far field damage.

In the quasi-isotropic systems, there was an apparent initiation fracture toughness which was lower than the reported propagation toughness. There was then a region where the value of G_{Ic} did not change with respect to crack length. This is the value of G_{Ic} reported in this study. For long crack lengths (on the order of 9-10 cm) there was then an increase in fracture toughness. This may be related to permanent damage occurring within the laminate as the crack grows.

For the AS4/3501-6 system the first two layups listed had significant permanent deformation after the cracks had grown to a length of 11.4 cm. For the AS4/APC2 system only the first layup listed had significant permanent deformation after the crack growth. These results are reported as upper bound limits to G_{Ic}. The remaining layups all had small amounts of permanent deformation and valid G_{Ic} results. The systems that had a large permanent deformation also had some twisting. This was not considered large enough to effect the mode I results.

For the AS4/3501-6 system there was an increase in fracture toughness as the number of splices increased. Grouping plies of the same orientation together increased the amount of permanent deformation as well as decreasing the fracture toughness. Splicing incresed the fracture toughness by at least 15%. (The increase may be larger than that for there is only an upper bound limit to the toughness for the system where 4 plies of each orientation were grouped together).

For the AS4/APC2 system, there was an increase in fracture toughness as the number of splices increased. The increase may be more than 70% (again it may be larger because of the uncertainty in the toughness for the unspliced system.) Both the AS4/3501-6 and AS4/APC2 systems had the biggest increase in toughness when the first round of splicing began. Grouping only 2 plies together was signicantly tougher than grouping 4 or 6 plies together, but further splicing did not have a significant effect upon fracture toughness (Compare the last two layups with each system with the first two layups for each system).

MODE II ANALYSIS--Two common methods of performing mode II delamination fracture toughness tests are the end notch flexure test (E.N.F.) and the end-loaded split laminate. Both methods have been used in this study. In a recent conference presentation Corleto and Bradley (3) reported a comparison of these two test methods for use with a brittle and a ductile resin system. Their conclusion was that these two methods give similar results.

The end-loaded split laminated has been previous used by the author (2,6). It was used on the T6T145/F155 and AS4/3502 systems. The E.N.F. test method was used in this study on the quasi-isotropic systems. There were nonlinearities in the load/unload curve for some of the systems. To deal with this issue, the example of Corleto and Bradley (3) was followed and G_{IIC} was calculated using the area within the load/unload curve as the energy requred to grow the crack. This was divided by the specimen width and the length of crack extension to get a value of G_{IIC}.

MODE II RESULTS--Mode II results are shown in Table 3. For three of the layups indicated in the table, there was significant permanent deformation and the mode II fracture toughnesses reported are upper bound limits to G_{IIC}. There was a small permanent opening displacement for these systems as well as a permanent bending of the laminate. In addition, for these same systems the outside layers (that were in tension) delaminated in a few locations prior to extension of the main crack.

When comparing the layps where a legitimate G_{IIC} could be obtained, splicing increase the fracture toughness of the system. In contrast to the mode I results, continuing to splice the laminate (after the first time) continued to increase the fracture toughness.

DISCUSSION OF RESULTS

In situ fractography of the T6T145/F155 off-axis layups indicated in Table 1 clearly indicates that significant resin deformation is occurring along the axis of the specimen (2,6). A specimen with all +-45 plies would certainly experience significant resin loading in the direction of the specimen axis as a result of bending stresses. While similar stresses would also be experienced by a unidirectional laminate specimen, the fibers would carry essentially all of the loading in the axial direction. It is this additional deformation, not only in the crack tip region, but presumably all along the specimen that is the far field damage previously mentioned as being responsible for giving an artificially high value of G_{IC} of 1333 J/m^2. It is worth noting that the specimen with +-45 degree plies at the interface where delamination is occurring, but mainly unidirectional plies otherwise had a delamination fracture toughness similar to that for the unidirectional laminate.

A J integral analysis developed at Texas A & M University has given a J_{IC} of 550 J/m^2 for the T6T145/F155 laminate with all +-45 degree plies (7). This suggests that the delamination fracture toughness in composite materials may be a material property independent of stacking sequence if the near and far field damage are properly separated. The large differences in toughness observed for G_{IC} in Table 1 is likely to be a result of not neglecting far field damage in the analysis. However it should be noted that the average value of G_{IC} for the first three laminates of T6T145/F155 (488 J/m^2) which contained mostly zero degree plies was slightly lower than the average of the J-integral for the last two laminates (525 J/m^2) which were composed of mostly plus/minus 45 degrre plies. This indicates that the fiber orientation may still play a role, but a much smaller role than would have been predicted by just observing the values of G_{IC}.

The nature of the effect of splicing on delamination fracture toughness is not surprising. If there was to be any effect at all, it would be expected that increased splicing would increase the delamination fracture toughness. When the laminate is cured, there will always be some residual stresses left within the laminate. These residual streses result from the different thermal properties of the resin and the fibers. If the laminate is heavily spliced then the tendency of one ply to cause the laminate to bend in one direction will be countered by its adjacent plies which may want the laminate to bend in a different direction.

If plies of the same direction are grouped together then different portions of the laminate will have very different residual stresses. These stresses will make it easier to delaminate the laminate, for there are now significant stresses present acting to pull the laminate apart at its weakest link (the ply interface). An illustration of this phenomenon is the observation that some of the less spliced laminate specimens opened up slightly (after removal from the autoclave) immediately upon being cut into their final dimensions.

It is interesting to note that splicing had a larger effect upon the more ductile AS4/APC2 system. This may be related to the poor interfacial bonding between plies in the AS4/3501-6 system. Previous work on this system indicated that delamination typically occurs at the resin/fiber interface rather than within the resin itself (5). This poor bonding may result in delamination occurring in the spliced layups before the intrinsic resin toughness could be obtained. The total increased toughness expected in the spliced system could not be obtained because of the fiber/resin interfacial failure.

In view of the results shown in Table 1, namely that the fiber orientation does not appear to make a large difference in the delamination fracture toughness if the far field damage is separated from the near damage, it is relevant to ask whether the differences observed in Tables 2 and 3 are "real differences" or only reflective of not accounting for the far field damage. The answer to that question is not a simple one.

The layups shown in Table 1 had very different moduli. This allowed for the more ductile ones to significantly microcrack and/or plastically deform. This process absorbed energy that must be accounted for in any attempt to analyze for an intrinsic fracture toughness.

For the quasi-isotropic materials shown in Tables 2 and 3, the situation is quite different. They all have about the same modulus which would indicate that any far field damage that might exist would be the approximately the same for each laminate. On the other hand, some of the unspliced specimens did have significantly more permanent deformation than did the spliced ones and this must be taken into account. If the material did have significant permanent deformation, then any estimated values of G_{IC} would be high, for it would assume all of the absorbed energy went into crack propagation which would not be true. The more highly spliced laminates did not have any permanent deformation and all fit within the linear elastic criteria established by Devitt (4). It therefore appears that varying the stacking sequence within the context of a quasi-isotropic system will, in fact, increase the toughness of the laminate.

This conclusion must be related to the conclusion that stacking sequence does not play a major role in toughness based on the data as reported in Table 1. The J-integral analysis that produced off-axis toughnesses that were approximately the same as those of a unidirectional layup still produced higher mode I toughnesses for systems with mostly 45 degree plies. This can be seen by comparing the results for the first three T6T145/F155 layups (which averaged 488 J/m^2) and the last two layups in that series (whose J-integral averaged 525 J/m^2). This may mean that the very high G_{IC} values for off axis laminates shown in Table 1 was caused by far field damage, but that there may still be some smaller effect caused by the stacking sequence itself. [It should be noted in Table 1 that the laminates with more like plies adjacent to each other had a lower toughness than the ones that were more highy spliced.]

SUMMARY AND CONCLUSIONS

The main conclusion of this work is that increased splicing will increase the delamination fracture toughness under both mode I and mode II loading conditions. This is useful to the designer who frequently may be limited to quasi-isotropic laminate layups, but would like to improve the fracture toughness. This shows that fracture toughness can be improved by a proper choice of stacking sequence. It must be noted that this improvement is not as great as might be obtained if the material system could be changed. (For example, there is a geater difference between the AS4/3501-6 and AS4/APC2 toughnesses than there is between the spliced and unspliced layups within each laminate system.)

This could prove useful to the designer for it provides a way where he can make some improvements in the toughness of a structure without sacrificing stiffness at the same time.

ACKNOWLEDGEMENTS

Some of the experimental work was completed while the author was a graduate student at Texas A & M University working on a project sponsored by the Air Force Office of Scientific Research. The remaining experimental work was done while the author was a Summer Faculty Research Fellow at the Air Force Materials Laboratory. The financial support of these Air Force agencies is gratefully acknowledged.

REFERENCES

1. Chai H., Composites, 15, No. 4, 277-290 (1984).

2. Jordan, W.M., Ph.D. Dissertation, Texas A & M University, 1985.

3. Corleto, C.R. and Bradley, W.L, "Mode II Delamination Fracture Toughness of Unidirectional Graphite/Epoxy Composites" presented at an A.S.T.M. Symposium, Cincinnati, Ohio, April 1987.

4. Devitt, D.F., Schapery, R.A., and Bradley, W.L., Journal of Composite Materials, 14, 270-285 (1980).

5. Cohen, R.N., M.S. Thesis, Texas A & M University, 1982.

6. Jordan, W.M., and Bradley, W.L., "Micromechanisms of Fracture in Toughened Graphite/Epoxy Laminates", in **Toughened Composites ASTM STP 937**, Norman Johnston, Ed., American Society for Testing and Materials, Philadelphia, 1987, pp. 95-114.

7. Schapery, R.A., Jordan, W.M., and Goetz, D.P., Proceedings of International Symposium on Composite Materials and Structures, Beijing, China, 1986, pp. 543-548.

Table 1*

EFFECT OF FIBER ORIENTATION UPON MODE I G_{Ic}

Composite System	Layup	G_{Ic} (J/m^2)	J_{Ic} (J/m^2)	Modulus (GPa)
T6T145/F155				
	[O(24)]	520		107
	[O(11)/10/-10/O(11)]	470		107
	[O11)/45/-45/O(11)]	474		107
	[45/-45/O(8)/-45/45/-45/45/O(8)/45/-45]	589		49
	[45/-45/O(4)/-45/45/-45/45/O(4)/45/-45]	848	500	52
	[45/-45(2)/45/-45/45(2)/-45/45/-45(2)/45/-45/45(2)/-45]	1333	550	26
	Average for first three layups that are mostly zero degree plies	488		
	Average for last two layups that are mostly plus/minus 45 plies		525	
AS4/3502				
	[O(24)]	190		134
	[45/-45/O(8)/-45/45/-45/45/O(8)/45/-45]	418	180	59
	[45/-45/O(4)/-45/45/-45/45/O(4)/45/-45]	383	200	49
	[45/-45(2)/45/-45/45(2)/-45/45/-45(2)/45/-45/45(2)/-45]	713		24

* The data in this table is from references (2), (6) and (7) by the author.

TABLE 2
EFFECT OF STACKING SEQUENCE IN QUASI-ISOTROPIC LAMINATES ON THEIR MODE I DELAMINATION TOUGHNESS

Material	Layup*	Number Plies	G_{Ic} (J/m^2)	Permanent Opening Displacement (mm)**
AS4/3501-6				
	[-45(4)/0(4)/45(4)/90(4)] S	32	133***	22.2
	[-45(3)/0(3)/45(3)/90(3)] S	24	154***	30.2
	[-45(2)/0(2)/45(2)/90(2)] 2S	32	158	6.4
	[-45/0/45/90] 4S	32	152	2.4
AS4/APC2				
	[-45(6)/0(6)/45(6)/90(6)] S	48	877***	27.8
	[-45(3)/0(3)/45(3)/90(3)] 2S	48	1454	7.1
	[-45(2)/0(2)/45(2)/90(2)] 3S	48	1523	5.6
	[-45/0/45/90] 6S	48	1509	4.0

* Systems delaminated at midplane in 90 degree direction.

** Permanent deformation after completion of mode I tests (with final crack length of 11.4 cm)

*** Upper bound estimate because of large degree of permanent deformation

TABLE 3
EFFECT OF STACKING SEQUENCE IN QUASI-ISOTROPIC LAMINATES ON THEIR MODE II DELAMINATION TOUGHNESS

Material	Layup*	Number Plies	G_{IIc} (J/m^2)	Upper Bound Estimate to G_{IIc} (J/m^2)
AS4/3501-6				
[-45(4)/0(4)/45(4)/90(4)] S		32		853
[-45(3)/0(3)/45(3)/90(3)] S		24		483
[-45(2)/0(2)/45(2)/90(2)] 2S		32	434	
[-45/0/45/90] 4S		32	564	
AS4/APC2				
[-45(6)/0(6)/45(6)/90(6)] S		48		2422
[-45(3)/0(3)/45(3)/90(3)] 2S		48	1687	
[-45(2)/0(2)/45(2)/90(2)] 3S		48	1692	
[-45/0/45/90] 6S		48	2077	

* Systems delaminated at midplane in 90 degree direction.

STRAIN CONCENTRATIONS AROUND EMBEDDED OPTICAL FIBERS BY FEM AND MOIRE' INTERFEROMETRY

Abraham Salehi, Andrew Tay
Dale A. Wilson, Dallas G. Smith
Tennessee Technological University
Cookeville, Tennessee, USA

Abstract

Moire' interferometry and the finite element method (FEM) were both applied to gain knowledge of mechanical behavior of an optical fiber embedded in a graphite-epoxy laminate. Five layups were considered -- one unidirectional case and four 0/90 cases. In each case the optical fiber was located at the laminate mid-plane. A uniaxial load was used.

Moire' interferometry provided full-field information for the in-plane surface displacements around the fiber. The displacement sensitivity was 0.417 μm per fringe. From these displacements, strains around the fiber were estimated.

The fiber caused a resin-rich discontinuity in the laminate. The shape and size of the discontinuity depended on the orientation of the plys surrounding the fiber. When the angle between the optical fiber and each adjacent ply was 90 degrees the resin-rich area was large and elongated.

The finite element model included the shape of the resin-rich area. The separate material properties of the optical fiber core, cladding and coating as well as those of the resin and ply were employed in the analysis.

The normal strain concentration factors were determined from both moire' and FEM. Strain concentrations associated with the normal strain in the load directions generally ranged from 2 to 4. Strain concentrations associated with the normal transverse strain were determined by FEM to be high for the case where the optical fiber was 90 degrees to the adjacent plys. FEM values for those cases ranged from 12.3 to 17.8. The moire' method failed to confirm these high values.

RECENT INTEREST IN MANUFACTURING AND DESIGNING self-health monitoring lightweight materials for aerospace applications has focused considerable attention on laminated composites containing embedded optical fibers -- so-called "smart skins". Small-diameter, lightweight optical fibers are included in the laminate during fabrication layup. These fibers have the potential to enhance the use of composites in aerospace structures.

In "smart-skin" applications many optical fibers are embedded and interconnected. The resulting network can be used to monitor and assess damage occurring during manufacturing and handling and also to monitor structural integrity of the vehicle during service life.

Presence of the optical fibers adds complexity to the design and analysis of the laminates. The fiber may affect the strength of the host laminate and cause a degree of structural degradation. Consequently, stress-strain analysis of "smart skins" is appropriate to ensure the benefits of embedded fibers without significant loss of laminate performance. Also of interest is the strain induced in the fiber by interaction with the host laminate. It is this strain and deformation which affects light transmission through the fiber. Recently a rather large concentration of strain around an optical fiber in a graphite-epoxy was reported by Czarnek et al. [1]. The results in reference 1 were obtained from moire' interferometry.

GEOMETRIC DISCONTINUITY - In the present work, moire' interferometry and finite element together are used to investigate the strain-displacement distribution around an optical fiber in graphite-epoxy laminates of five different layups. Uniaxial tension test specimens were fabricated with the layups shown in Table 1. Each specimen has an optical fiber embedded at the mid-surface oriented 90 degrees to the direction of applied load. Each ply in the laminates is approximately 0.127 mm (0.0050 inch) thick and the optical fiber has a diameter of approximately 0.142 mm (0.0056 inch). A cross-section of the fiber is shown in Fig. 1.

Table 1 - Laminate Layups

Case	Layup[1]
1	$[0_{12}(fiber)0_{12}]$
2	$[0_4/90_4(fiber)90_4/0_4]$
3	$[90_4/0_4(fiber)0_4/90_4]$
4	$[0_2/90_2(fiber)90_2/0_2]$
5	$[90_2/0_2(fiber)0_2/90_2]$

[1] Ply Thickness = 0.127 mm (0.0050 inch)

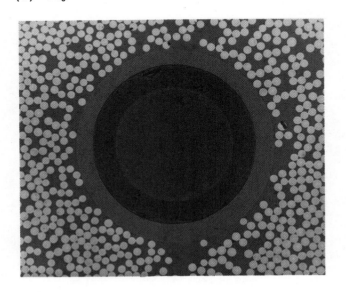

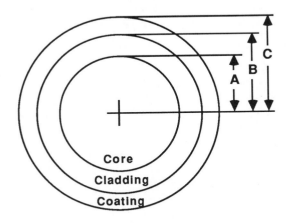

A = 0.0432 mm (0.0017 inch)
B = 0.0584 mm (0.0023 inch)
C = 0.0711 mm (0.0028 inch)

Fig. 1 - Properties of the optical fiber

The embedded optical fibers in various layups have distinctly different resin-rich regions around them. The shape and size of the resin-rich region around the optical fiber depends on the orientation of plys surrounding it. For the present layups the geometric discontinuity induced by the optical fiber is shown by the photographs in Fig. 2. The smallest resin-rich region occurs when the optical fiber and the adjacent plys have the same orientation. The largest resin-rich area is observed when the angle between the optical fiber and each adjacent ply is 90 degrees. For that case the parting of the plys leaves a wedge-shaped, resin-filled region that extends in each direction a distance of nine diameters of the optical fiber. Strain distribution and concentration around the optical fiber is, of course, strongly influenced by the geometry of the resin-rich region.

FINITE ELEMENT MODEL

Strains induced in the embedded fiber and graphite-epoxy host material were analyzed using I-DEAS SUBERTAB, a finite element software package developed by Structural Dynamics Research Corporation [1]. The investigation primarily focused on a small region having dimensions of 2.03 mm (.080 inch) high, 1.02 mm (0.040 inch) wide by 1.27 mm (0.050 inch) deep. Three-node, linear triangular elements were used.

Figs. 3 and 4 illustrate the finite element model for the optical fiber oriented parallel and perpendicular, respectively, to the adjacent plys. Only that portion of the mesh in the fiber region is shown. Because of symmetry only one-half of the region is pictured. The model created reflects the dimensions and properties of the optical fiber core, cladding and coating as well as the properties of the surrounding plys and the embedded resin-rich region. Material properties used for the various component parts are given by Table 2.

Table 2 - Material Properties

Material	E(GPa, MPsi)	ν
Core	71.0 (10.3)	0.14
Cladding	71.0 (10.3)	0.14
Coating	1.72 (0.25)	0.35
Longitudinal graphite-epoxy	149 (21.3)	0.30
Transverse graphite-epoxy	11.4 (1.65)	0.023
Resin	3.44 (0.500)	0.34

Figs. 3 and 4 show typical mesh patterns (for cases 4 and 5); the actual mesh pattern was varied to accommodate the various layup cases. A total of 730 elements and 390 nodes was created for the case 4 layup. Comparable numbers were used for the other cases. In order to directly compare the moire' interferometry and finite element results the actual remote normal test strain was used to compute a remote displacement. This remote displacement was then used as input for the finite element model.

The inside radii of the optical fiber used in the model, Fig. 1, are slightly different from those of the test coupons. The outside radius, however, was the same. The main difference was in the radius of the core. But the core and cladding had identical stiffness properties so that a different core radius should have no affect on strain distribution.

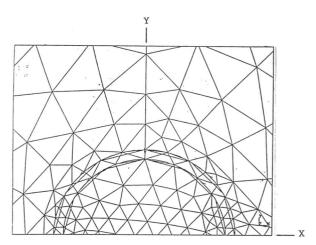

Fig. 3 - Finite element mesh for the optical fiber parallel to adjacent ply fibers, case 4.

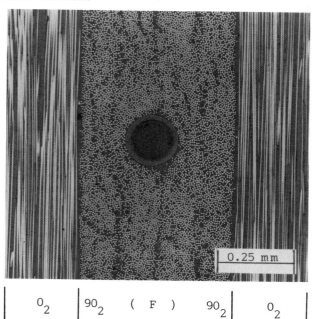

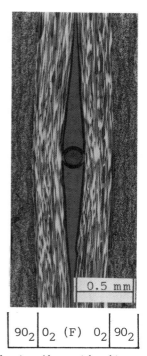

Fig. 2 - Laminate discontinuity produced by optical fiber -- embedded parallel and perpendicular to adjacent ply, respectively.

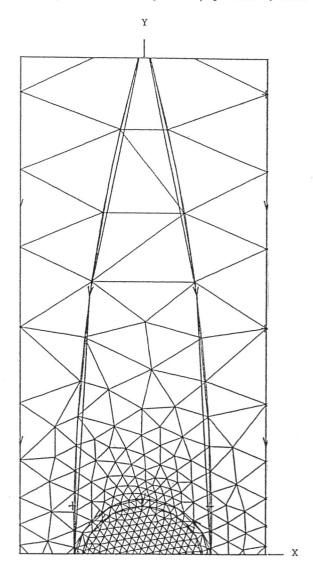

Fig. 4 - Finite element mesh for the optical fiber perpendicular to ply fibers, case 5.

THE MOIRE' INTERFEROMETRY METHOD

Moire' interferometry is a highly sensitive optical method that produces full-field values for in-plane surface displacements. The development of the method is due primarily to the work of Post and his associates [2, 3]. The method requires a high frequency diffraction grating on the specimen surface illuminated with coherent laser light. Interference fringes, representing surface displacements, are produced. From these displacements strain can be determined. Post has applied the method to a variety of composite materials problems [4, 5, 6].

The diffraction grating is attached to the surface of the specimen with an epoxy adhesive using a method developed by Post [2]. The grating is a cross-line reflective phase type. The grating is thin, on the order of 0.1 µm; its epoxy adhesive is much thicker, on the order of 25 µm. In the present work the frequency of the grating is 1200 lines/mm (30,000 lines/inch).

APPLYING THE METHOD - Fig. 5 shows a basic schematic of the method. Since the optical fiber intersects the edge, the diffraction grating is placed on the edge of a tensile test coupon. Two coherent collimated beams of light lying in the x-z plane, represented by rays A and B, impinge on the diffraction grating. The two beams make equal angles with the normal, z. When the test coupon is strained the diffraction of the two beams produces an interference fringe pattern which can be collected by a camera lens and focused on a screen. The physics of the interference has been explained by Post [7]. The fringe pattern represents a contour map of displacements, u, in the x-direction. To obtain displacements, v, in the y-direction, instead of A and B, two beams C and D, lying in the y-z plane, are used.

The two fringe patterns are simply converted to displacements by:

$$u = N_x/f$$
$$v = N_y/f \quad (1)$$

where u and v are displacements in the x and y directions, respectively; N_x and N_y are fringe order numbers for the x and y patterns respectively; and f is twice the frequency of the grating. For the present f is 2 x 1200 lines/mm or 2400 lines/mm. The sensitivity of the method is represented by the displacement corresponding to one fringe -- in this case (1/2400) mm or 0.417 x 10^{-6}m (16.4 x 10^{-6} inch).

Normal strains, ε_x and ε_y, can be found by numerical differentiation of the displacements. The strain-displacement equations are used, written in finite form;

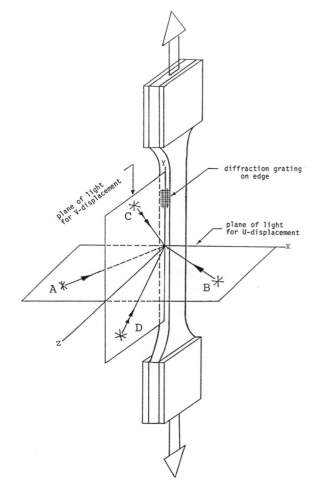

Fig. 5 - Application of the moire' interferometry method to obtain laminate edge deformation around the optical fiber.

namely,

$$\varepsilon_x = \frac{\Delta U}{\Delta X} = \frac{1}{f}\frac{\Delta N_x}{\Delta X}$$
$$\varepsilon_y = \frac{\Delta V}{\Delta Y} = \frac{1}{f}\frac{\Delta N_y}{\Delta Y} \quad (2)$$

Thus normal strains were determined along lines of interest through and near the optical fiber.

RESULTS AND DISCUSSION

FRINGE PATTERNS - The moire' fringe patterns for N_x and N_y for all five layups are shown in Figs. 6-10. Those patterns exhibit several features. The view for each one is of the specimen edge. The specimens are thin, varying from 1.02 mm (0.04 inch) for the 8-ply layup to 3.05 mm (0.12 inch) for the 24-ply layup. The optical fiber comprises a very small portion of the total field. Large magnification was necessary. The negative image was 4X. Further enlargement -- to a total of 45X -- was necessary before digitizing the fringe patterns.

While image enlargement aids fringe measurements it does not increase the amount of fringe data present. A relative paucity of fringe data exists in the immediate neighborhood of the fiber such that only a few fringes (data points) cross the actual fiber. Strain obtained represents average strain between successive fringes. Thus in small local regions of high strain gradient the average strain is only an estimate of the actual strain at a point. The FEM on the other hand permits many strain values in the small distance of only one fiber diameter.

Loading a thin specimen without significant accidental bending represents a challenge. For a perfect uniaxial load the N_x fringes are straight and parallel to the y-direction. Curving or sloping N_x fringes indicate bending and rotation. An elementary bending analysis for a hypothetical load eccentricity of only 0.0254 mm (0.001 inch) for the cases 4 and 5 specimens indicates a through-the-thickness stress variation of +15 percent and -15 percent from the average uniaxial value. An attempt was made to minimize bending, especially in the region near the fiber.

The kinks in the N_y patterns of Figs. 7-10 mark the layer between the 0 degree and 90 degree plys. Each kink indicates a v-displacement which is rapidly changing in the x-direction. This requires shear strain. Thus a degree of interlaminar shear strain is indicated along the boundary.

NUMERICAL RESULTS - The fringe patterns were digitized along certain lines. Distributions of v-displacements along vertical lines (x = constant) and u-displacements along horizontal lines (y = constant) were determined and plotted. Fig. 11 shows, for example, the v-displacement along the two lines x = 0.0 and x = 0.427 mm for the case 4 layup.

From plots such as Fig. 11 for the y-direction and similar plots for the x-direction, normal strains ε_y and ε_x, respectively, were determined. Fig. 12 shows the normal strain ε_y along x = 0 for for the case 4 layup. The strain along x = 0, a line through the center of the optical fiber, is of primary interest. Approaching the fiber, this strain gradually decreases to a value of about two-thirds the remote strain. Then, at the edge of the fiber the strain rises sharply, displaying peaks of 9200 $\mu\varepsilon$ and 10,600 $\mu\varepsilon$ at the two fiber edges. In the fiber core region, between the two peaks, the strain is somewhat less.

The strain peaks occur in the region of the fiber coating. Larger strains can be expected there since the stiffness of the coating is considerably less than that of the core-cladding and also less than that of the surrounding graphite-epoxy -- see Table 2.

The results of the finite element model (FEM) along x = 0 are also shown on Fig. 12. The trends of the FEM and the moire' results agree. The strain from the FEM, of course, exhibits symmetry about y = 0. The FEM permits strain values of many more points along a given line than moire', thus producing a smoother distribution. The FEM resolution is much better. Thus FEM can find local maxima and minima which may occur between data points of the moire' method.

The strain peaks from the FEM results, rising to 13,900, are somewhat higher than those of the moire'. Yet results are fairly close. This is somewhat fortuitous. As previously mentioned the moire' results must be regarded as estimates because of the averaging between fringes and because of the difficulty of obtaining data in such a small region. Indeed moire' data were insufficient for even an estimate for case 5 and also for ε_x for case 4.

STRAIN CONCENTRATION - Of considerable interest is the strain concentration factor, the ratio of the local maximum strain to that

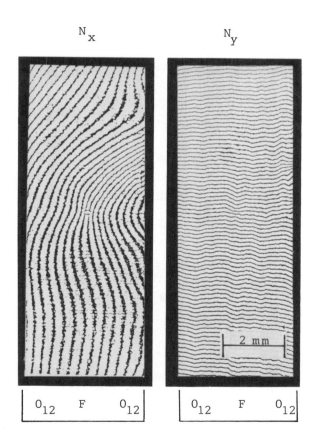

Fig. 6 - Moire' fringe patterns for case 1 layup, $[0_{12}$ (fiber)$0_{12}]$.

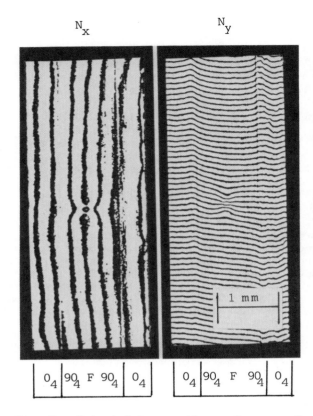

Fig. 7 - Moire' fringe patterns for case 2 layup, [$0_4/90_4$ (fiber) $90_4/0_4$].

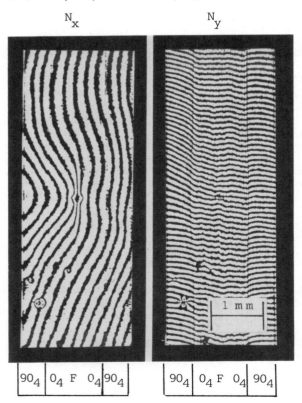

Fig. 8 - Moire' fringe patterns for case 3 layup, [$90_4/0_4$ (fiber) $0_4/90_4$].

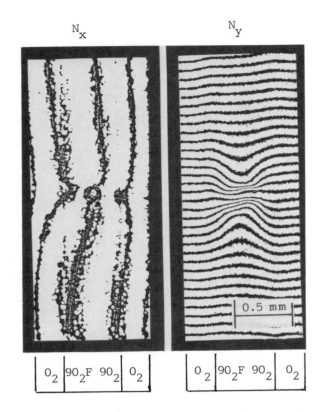

Fig. 9 - Moire' fringe patterns for case 4 layup, [$0_2/90_2$ (fiber) $90_2/0_2$].

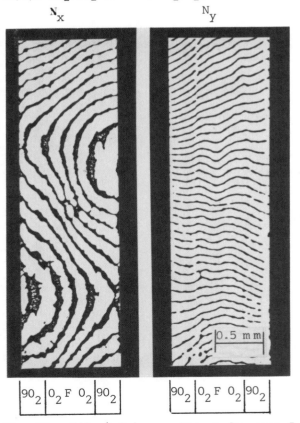

Fig. 10 - Moire' fringe patterns for case 5 layup, [$90_2/0_2$ (fiber) $0_2/90_2$].

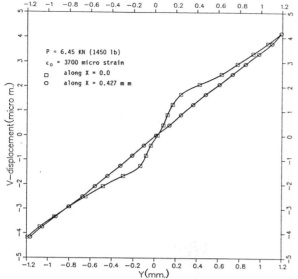

Fig. 11 - V-displacement in region of optical fiber for case 4 layup.

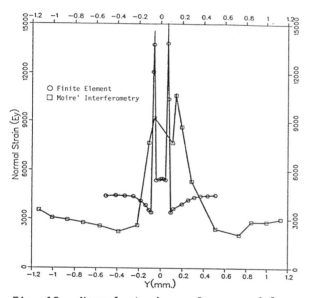

Fig. 12 - Normal strain ε_y for case 4 layup.

of the remote strain. The results may be summarized by defining strain concentration factors K_x and K_y for both the x- and y-directions. Namely,

$$K_x = \frac{\varepsilon_{xmax}}{\varepsilon_{xo}}$$
$$K_y = \frac{\varepsilon_{ymax}}{\varepsilon_{yo}} \qquad (3)$$

where ε_{xo} and ε_{yo} are normal strains remote to the fiber region. The values of K_x and K_y for the five layups from both moire' and FEM are shown in Table 3. For brevity, strain values are omitted and only strain concentration factor values are given.

The greatest strains occurred in the coating region of the fiber, on the y-axis for ε_y and on the x-axis for ε_x. For cases 2 and 4 strain concentrations for either ε_x and ε_y are moderate, falling generally in the 2 to 5 range. For these two cases the optical fiber is parallel to the adjacent plys and the fiber represents only a moderate disturbance of the laminate. The moire' values, while not agreeing precisely with FEM, support a moderate value for both K_x and K_y.

Cases 1, 3, and 5 are more severe according to FEM, especially when considering K_x. For all those cases the optical fiber was oriented 90 degrees to the adjacent plys resulting in a large geometric discontinuity. K_x values from FEM range from 12.3 to 17.8. These high values were not confirmed by the present moire' data. Indeed values of 4.04 and 3.05 are given for cases 1 and 3, respectively (estimates were not possible for case 5).

Significantly, the high FEM values for K_x support high values previously reported by Czarnek [8] for the layup of case 5. The reported value [8] was 14.3 while the present value is 16.5. The value in reference [8] was determined from moire' interferometry. It should be noted that while the layups were the same there may be material differences in the specimen of reference [8] and the present. While the present moire' method was not capable of finding a K_x value for case 5 it did determine a value for a similar case, case 3; and that value was only 3.05 -- far less than the FEM results. Perhaps the large concentrations occur in an area too small to be detected by the present moire' system.

CRACKS - In preparing specimen edges for the diffraction grating, pre-existing cracks were noted in some optical fibers. In all cases noted the cracks were in fibers oriented at 90 degrees to adjacent plys. Presumably, these cracks occurred during fabrication. Fig. 13 shows a fiber with two such cracks. Two views are given -- an optical micrograph and a scanning electron micrograph. The cause of these cracks is not known. If they occur frequently they represent a problem which deserves attention in smart-skin technology.

FAILURE CRACKS - When sufficiently loaded cracks occur in the 90 degree plys. These cracks sometimes occurred remote to the optical fiber. Other times the crack occurred at the fiber. Fig. 14 shows one such case, for layup case 2. The crack there extended through the optical fiber coating and around the cladding. Such cracks become readily apparent in the application of the moire' interferometry method. Fig. 15 shows the fringe disturbance caused by the crack. The crack stopped at the 0 degree plys and the 90 degree ply load was distributed to the 0 degree plys.

Table 3 - Strain Concentration Factors

Case		FEM $K_x^{(1)}$	$K_x^{(1)}$	MOIRE' $K_x^{(1)}$	$K_y^{(1)}$
1	$[0_{12}(F)0_{12}]$	12.3	3.17	4.04	3.46
2	$[0_4/90_4(F)90_4/0_4]$	2.06	4.70	1.92	3.34
3	$[90_4/0_4(F)0_4/90_4]$	17.8	2.23	3.05	2.43
4	$[0_2/90_2(F)90_2/0_2]$	4.06	3.16	---(2)	4.18
5	$[90_2/0_2(F)0_2/90_2]$	16.5	3.55	---(2)	---(2)

(1) Equation (3)
(2) Insufficient data for an estimate

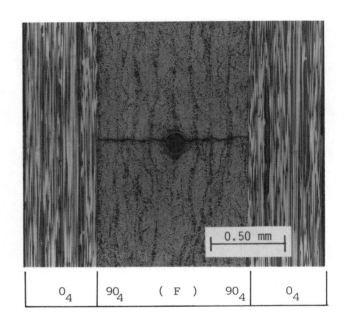

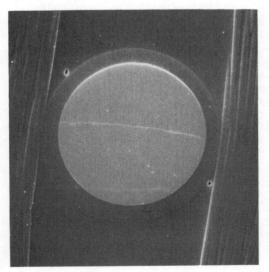

Fig. 13 - Optical and SEM images, respectively, of pre-existing crack in optical fiber, layup case 3.

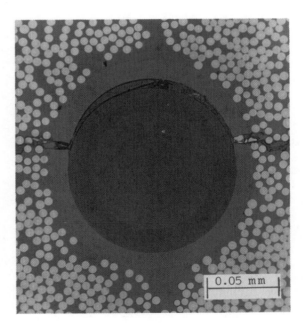

Fig. 14 - Two views of transverse ply failure in region of optical fiber, layup case 2.

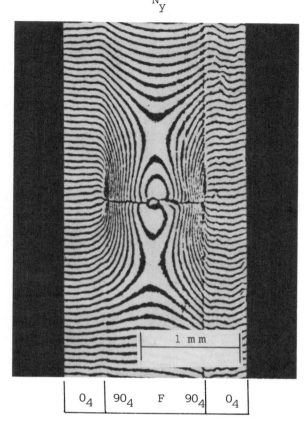

Fig. 15 - V-displacement fringes following transverse ply failure, layup case 2.

SUMMARY

Moire' interferometry and finite element strain analysis were obtained for an optical fiber in five graphite-epoxy laminates. The following are noted:
1. When the optical fiber is parallel to adjacent plys the laminate discontinuity is small. Strain concentration factors are moderate for both K_x and K_y, ranging generally from 2 to 5.
2. When the optical fiber is 90 degrees to adjacent plys the laminate discontinuity consists of a relatively large, elongated discontinuity. Strain concentration factors based on ε_y remain moderate but values based on ε_x are high, ranging from 12.3 to 17.8, according to FEM. These high values support a high value reported previously [8], determined from moire' interferometry. The present moire' data are insufficient to support these elevated values.
3. The greatest strain occurs in the coating of the optical fiber.
4. Pre-existing cracks in the optical fiber were noted occasionally, but only when the optical fiber was at 90 degrees to adjacent plys.

ACKNOWLEDGMENT

The financial support for this work was provided by the Center for Manufacturing Research and Technology Utilization, Tennessee Technological University and by Textron Aerostructures, Nashville, Tennessee. The authors gratefully acknowledge their combined support. Additionally, the authors gratefully acknowledge the contribution of Lee Wood of Textron, who provided essential fabrication of the test specimens. Professor Daniel Post of Virginia Tech provided extensive help and advice in the initial assembly of the moire' system. The authors are most grateful for his scholarship and generosity.

REFERENCES

1. I-DEAS™ Supertab Pre/Post Processing™ Engineering Analysis User's Guide," Structural Dynamics Research Corporation, Milford, Ohio, 1988.
2. D. Post, "Moire' Interferometry at VPI and SU," Experimental Mechanics, 23, 203-210 (1983).
3. D. Post, "Moire' Interferometry for Deformation and Strain Studies," Optical Engineering, 24, 663-667 (1985).
4. Czarnek, R., Post, D. and C. T. Herakovich, "Edge Effects in Composites by Moire' Interferometry," Experimental Techniques, 1, 18-21 (1983).
5. D. Post, "Moire' Interferometry for Damage Analysis of Composites," Experimental Techniques, 7, 17-20 (1983).
6. Post, D., Czarnek, R., Joh, D., and J. D. Wood "Deformation Measurements of Composite Multi-Span Beam Shear Specimens by Moire' Interferometry," NASA Contractor Report No. 3844 (1984).
7. Ruiz, C., Post, D., and Czarnek, R., "Moire' Interferometric Study at Dovetail Joints," J. Applied Mechanics, 109-114 (March 1985).
8. Czarnek, R., Guo, Y.F., Bennett, K. D., and R. O. Claus, "Interferometric Measurements of Strain Concentrations Induced by an Optical Fiber Embedded in a Fiber Reinforced Composite," SPIE, 986, Fiber Optic Smart Structures and Skins, 43-54 (1988).

HIGH PERFORMANCE COMPOSITES FOR THE AUTOMOTIVE INDUSTRY

Louis Dodyk
GenCorp Automotive
Marion, IN 46952 USA

INTRODUCTION

The use of plastics is widespread in today's automobiles. One cannot imagine a car or truck that does not contain any plastics material. With all the different materials available, choosing the right material is a very difficult decision. Add the different processes used to manufacture plastics parts and the decision becomes even more difficult. The development of Sheet Molding Compound (SMC) has allowed design engineers freedom to use reinforced plastics extensively for exterior parts such as the new G.M. Minivan, the Mack Truck Ultra Liner, the Fiero and of course the Corvette. In addition, Ford, Chrysler and other truck manufacturers make liberal use of SMC from grille opening panels in the front to taillight panels and liftgates in the rear.

Although SMC is used sporadically in the European auto industry, it has conquered an application which in this country has been traditionally steel or aluminum; bumpers! In 1972, Renault assembled the R5 with SMC bumpers. Audi, BMW, Mercedes, Peugeot, and Volkswagon utilize SMC bumpers in their production vehicles. However high performance composites for structural applications has seen only limited use in the U.S. automotive industry. GenCorp Automotive has developed an SMC specifically formulated for bumpers to meet the performance require- ments of a major European automobile manufacturer. Based on this success, GenCorp is now working with American automotive manufacturers to develop high performance structural SMC for use in the domestic market. It is anticipated that the use of high performance SMC will extend to other automotive structural applications as industry gains experience in designing and working with these materials.

SMC PROCESS

GenCorp Automotive's 7150 SMC is made on a conventional SMC machine. A paste containing the resin, filler, catalyst, mold release and thickening agents is uniformly metered onto the carrier film.

Fiberglass roving is chopped to 1" length and dropped randomly on the paste on the carrier film. Additionally continuous strands are added to the SMC, bypassing the chopper unit. The second carrier film, coated with a metered layer of resin paste, is applied on top of the layer of glass forming a sandwich of resin, glass fiber and resin.

The SMC then goes into the compactor unit where uniform pressure is applied to wet out the glass and produce a uniform SMC sheet weight and thickness. The SMC sheet is then packaged in boxes or on rolls as desired and then put into a climate controlled room for maturation to a specified viscosity prior to molding.

Once the molding process is defined for a particular material, it is very important that the material is made consistently to insure properly molded parts which exceed the customer's requirements.

PHYSICAL PROPERTIES

The mechanical properties of SMC are influenced by the quality of resin, the type and amount of fillers, the type and amount of glass fiber and the SMC manufacturing process. Table I shows a comparison of GenCorp Automotive's 7150 SMC versus high strength compounds reinforced with 50-65% random glass fibers. The ability to develop SMC systems with higher mechanical properties increases its competitiveness with steel and aluminum based on cost and weight. Through the combined use of continuous glass and chopped rovings, 7150 out performs SMC systems with higher glass loadings. Depending on the application, the amount of continuous roving can be altered to either increase or decrease properties. (Table II) These increases are obtained in the direction of the continuous rovings. Table III illustrates the high performance properties of 7150 as reported by Dow Chemical.

In a study conducted by Owens-Corning Fiberglas, they determined that to test bumpers or bumper systems one has to select the appropriate test method, such as barrier or pendulum. In a barrier test, the impacting forces are directly transferred through the bumper supports into the chassis when a straight bumper is used. This test is very useful to determine the performance of bumper systems: e.g., the function of shock-absorbers or energy-absorbing materials which are placed directly on the supports or on the bumper itself. In a pendulum test, the cross-member (bumper) between the supports is submitted to impact forces. This test determines the impact resistance of the bumper material and the stiffness of the bumper.

To test our bumpers, GenCorp Automotive uses a static bending test as a mandatory quality requirement. Equipment required include (Fig 1)
- Instron Testing machine with HP86 Data System
- Chart Recorder
- Test Fixture Anchor Plate
- Test Fixture I-Beam/Hinge Mounting Assembly
- Mounting Brackets and Bolts

TEST PROCEDURE - After calibration of the Instron
- Attach the bumper mounting brackets to the bumper with the attachment bolts.
- Attach the bumper to the test fixture with the long hex socket bolts and tighten.
- Lower the cross-head and pressure hammer with the rubber compression pad between the hammer and the bumper at centerline.
- Preload the bumper to 0.15 Kn. and reset the gage length to the 1 cm. mark on the chart.
- Start the test by pushing the down key on the control console. The Instron is now loading the bumper and the recorder should be recording the applied load and deflection.
- Take the bumper to total failure (Fig 2 and Fig 3) - Maximum load and maximum deflection.
- Stop the test at failure and reverse the cross-head to a predetermined set point.
- Remove the bumper and prepare for the next test.

RECORDING TEST DATA - Record the millimeters of deflection at 26 Kn. (5845 lbs.). Record the maximum load at failure and the maximum deflection at failure.
- Record all molding parameter data according to the Quality Assurance Program requirements.
- Figure 4 is a graphical explanation of force-vs-deflection.

EVALUATING THE DATA - Our customer has determined that to qualify the material and the process used to manufacture the bumpers, the deflection of the beam tested must fall within 25-33 mm at a load of 26 Kn. Table IV shows our 7150 SMC capable of meeting this specification statistically. Table V shows that this same material not only meets the requirements of 26 Kn. statistically but exceeds it significantly.

SUMMARY

The performance/cost ratio for SMC compared to steel and aluminum is greatly enhanced by the use of continuous glass fiber rovings. GenCorp's 7150 SMC was designed exclusively to meet the high performance requirements of bumper beam systems. Appropriate design considerations are necessary to utilize these composites for structural applications in the domestic automotive industry. Hopefully, the increased in strength and economics with lower glass content is interesting enough to consider future applications.

References:

JOURNAL ARTICLE -
OCF Status Report: SMC in Automotive Bumpers by Gerd Ehnert

JOURNAL ARTICLE -
OCF High Performance Fiberglas Reinforced Composite Bumpers
by Gerd Ehnert

Table I - Typical High Strength Molding Compound Properties

Formulation					GenCorp Automotive 7150 SMC
Glass %	65	60	55	50	50
Resin %	27.7	24.3	23.6	24.5	-
Filler %	7.3	15.7	21.4	25.5	-
Tensile Strength (MPa)	214	200	180	165	340
Tensile Modulus (MPa)	15,870	- -	- -	15,870	21,400
Flexural Strength (MPa)	407	384	362	333	600
Flexural Modulus (MPa)	15,180	15,663	15,732	15,387	20,240
% Elongation	1.5	1.5	1.5	1.5	4.0

Table II - Effect of Continuous Strands on Mechanical Properties

	C60R6	C25R25	C10R20
Ten. Str. (MPa)	380	340	150
Flex. Str. (MPa)	650	600	318
Flex. Mod. (MPa)	38,300	20,240	12,700

Table III - GenCorp Automotive 7150 SMC

Tensile Strength (ultimate, MPa)	407 ± 28
Tensile Strength (.2% offset yield, MPa)	307 ± 58
Youngs Modulus (MPa)	21854 ± 1072
Flexural Stength (MPa)	635 ± 19
Flexural Modulus (Tangent, MPa)	21305 ± 720
Barcol Hardness (934-1 Impressor)	57 - 60
Poisson's Ratio	.2738 ±.0165
Shear Strength (Iosipescu Method, MPa)	91.2 ± 5.5
Shear Modulus (Iosipescu Method, MPa)	4522 ± 227

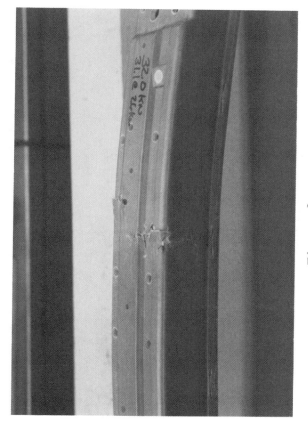

Figure 2

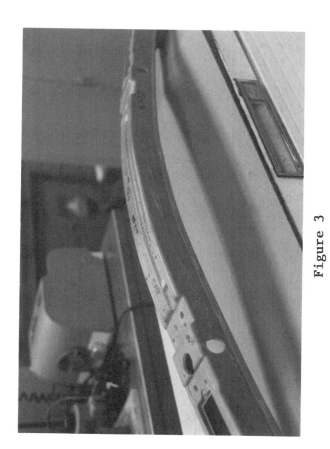

Figure 3

Figure 1

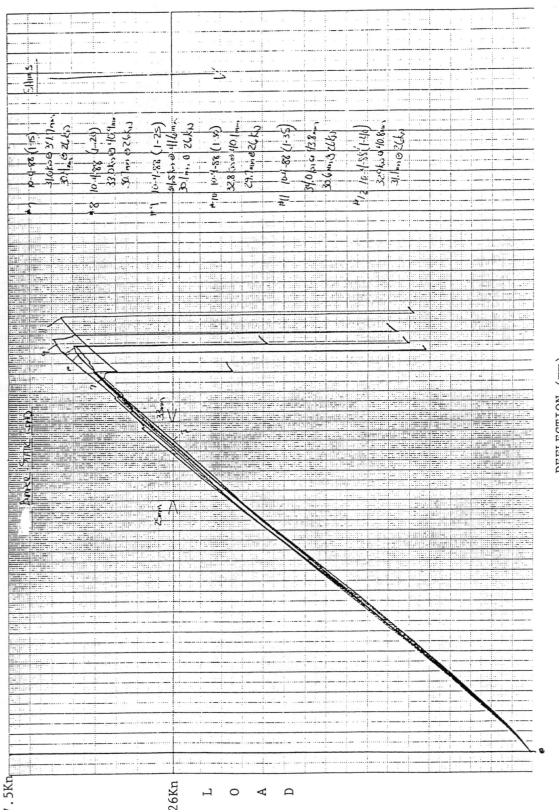

DEFLECTION (mm)

Figure 4

Table 4 - Deflection @ 26 Kn.

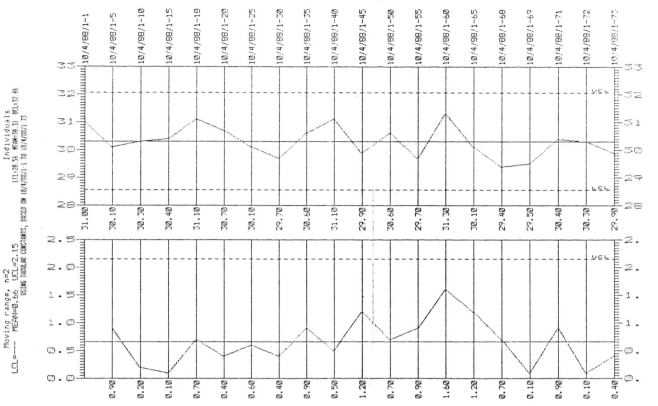

Table 5 - Peak Load

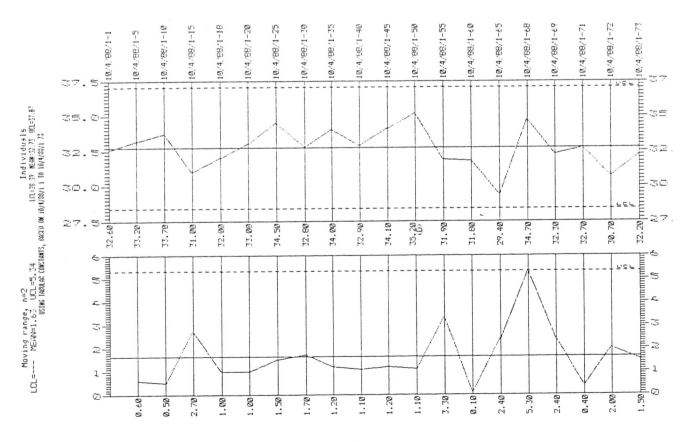

PROCESS DRIVEN DESIGN OF A PLASTIC BUMPER BEAM

Robert G. Dubensky
Chrysler Motors
Highland Park, MI, USA

Donald E. Jay
Chrysler Motors
New Mack Avenue Process
Detroit, MI, USA

Ralph K. Salansky
Creative Industries Group©
Auburn Hills, MI, USA

ABSTRACT

Plastic bumper beams are proposed to save vehicle weight. This work describes process driven design and the benefits of this technique. Its application to a generic bumper beam design is described in detail.

KEYWORDS

Plastics, Composites, Bumper, Beams, Design, CAE, Process, Process Driven

OBJECTIVE

The objective of this paper is to describe the steps in the process driven design of a plastic bumper beam. It is presented to provide information on this new engineering technique, along with the advantages and limitations. It includes the problem description, background material, and the needs for cost and weight saving. A list of bumper design requirements are included, both product and process.

THE MANUFACTURING PROCESS to be used for making an item of plastic should ideally be selected when the design and material are being determined. Design, material selection, and processing should be considered simultaneously because they are so closely interrelated. Making decisions on one of these factors without adequate consideration of the others will lead to unsatisfactory results.

In many cases, the engineer does not have the ability to choose freely from all of the design, material, and process alternatives. For example, the design is often heavily constrained by the need to fit an existing assembly, and the material and process may be largely determined by the need to use existing facilities. However, to optimize the results, the engineer should establish the extent of his design freedom early in the design process and should explore the design, material, and process alternatives within these bounds.

Before selecting the manufacturing process, the entire product processing should be considered, including secondary operations, such as painting and decorating. It is still common practice to pay little attention to these secondary operations until after the basic design, material, and process are finalized, despite the fact that poor execution of these secondary operations can often significantly affect the cost and performance of the application.

Other features to be considered will be the method of assembly, repair, and/or servicing. With plastic materials, the method of recycling must also be considered.

CURRENT MANUFACTURING TECHNIQUES

In the current bumper design procedure, the product and process design personnel examine the clay model in the design studio and obtain information about the front and rear fascia dimensions.

In fact, the real truth is "that the plastic processing expert may spread his hands in the horizontal and vertical direction, looks up at the ceiling with a pensive look and pronounces, that will take a 4,000 ton press." The analyst who was attending this clay model review said, "If you had an accurate estimate of the projected area, wouldn't a 3,500 ton press work?" The processing expert answered, "Yes, possibly!" The analyst asked, "How much money would be saved?" "Approximately $500,000 or $1,000 per ton of press capacity", replied the plastic expert. "How many presses will be used? One or two?," asked the analyst. "Eighteen for the rear only," replied the plastic expert. "That's real money," replied the analyst. This scenario provided the impetus to develop the techniques described in this paper.

With a knowledge of these outside dimensions, various bumper beam proposals are made (usually as design sketches), see figure 1 for a cross section sketch of one beam design. The proposed material is discussed and then a simple stress analysis is made using the cross section centerline data for the proposed material thickness. The material thickness is adjusted, and various factors are added until the stresses are within expected values. Figure 2 shows the stresses and the calculated deflection.

Many assumptions are made in this analysis procedure. With the press of day to day business, the total spectrum of new materials cannot be analyzed. Hence the designs tend to be of conventional types, while the competition could have designs in composites and plastics on the road today.

DISADVANTAGES OF THIS METHOD

A great deal of assumptions are required, and the results are not reported or documented, let alone stored in the computer. No accurate weights or volumes are calculated. This work (creating dimensions) to calculate these values must essentially be redone by process personnel, thereby losing the time invested by the original designer.

PROCESS DRIVEN DESIGN

TRANSITION FROM CURRENT TO SIMULTANEOUS ENGINEERING - Coming to grips with the subject is difficult because a clear definition or even a name that is accepted by all cannot be agreed upon. Some call it the "team approach," or "concurrent engineering," or "integrated engineering," or "life-cycle engineering," or "design for manufacturing," or . . .

One definition of simultaneous engineering is an approach to product development where everyone participates as early in the process as possible and gets involved in every step, including concept design.

This involves the process engineer, the manufacturing engineer, the product designer and development engineer, the financial analyst, and the product planners. Both marketing and sales personnel, the stylist, and senior management are all involved in the process right from the start. The old style over-the-wall "hand-off" described above now becomes an inter-disciplinary "reaching ahead" for information rather than waiting for it, and all individuals act on it together as an integrated team with a common focus.

The Japanese are known to be strong proponents and users of simultaneous engineering. An intriguing question is how they use these procedures. It is known that they stress cooperation, teamwork, and discipline. Reference 1 gives a list of the

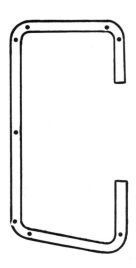

Figure 1 Cross Section of Generic Bumper Beam

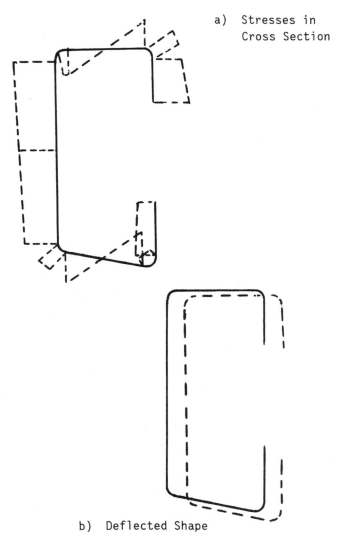

a) Stresses in Cross Section

b) Deflected Shape

Figure 2 Stresses and Deflections for a Bumper Beam

products designed by these methods ranging from cameras to personal computers and automobiles. The utilization of these techniques in Japanese industry is not the primary objective of this paper and will not be discussed in further detail.

This section will conclude with a short statement of the objectives of simultaneous engineering. These include a process to:

- reduce the time required to develop new products
- adhere to the product timing schedule
- improve manufacturing feasibility
- reduce the number of engineering changes
- place emphasis on early outside supplier involvement and development.

Meeting all of these objectives will help improve product quality and provide a low cost product that will truly lead to a "world class" product for the corporation.

PROCESS DRIVEN DESIGN TECHNIQUE - A natural evolution of simultaneous engineering is to modify it by bringing manufacturing processes as the first step in the new product development process. Naturally, the objectives are defined, and then manufacturing leads with the preferred process for constructing the product system and subsystems (often called modules), which will achieve the goals for quality, reliability, productivity, and cost. This process provides the basis for generating the product design guidelines.

Then product engineers, product designers, and stylists decide on the best ways to design the product and its components to take advantage of the process or processes. All of these individuals develop the product in a synergistic team approach with manufacturing using the process-driven guidelines.

INFORMATION FLOW - The driving force in all methods described is information. A key aspect in both simultaneous engineering and process-driven design is that this information is pushed to the beginning of the process. This can minimize change.

According to Henry Stoll (ITI), design is an interactive process; change is a part of it. "But we want to minimize the number of changes, and make them early on, when its nothing more than erasing a line on a piece of paper, not obsoleting a piece of equipment that has already been contracted." (Reference 1) Today the lines would be changed on a Computer-Aided-Design (CAD) system.

If information is the driving force behind this process, the flow of information is the basis for success of the process. Before any procedure, or portion of it, can be computerized, the detailed flow of information must be known and documented. The information flow has been studied in a small team of highly skilled product engineers, manufacturing engineers, procurement and supply experts, product planners, finance people, and other professionals that were teamed to develop new products via process driven design techniques described above. (Reference 2) This has been completed and documented in a previous publication (Reference 5) and will not be repeated. The key is to apply Computer-Aided-Engineering (CAE) procedures to this process.

CAE BASED PROCESS-DRIVEN DESIGN

The product designer needs to determine various items from the product design, including:
  a. Product Volume
  b. Product Weight
  c. Strength of the Product
  d. Rigidity of the Product
  e. Vibrational Characteristics of the Product

The process engineer usually is concerned with the following factors of the designed parts:
  a. Amount of material used to make the part (determines product cost).
  b. Surface area (to determine any coating cost).
  c. Projected part area (to determine the size of the presses required to form the part).
  d. Clearance between adjacent parts for assembly.
  e. Type of assembly process and fasteners to be used.
  f. Part inspection techniques and material handling process.

Additional items include the production rate, number of presses required, square footage required in the factory for these presses, and the cost associated with this equipment (i.e. capital equipment cost and capital facilities cost). Tooling questions also come up; such as, Will the material flow in the die cavity? and what is the tool temperature and life (especially if its subject to cyclic heating and cooling)?

All of this information can be determine by conventional methods. This process is long and involved, and with timing commitments, only a few of the many alternative design and material being generated can be analyzed. Thus, a number of important design options can be missed. This can be alleviated by using a CAE technique, i.e. solid modeling to define the initial part geometry at the concept design stage.

PLASTIC BUMPER BEAM - PROCESS DRIVEN DESIGN

The numerous bumper beam concepts were investigated by a group of individuals (all experts in their field of specialization, including product designers, process engineers, cost analyst personnel, analysts, and stylists). Constant interaction by these personnel would enhance the flow of information in the process-driven design cycle.

The stylist would determine the outer geometry of the fascia and the product designer would determine a preliminary beam section and energy absorbing system and attachments. These preliminary designs would be given to process engineer who would look at them and recommend changes to facilitate manufacturing and assembly of the bumper beam into the bumper system.

The analyst would then create a solid model of the proposed design. By using the solid modeler, which is a part of a vertically integrated CAE package, the material can be attached directly to this to obtain part weight from the part volumes. Section and inertial properties can be obtained from this model also.

Next a finite element analysis mesh is generated on this solid, and the analysis conducted to simulate the type of testing desired. All values are stored in the system. Projected area of this bumper beam is obtained and used to obtain the press capacity required to form this bumper beam. Table 1 shows the processing plan for a composite plastic beam. Table 2 lists the bumper design requirements.

All values are stored within a computer spread sheet and can be sorted by any parameter desired. These parameters will be detailed in another paper. The procedure has been developed and tested on five separate bumper beams. It is estimated that eight concepts can be analyzed within one week.

The advantages of this procedure are:
- All functions are considered early in the process (no over-the-wall hand off).
- Information available for everyone.
- Saves time (which is money).
- Reduces development time.
- Better documentation.
- Allow more alternatives to be considered.
- Results in optimum manufacturing.

SUMMARY AND FURTHER WORK

This paper has presented the aspects of process-driven design as implemented by CAE techniques. The techniques have been used to accelerate the process to complete a bumper beam concept analysis of a proposed design in approximately six hours. This will allow the investigation of new materials and design, never possible in the past. This will provide better designs via these CAE techniques.

Further work is required to incorporate some expert system programs to calculate the capital costs. Additional work will be required to integrate this model directly to a moldflow type analysis program.

REFERENCES

1) Vasilash, G. S., "Simultaneous Engineering - Management's New Competitiveness Tool", Production, July, 1987, pp. 36-41.
2) Hinckley, Jr., J. P., "Simultaneous Engineering - A New Approach", SAE, March, 1986.
3) Damitio, A., Davis, C., Lechner, J., Ziola, B., "Management of CAD Implementation in a Power Train Design Activity at Chrysler Corporation", SAE 830262, SAE, Warrendale, Pa., 15096, 1983.
4) Dubensky, R. G., "What Every Engineer Should Know About Finite Element Analysis Methods", SAE, Warrendale, Pa., 1986.
5) Dubensky, R. G., "CAE Aspects of Process-Driven Design", SAE Paper 871680, SAE, Warrendale, Pa., 1987.

Table 1 Processing Plan For Composite Plastic Material

- Supply plastic sheets to the fab line
- Cut raw material into blanks
- Place cut blanks into (3) stage infra red oven for preheating of material
- Place heated material into mold
- Close compression molding press
- Mold bumper beam
- Open mold and remove part
- Trim flash from part
- Cut holes in part
- Inspect

Table 2 Bumper Design Requirements

- Meet Styling Requirements
- Fit Into The Design Package
- Manage Energy Absorption
- Meet All Government(s) Requirements
- Meet Corporate Weights & Costs
- Meet Manufacturing Requirements
- Attempt To Meet Z-Axis (Vertical) Loading
- Be Completely Servicable
- Must Be Recyclable

DESIGN SYNTHESIS AND ASSESSMENT OF ENERGY MANAGEMENT IN A COMPOSITE FRONT END VEHICLE STRUCTURE

Robert L. Frutiger, S. Baskar
CPC Group/General Motors
Pontiac, MI USA

King H. Lo, Robert Farris
Shell Development Company
Houston, TX USA

ABSTRACT

Composite applications in automotive vehicle structure has been a topic of increased interest in the past few years. One of the major product issues that needs to be addressed prior to the implementation of this technology is energy management in a composite vehicle structure. In this paper, the design synthesis and assessment of a composite front structure to demonstrate stable, progressive crush is described.

The design synthesis approach is used to design and analyze a fully composite front structure, based on packaging constraints representative of a production vehicle, to assess 0 and 30 degree front barrier performance. Issues discussed include the design and analysis, design features, composite material tailoring, prototype fabrication and component and vehicle crush tests. An RTM process with glass fiber reinforcement and foam cores was used to make the prototype structure. Static crush tests on various components of the structure were successfully conducted. Both static and dynamic barrier tests were performed on a trimmed out front structure. This paper summarizes the design synthesis process used to engineer the front structure to achieve a stable crush performance, component and barrier test results and remaining technical areas that need to be addressed prior to application of this technology to a vehicle program.

IN TODAY'S AUTOMOTIVE DESIGN ENVIRONMENT, composite exterior panels are competitive from a product performance point of view and also from a cost point of view for medium to low program volumes. These panels are designed to meet local stiffness and strength requirements in addition to class A surface appearance requirements. Semi-structural load bearing panels for interior non-appearance applications are also being considered (1).

The use of composites in load bearing primary structures represents a significant challenge. Recent studies have used structural composites to achieve major parts consolidation in van crossmembers (2,3). Part performance and durability were also demonstrated. A hybrid vehicle structure, consisting of a composite passenger module on a steel frame, has also been proposed (4). In this concept composites were used in the noncrush portion of the structure, with steel in the front and rear crush zones.

A major product issue that must be addressed before composites can be used in a total vehicle structure is energy management. The vehicle must be capable of meeting stringent safety objectives, which places severe performance requirements on the structure. Some composites have been shown to be more efficient per unit mass in absorbing energy than steel in simple components under axial loading (5,6). Applying these concepts to the complex geometry and loading in a vehicle structure, however, is a major challenge.

Some progress has been made in this area. The impact performance of a Kevlar/epoxy filament wound space frame structure has been demonstrated (7). A conventional steel front structure of a Ford Escort was replaced by a glass/vinylester composite made by a modified resin transfer molding process (8). Stable progressive crush was reported for both 0 and 30 degree dynamic tests (9).

The objective of this work was to use design synthesis to engineer a glass fiber reinforced composite front structure to achieve a stable, progressive crush for realistic vehicle packaging constraints. A mainstream, mid-size vehicle packaging envelope was choosen as the design base. The parts were made by a prototype version of the Structural Reaction Injection Molding (SRIM) process. Frontal barrier impacts were performed to assess the crashworthiness of the composite for both 0 and 30 degree impacts.

The scope of this program was limited to energy management. Other design requirements such as in-service road input, fatigue, abuse loads, environmental effects, damage assessment and repairability were not considered. These requirements would be met in later design iterations.

This program was part of a joint development effort between CPC Advanced Vehicle Engineering and Shell Development Company. A synthesis team was established to perform the analysis, design, prototype manufacture, and part assessment.

DESIGN SYNTHESIS

A synthesis team consisting of GM/Advanced Vehicle Engineering (AVE) and Shell Development Company was formed to accomplish the design, analysis, and prototype manufacture of the composite front end vehicle structure. Vehicle design and packaging requirements, finite element analysis, material tailoring, and processing input were combined to develop the feasible design.

DESIGN CRITERIA - For the given vehicle packaging constraints, the composite front end structure should meet the following requirements:

1. Provide progressive crush in a controlled manner to absorb the impact energy for the given crush loading experienced during both the 0 and 30 degree impacts;

2. The collapsed structure should retain the backup structure integrity during 0 and 30 degs. impacts, i.e., the backup structure should bear the simultaneous axial and bending loadings without catastrophic failure; and

3. The rear cradle joint mount should remain integral with the structure.

The design is mainly to assess the crashworthiness of the composites for energy management. The other types of loadings such as service, fatigue and abuse loadings were not accounted in this design. Further, other design issues of long and short term environmental effects, damagability, and repairability were not considered in this early part of the prototype design.

SYNTHESIS APPROACH - Based on structural RIM composites, a preliminary design was developed to establish the structural configuration. This initial design was governed by the vehicle packaging constraints, structural dimensions and impact loadings. The impact forces transmitted through the upper and lower paths during 0 degree collision at 30mph provided the design requirements for the upper and mid rail cross-sections.

FEM analysis was used to lead the design. The front structure was first divided into crush and non-crush portions as shown in Figure 1. The rear portion of the structure from the shocktowers rearward was designed to carry the loads developed during the crush without failure. This included the rear cradle attachments which carried significant loads. The front crush portion of the upper and lower rails were designed empirically by tube and rail crush tests.

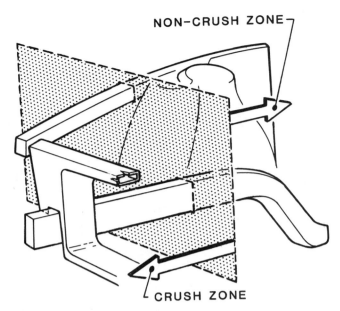

Fig. 1 - Crush and non crush portion of front structure

The initial design of the noncrush portion of the structure was determined by applying the barrier loads to the upper and lower rails and modeling the stresses developed in the composite. The gages and geometry were selected, based upon the packaging constraints of the structure, to keep the stresses below failure. Initial analysis was performed for an all random mat material. In locations where stresses were high, and large gages would be required, other glass fiber preforms were used in combination with the random material, such as woven rovings, biaxial and triaxial fabrics.

Many glass reinforced composite tubes were investigated for various parameters such as, geometry effect, laminate configuration, fiber orientation angle, and different weave pattern of knit and braid. Under the design envelope, these studies revealed that tube cross-sections of tapered thickness with 90 deg. fiber wrapped outside yielded more energy absorption than the steel design. A typical curve of the load developed during crush is shown in Fig. 2 for a tapered glass vinylester rectangular tube. This demonstrates the stable crush mode and efficient energy absorption possible with composite tubes.

DESIGN FEATURES - Several interesting design features were developed as a result of the synthesis process. These are described below.

Mid and Upper Rails - In order to have a progressive and controlled crush, the rails were designed with tubular cross-section and tapered wall thickness. The rails were fabricated with the different stacking sequence of various glass cloths according to the thickness in the rails. Foam cores were used as mandrels around which the fiber preforms were wrapped. The foam core was removed in the crush portion of the rails because crush tube tests showed that the foam core caused an instability in the crush mode of the tube wall and reduced the energy absorption efficiency.

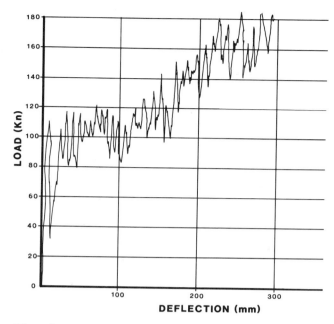

Fig. 2 - Static crush load - deformation on a tappered glass vinylester rectangular tube.

Shear Panel - The shear panels as shown in Fig. 3 were used to tie both rails. They served as flanges and provided stability for the rails. Six layers of 0/90 glass cloth were used in the design.

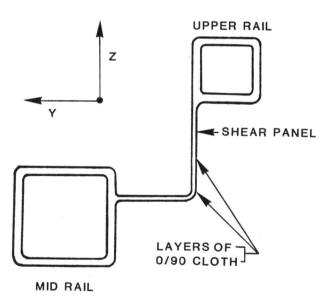

Fig. 3 - Shear panel

Shock Tower Double Wall - The shock tower structurally integrated the upper and mid rails and provided lateral stiffness for the corner impact. A double walled construction (Figure 4) was used to enhance the overall stiffness and strength of the backup structure for the case of 30 deg. corner impact. A shear strip bridged the thickness variations between the rail and the wall and helped in the transition of stresses. The woven rovings cloth was wrapped around the core to increase the structural stiffness in the shock tower region.

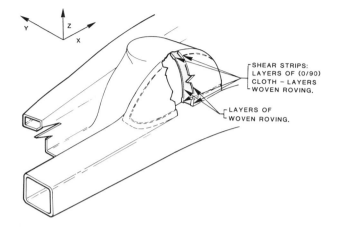

Fig. 4 - Double walled shock tower region

S Bend in the Mid Rail - The S bend of the mid rail was reinforced to 5mm thick to avert collapse of the backup structure subjected to out of plane bending. Several analytical iterations were performed to evaluate the fiber orientation and thickness of the composite in that region. Considering the structural and the manufacturing requirements, various layers of glass reinforcements such as, biaxial, triaxial and woven rovings cloths and were used.

Rear Cradle Attachment - The rear cradle attachment mount should withstand the axial force and the moment due to the offset of the crush load path through the cradle and remain attached after barrier impact. The steel design showed significant reinforcements. The composite plate was also stiffened by steel reinforcements. A sandwich construction consisting of an upper and lower steel plate with the composite between was used in the design. In order to enhance the integrity of the cradle mount and remain attached, the part of the bottom steel reinforcement plate in the above sandwich construction was extended to connect the inboard of the mid rail. The composite layer in the middle of the sandwich construction was designed with woven rovings and triaxial cloths.

MATERIALS

The composite front end vehicle structure consisted of various glass reinforced laminates. Several preforms such as random mat, rovings cloth, biaxial and triaxial cloths were used. The triaxial fabric is made of three unidirectional layers with the orientation, $0_2/-45/+45$, and provides the best combination properties. A low viscosity vinylester resin, Ashland Hetron D1079, was selected because it provided low injection pressure, and higher temperature resistance when cured. The various glass reinforced composite properties are shown in Table 1.

Table 1. Various Glass Reinforced Composite Properties:

Volume of Fiber	Glass Reinf.	E11 N/Cm² 10⁵	E22 N/Cm² 10⁵	G12 N/Cm² 10⁵	Nu12	ST N/Cm² 10³	SC N/Cm² 10³
0.4	Rovings cloth	19.31	19.31	2.41	0.125	42.75	29.70
0.3	Random Mat	11.58	11.58	4.27	0.341	27.59	20.00
0.4	Random Mat	14.68	14.68	5.52	0.335	33.10	22.75
0.4	Triaxial Cloth	19.66	9.24	5.51	0.246	44.83	31.03

Biaxial cloth properties are the same as rovings (0/90) cloth with orientation of 45 deg.

Density (assumed) = 1.8×10^{-5} N-S²/Cm⁴

The foam cores were used primarily as tooling aids to form the three dimensional shape of the part. Various glass reinforcements and core were assembled together as preforms for molding the composite parts. The cores were molded separately from rigid polyurethane foam. A structural adhesive, Lord Fusor 320, was used to join the steel reinforcements and the composites. In addition, some mechanical fasteners were used to increase the failure strength in the highly loaded joints.

COMPOSITE MATERIAL TAILORING

Several analyses such as force capacity, composite material tailoring and rear cradle attachment were conducted to achieve a successful design.

Force Capacity - Initially the structure was assumed to be made of all random composite (Table 2). The objective of this analysis was to determine a graduated force capacity at various locations of the structure for impact loads at 0 and 30 degs. This enabled the cross-sections and gages to be determined so that the load bearing capacity increased along the car (x) direction. This aided in obtaining a progressive crush. Several levels of calculation were performed using Nastran to arrive at the thickness of the cross sections in the various structural components. The force capacity curves for several design iterations are shown in Fig. 5. The curve marked level 4 gave the best performance. The dip in the curve at the S bend location (202-212 cm) is due to reduced force carrying capacity in the midrail due to out of plane loading.

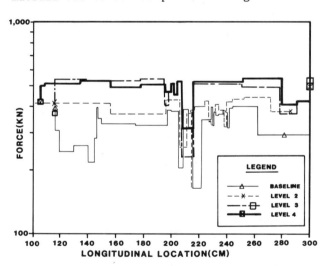

Fig. 5 - Force capacity analysis

Table 2. Random Composite Properties (Isotropic):

E (Modulus) = 0.8276×10^6 N/Cm²

S (Tensile Strength) = 18620 N/Cm²

Nu (Poisson's Ratio) = 0.3

Rho (Density) = 1.55×10^{-5} N-S²/Cm⁴

V (Fiber Volume) = 0.40

Composite Material Tailoring - After the preliminary design, a detailed finite element analysis was performed using the FE2000 code (Numerics Corp.). The FEM analysis derived the specific geometry and gages and was mainly used to identify the critical stress levels in the non-crush portion to avoid failure during impact at 0 and 30 degs. Several candidate alternate materials and laminate constructions were considered. The laminate plies were modeled as composite shell elements with orthotropic material properties, as given in Table 1. Various material tailoring such as stacking sequence and orientations of plies ($\pm\theta/\pm\theta$) had been analyzed and suggestions were made for the final design. Figs. 6 and 7 show the stress levels due to 100 KN at 0 deg. and 80 KN at 30 deg. applied on mid rail only.

The Von-Mises' stress levels in Fig. 6 for 100 KN impact load at 0 deg. showed all the structural components below the design stress level, which was taken to be 12000 N/Cm^2 in the non-crush portion. The Von-Mises' stress levels for 80 KN impact load at 30 deg. were well below the safe stress level (Fig. 7) at the backup structure and was ensured avoidance of potential bending failure in the rails. All stress components were examined during the analysis, although only the Von-Mises' stress plots are shown here.

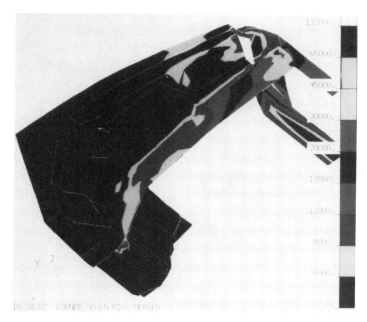

Fig. 7 - Von Mises' stress plot for 80 kn force at 30 degree corner

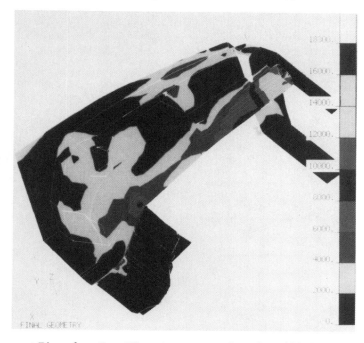

Fig. 6 - Von Mises' stress plot for 100 kn force at 0 degree axial

Rear Cradle Attachment - The rear cradle attachment was designed as a 3 layer construction and the composite was sandwiched as a core in between the steel plates. The steel plates had taken the major distribution of stresses to avoid the bearing load on the composites. Several iterative studies were performed to optimize the plate thickness and width. This sandwich design concept was verified on a test plaque with cradle attachments. The rear cradle joint was analyzed for an axial load of 70 kn and Von-Mises' stress levels shown in Fig. 8. The maximum stresses were found to be above the safe stress level (steel) 30000 N/Cm^2 and occurred at the small portion of the inner(top) steel plate. Since, the stresses were calculated using static linear analysis the results will be conservative.

BARRIER SIMULATION

The barrier impact analysis was performed to evaluate the vehicle's structural crashworthiness. The general purpose program, Language for Structural Dynamics (LSD), lumped mass and

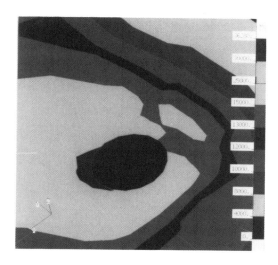

Fig. 8 - Von Mises' stress plot for 70 kn cradle load

nonlinear spring model, shown in Fig. 9 was used to simulate the vehicle barrier performance for 0 deg. impact at 30 mph and guide the design. The empirical static force deformation characteristics of the crush rails (design curves), packaging clearances, and concentrated masses of the vehicle structural components were used as inputs to the simulation model. The LSD program provided outputs of force deformation plots, and time history plots of force, velocity and acceleration.

LSD predicted the peak deceleration, calculated over the most severe 20 millisec. span of the velocity vs time plot, for the composite baseline design. This was then compared to the test values achieved from the steel design and used to guide the analysis and design process.

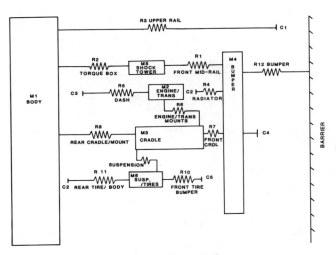

Fig. 9 - LSD model

PROTOTYPE MANUFACTURING

THE STRUCTURAL RIM PROCESS - A prototype version of the Structural Reaction Injection Molding (SRIM) process was used to fabricate the composite front structure. This process allows the fabrication of large, integrated high-performance structures at low to moderate pressures. Fig.10 illustrates the SRIM process as applied to a structural floorpan concept. Essentially, a dry fiber sheet consisting of multiple layers of random and/or oriented fabrics are made into three dimensional preforms. Light weight cores, such as rigid urethane or balsa wood, are often used as an integral part of the preforms. The dry preform is loaded into a heated mold and resin is injected as soon as the mold is closed and sealed. The resin flows through the preform and subsequently cures to form a rigid, load bearing structure. As soon as the cure is complete the component is removed from the mold for post-cure (if required), trimming and secondary machining. High Speed Resin Transfer Molding (HSRTM) is another name for this process (3).

FOAM CORES - A total of ten different foam cores were molded for each composite front structure, Fig. 11. With the exception of the wheelhouse construction, the cores were primarily used as tooling aids to locate metal inserts and achieve the three dimensional shape of the structure. The cores were molded in two piece epoxy molds using rigid urethane foam blown to a density of five to ten pounds per cubic foot. Metal inserts were used for the two rear engine cradle attachments and were placed in the cross bar located beneath the toepan and mid rail foam cores.

A two part foam system was injected into closed molds which were held together by simple perimeter clamps. Cream time for the foam was approximately 90 seconds. Total cycle time for the prototype foam varied from 20 minutes to one hour depending primarily on part thickness. Thicker parts required longer cool down times to minimize dimensional instability.

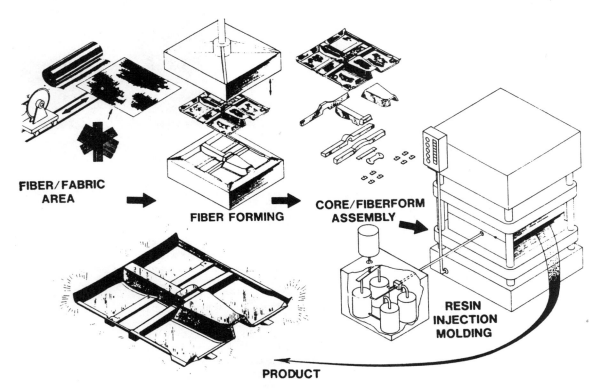

Fig. 10 - The high speed resin transfer process

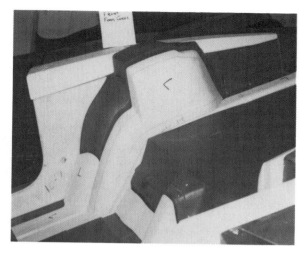

Fig. 11 - Left side foam cores located in bottom mold half

PREFORM - The preform necessary to meet the structural and crush requirements for the front third of a body-in-white is relatively complex. Upper rails, mid rails, shock towers, A pillars, #1 bar, etc, had to be constructed and integrated into one large preform. Chopped strand random, woven roving, woven cloth, non-woven biaxial and non-woven triaxial E-glass fabrics were used in the preform.

The main structures of interest, the upper and mid rails in the energy management portion of the structure had the typical construction shown in Fig. 12. The main construction consisted of varying numbers of 0/90 deg glass fabrics wrapped around the rails (oriented with respect to the major axis of the rails) which were connected by a thin shear panel. This shear panel could have been made with random mat (as shown in Fig 12), however, light weight glass fabric was used to facilitate prototype fabrication. This design facilitates anticipated mass production techniques in the future. For example, the beams could be made by a continuous pultrusion/winding process and the shear panel could be made by either stamping formable random mat or automated spray up.

A "cut and sew" technique was used to form the prototype front structure preform. Patterns for orienting and cutting the glass materials to conform to the shape of the part were developed using trial and error manual methods. Local patches were used at high stress locations and were major subsections were combined. For example, extra layers of woven roving were added to the #1 bar-mid rail joint at the rear cradle attachments. The preform was constructed by fabricating subassemblies of major components such as rails, #1 bar, shock towers, etc., then combining these components into larger subassemblies and finally combining these assemblies into the final preform, Figs. 13-14. Glass materials were pulled tightly over the foam cores and stapled into place.

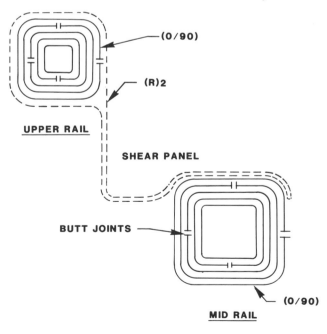

Fig. 12 - Typical preform design in energy management portion of front structure

Fig. 13- Left midrail preform

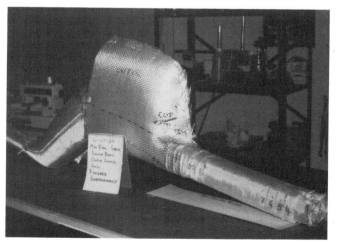

Fig. 14 - Left midrail shock tower preform assembly

Figures 15 and 16 show the finished preform in the bottom mold half.

Fig. 15 - Completed left side preform

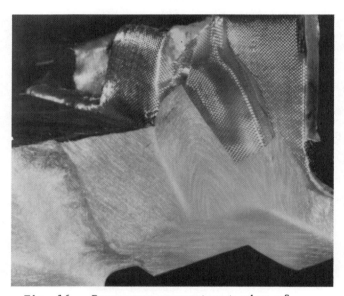

Fig. 16 - Passenger compartment view of completed front structure preform

COMPOSITE FRONT STRUCTURE MOLDING PROCESS - A large two piece high temperature epoxy mold with an "o" ring seal was used to mold the one piece composite front structure. After the preform was loaded into the bottom mold half, vent tubes and injection lines were prepared and the mold was closed. The part was oriented at approximately 13 deg from vertical to allow the front tips of the upper rails to be the highest locations of the part (except for the shock towers) for venting purposes. Resin was then injected into the room temperature mold and the vent tubes were successively closed as air and volatiles were vented and pure resin began to flow through them. Resin was injected at the center of the #1 bar through the upper mold half. Four vents were located along both sides of the part at the parting line, Fig. 17. Injection times ranged from 15 to 30 minutes. The part was allowed to cure in the mold for at least three hours before removing it from the press. After the part was removed from the mold, trimmed and prepared, it was post cured at 120 deg C for 25 minutes.

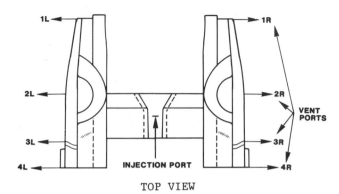

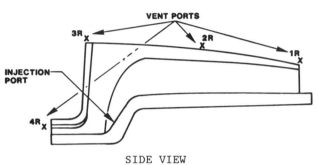

Fig. 17 - Injection and vent locations

The working surface of the mold was reinforced to withstand injection pressures as high as 150 psig (1.03 MPa). Normal injection pressures were not directly measured but were calculated to average between 15 and 35 psig (0.10 - 0.24 MPa) with a local area around the gate possibly seeing a maximum of 100 to 140 psig (0.69 - 0.96 MPa). Since the objective of this program was to design for energy management and not mass production, gate and vent designs were not optimized for maximum injection rates.

Figs. 18 and 19 show the completed composite front structure of front, and rear view, respectively. Final part weight was 150 pounds (68 Kg), approximately 25 pounds (11.3 Kg) of foam and metal inserts and 125 pounds (56.75 Kg) of resin and fiberglass. Fiber content varied from 45% wt to 65% wt depending on location and preform design.

VEHICLE BUILD

A prototype coupe was modified to install the composite front end structure. Items such as the bumper, radiator, engine, battery, front suspensions, tire and wheel, shock strut, and coil spring were removed from the steel front end and kept for reinstallation for the composite front end structure. The engine and suspension components were assembled separately and mounted at the cradle and shock tower locations of the front end.

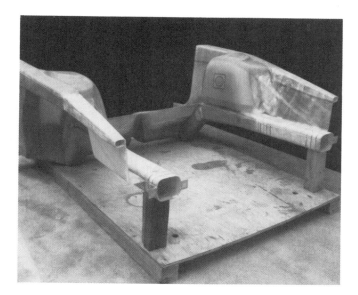

Fig. 18 - Completed composite front structure - front view

Fig. 19 - Completed composit front structure -rear view

The bumper was bolted to the mid rails and the ends of the radiator support were fastened and bonded to the composite structure. The preassembled composite front end was mated to the steel vehicle at the A pillar and rocker locations using specially fabricated fixtures. Lord Fusor 320 adhesive was used to bond the steel and the composite. In addition to the adhesive bonding, mechanical fasteners were used in the joints of the toe pan to the steel floor pan, as well as A-pillar and rocker joints to the steel fixtures. Fig. 20 shows the assembled view of the fully trimmed composite front end structure.

CRASHWORTHINESS VALIDATION PROGRAM

Various tests as shown in Table. 3 for the composite front end structure were conducted to verify the progressive crush of the rails for axial and corner impacts under static and dynamic loads. Further, these test results served as a validation of the design synthesis approach and verification of the analyses predictions.

Fig. 20 - Fully trimmed composite front end structure

Table 3. Composite Front End Crush Tests Program:

1. Component Crush Tests:
    a. Upper Rail
    b. Mid Rail

2. System Interaction Tests:
    a. Rear Cradle Attachment
    b. Back Up Structure at 0 Degree Load
    c. Back Up Structure at 30 Degree Load
    d. Complete Driver Side Static Crush

3. Dynamic Barrier Test:
    a. Complete Structure With Trim at 0 Degree

COMPONENT CRUSH TESTS - Crush tests were performed on the upper and mid rails cut from a prototype front structure. These tests were conducted to verify the progressive crush of the rails and the load carrying capability of the structure.

Both mid and upper rails were tested statically and the force-deflection curves were compared with the design load for the energy absorption characteristics. The agreement of these static crush results with the design curves was generally good, with some deviation due to inaccuracies in the molds and foam cores.

The cradle attachment was tested statically to insure that it could carry the crush load and remain attached to the #1 cross bar. Fig. 21 shows the force - deflection characteristics of the cradle attachment and the design load. The test ran successfully without any fractures in the composite and proved the attachment area could sustain the cradle reaction load and remain attached.

SYSTEM INTERACTION TEST - Backup structure stiffness tests for 0 and 30 degrees and a system crush test were performed on the static crusher.

Stiffness tests were conducted on the passenger side mid rail to verify that the rail can react the lower crush load without any failure in the backup structure. A proof load (mean crush load) of 100 KN was applied in the 0 degree direction. Due to bending of the test fixture, a lower proof load was acheived for the 30 degree angle test. No failure was observed. This also gave the stiffness of the reaction structure for LSD barrier simulation model to predict the deceleration pulse.

The system test was performed on the driver side of the front structure to study the interaction of the individual components such as cradle bolt, mid rail and upper rail. The structure was loaded slowly in the static crusher up to 500 mm. Fig. 22 compares the total force - deflection curves of the test and design. The good correlation proved the importance of the design synthesis to achieve a successful design. Further, this static test showed the interaction of the system and the load carrying capability of the reaction structure.

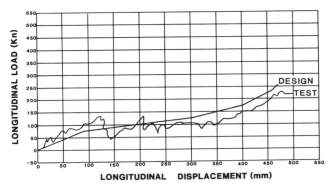

Fig. 22 - Total force vs. deflection on the static system interaction test

DYNAMIC BARRIER TEST - The dynamic test was performed on the fully trimmed vehicle fitted with the composite front end structure. The test was conducted at 30 mph at 0 degree frontal impact. The hood and front end sheet metal were removed to view the crushing behavior of the individual components. Load cells were attached on the rocker, the A-pillar, and door beam to verify the load path in the composite structure. Due to the modifications required to mount the composite to the steel structure, the windshield could not be mounted. The 50% male dummy was placed on the driver side and restrained to evaluate the occupant behavior. Vehicle and occupant data were recorded during the barrier test.

The total barrier load - rocker panel displacement is shown in Fig. 23. The deceleration pulse was calculated and correlated well with the predicted value. The

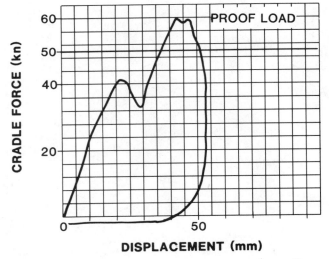

Fig. 21 - Force vs. displacement of cradle attachment

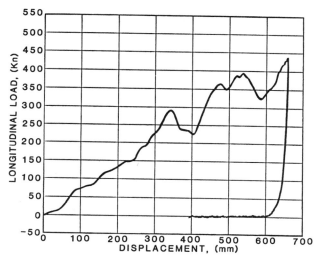

Fig. 23 - Total barrier load vs. displacement for 0 degree impact at 30 mph

rails were crushed progressively and yielded good energy absorption. However, the passenger side rails suffered some degradation of performance due to the interference of the radiator support. The support failed to break away as intended and caused local buckling of the rails. However, a mean crush load was still maintained. The dummy data met the federal requirements and the force levels showed that the composite front end structure can be crushed progressively. Finally, the test results served as a validation of the design synthesis approach and design verification of crashworthiness of the composite structure.

SUMMARY AND CONCLUSIONS

A structural composite vehicle front end has been designed, prototyped and evaluated for energy management as a joint effort between CPC Advanced Vehicle Engineering and Shell Development Company. Design synthesis and analysis was used by the team to design the geometry and composite material to achieve crush performance for axial (0 degree) and corner (30 degree) impact loads. The composite front structure was molded by the resin transfer molding process, using low cost epoxy tooling and E glass fabrics and polyurethane cores.

The synthesis approach, which took into account the performance requirements, design, material and manufacturing, has contributed to the confidence in being able to design structural composites for automotive applications. The static and dynamic component and trimmed vehicle crush tests have given encouraging results. It appears that with correct engineering and design synthesis, composites can be successfully utilized in vehicle energy management applications.

Additional development work is required in addressing issues of durability, environmental effects, and attachments. The area of manufacturing, and low cost glass preforming methods in particular, will require additional development to make these applications cost effective for the automotive industry.

ACKNOWLEDGEMENTS

The authors would like to thank the team members from CPC and Shell who contributed to the success of this technology assessment project. The technical contributions from the CPC engineering team, including S. Abbasspour, M. Bhatti, S. Longo, S. Niks, and H. Weiss, are greatly appreciated. Special thanks is due Steve Longo for the coordination of the design and tooling stage of the project. Also, the support of Robert Cavalier (Shell/Ardyne) is appreciated.

REFERENCES

1. Chavka, N.G. and C.F. Johnson, "A Composite Rear Floor Pan", Proc. of the 40th Annual Conf., Reinforced Plastics, CI/SPI, 1985, Session 14-D.
2. Farris, R.D., "Composite Front Crossmember for the Chrysler T-115 Mini-Van", Proc. of the Third Annual Conf. on Advanced Composites, ASM/ESD, Sept. 15-17, 1987, Detroit, Michigan, pp. 63-73.
3. Johnson, C.F., N.G. Chavka, R.A. Jeryan, C.J. Morris, and D.A. Babbington, "Design and Fabrication of a HSRIM Crossmember Module", Proc. of the Third Annual Conf. on Advanced Composites, ASM/ESD, Sept. 15-17, 1987, Detroit, Michigan, pp. 197-217.
4. Maier, E.K., T.T. Pierce, R.B. Freeman, and H.A. Jahnle, "Composites Intensive Vehicle: Rationale, Design, and Prototype Build", Proc. of the Third Annual Conf. on Advanced Composites, ASM/ESD, Sept. 15-17, 1987, Detroit, Michigan.
5. Thornton, P.H., J. Comp. Mat. 13, 247-62 (1979).
6. Farley, G.L., J. Comp. Mat. 17, 267-79 (1983).
7. Schmueser, D.W., L.E. Wickliffe, and G.T. Mase, "Front Impact Evaluation of Primary Structural Components of a Composite Space Frame", Proc. of the Seventh Intr. Conf. on Vehicle Structural Mechanics, SAE, April 11-13, 1988, Detroit, Michigan, pp. 67-75.
8. Johnson, C.F., N.G. Chavka, and R.A. Jeryan, "Resin Transfer Molding of Complex Automotive Structures", Proc. of the 41st Annual Conf., Reinforced Plastics, CI/SPI, Jan. 27-31, 1986, Atlanta, Georgia, Session 12-A, pp. 1-7.
9. Thornton, P.H. and R.A. Jeryan, "Composite Structures for Automotive Energy Management", Proc. of the Third Annual Conf. on Advanced Composites, ASM/ESD, Sept. 15-17, 1987, Detroit, Michigan, pp. 75-82.

A CAE METHODOLOGY FOR PLASTIC COMPONENT DESIGN

Robert G. Dubensky
Chrysler Motors
Highland Park, MI, USA

ABSTRACT

Plastic parts are proposed to save vehicle weight and cost. This work describes a CAE methodology for plastic component design utilizing the solid model of the concept part as the starting point. The application is general, rapid, and allows the investigation of a greater number of designs at the early concept stage.

KEYWORDS

Plastics, composites, computer-aided-engineering, CAE, design, design methodology.

PLASTIC PRODUCT creation and utilization is increasing in the automotive industry. (References 1 - 7)* Weight and cost savings provide the largest motivating factor. An additional factor is the ability with plastics to integrate a number of parts into a single unit. This part integration is an important feature and must be studied early in the concept stage for proper application. It cannot be used effectively, at the last minute, to substitute one plastic part for a multiplicity of other parts of different metallic materials. This technique must be studies at the earliest portion of the concept design stage.

No easy method exists to consider alternative concepts at the early concept stage and provide quantitative values upon which to make effective product and process decisions which will affect the product performance and profitability during the life of the product. The British Aerospace Industry has studied the design and production cost of the process and have shown that 80% of the product and capitol costs are committed over the life of the product, within the first 5% of the product design cycle. (Reference 9)

* Numbers in parenthesis refer to references.

Computer-aided-engineering (CAE) techniques have assisted in the product design process for metal part design. Their use is now expanding to plastic part design. A CAE methodology has been developed using the vertical integrated CAE capabilities. This is based upon creating the initial part as a solid model. Volume, weights, and other values can be determined directly and used in the concept studies of plastic parts. This type of tool permits asking and answering a series of "what if the parts were longer ..., thicker ..., deeper ...," or "what if it was molded from material A ..., molded from material B ..., etc.".

OBJECTIVE

A methodology has been developed which provides quantitative values to product and process variables (such as weight and material cost) for plastic components. This can be accomplished rapidly and allows the expanded study of more concept very early in the concept design cycle. The objective of this paper is to describe a CAE methodology (based on solid modeling) which gives rapid quantitative values to product and process variables for plastic part design.

SOLIDS AS THE STARTING POINT FOR CAE

Process-driven design techniques will have to be used for proper plastic product design at the concept design stages. At this stage, various alternatives are proposed and evaluated for hands-off to other stages in the design process. This section describes concept design and solid modeling.

WHAT IS CONCEPT DESIGN? - Concept design, as used in this paper, means the initial creation of a product idea and/or geometric arrangement of materials and systems to achieve the desired product goal. How the request for the product idea originated is beyond the scope of this work and is contained in books on industrial design and product development.

The flow of information in this concept stage is shown in Figure 1. Not shown on this figure are the countless interations (or repetitive loops required to meet all of the product goals. It may be slightly different for various industries, but the flow of information is similar for most industries.

The two main aspects of concept design for most products are styling and spatial allocation, referred to as "packaging" in the automotive industry. The method of accomplishing these tasks today is the traditional one of the artist and stylist preparation of art work or "renderings", then the creation of a physical model (ranging from paper, clay, wood, plaster, and plastic model), selection of a final design, and then the preparation of a prototype part. The newest technique is the computer workstation with a color, solid model representation of the proposed part.

In the current bumper and bumper beam concept design procedure, the product and process design personnel examine the clay model in the design studio and obtain information about the front and rear fascia dimensions.

In fact, the real truth is "that the plastic processing expert spreads his hands in the horizontal and vertical direction, looks up at the ceiling with a pensive look and says that will take a 4,000 ton press." The analyst who was attending this clay model review said, "If you had an accurate estimate of the projected area, wouldn't a 3,500 ton press work?" The processing expert answered, "Yes, possible!" The analyst asked, "How much money would be saved?" "Approximately $500,000 or $1,000 per ton of press capacity," replied the plastic expert. "How many presses will be used? One or two?," asked the analyst. "Eighteen for the rear only," replied the plastic expert. "That's real money," replied the analyst. This scenario provided the impetus to develop the techniques described in this paper.

WHAT IS SOLID MODELING? - Solid modeling is the creation of a computer representation of a part which posses true characteristics of solids, including volume and weight calculations, mass moments of inertia and clearance and interference calculations.

Solid modeling computer software and hardware have both progressed to the point where the graphic representation on the screen can compare with a high quality color photograph. The solid modeling software selected and the computer size control the styling and packaging aspects of the proposed design.

Styling of the part via solids relates to the combination of basic shapes (called primitives) to create the product and the color representation of the solid. The part is created by combining (both adding and subtracting) the primitives to obtain the desired geometry. Some solid modeling systems must also display the solid part on the computer screen as a series of faceted surfaces (this is shown as a series of small triangular area to speed up the display process only). Some solid modeling software packages also have the ability to add light sources, and various surface textures and reflectivity to obtain realistic part appearance.

Spacial allocation or packaging features refer to the ability of the software to position the proposed object in a coordinate system and the capability of the software to determine the clearance and/or in concept design with a solid modeler. These capabilities are all being utilized, both by product and process individuals, for the design of solid parts in the aircraft, computer, consumer goods, and automotive industries.

ADVANTAGES AND DISADVANTAGES OF SOLID MODELING - A few of the benefits of the solid modeling have been described above. Other advantages include:
- COLOR DESIGNATION OF SPECIFIC PARTS
- PICTORIAL REPRESENTATION OF PART (ALONG WITH ANALYTICAL DATA)
- RAPID IMPLEMENTATION
- PART SCALING
- ASSEMBLY CLEARANCE DETERMINATION
- POWERFUL ANALYTICAL CAPABILITIES (AREA, SURFACE AREA, VOLUME)
- FOUNDATION FOR OTHER ASPECTS OF CAE PROCESS-DRIVEN DESIGN

Although not a disadvantage, the item that must be overcome is the individuals resistance to use the new tool to accomplish the task. One past this hurdle, and after receiving proper training on the system, the individual will have to rethink how the part must be formed. Successful work in solid modeling requires thinking in the same way as the clay modeler; ie, start with a solid part and subtract or remove material.

Other disadvantages include:
- LONG LEARNING CURVE (FOR THE FIRST SOLID MODELING PACKAGE)
- COMPUTER MEMORY LIMITATIONS (REQUIRES SIMPLIFICATION OF PART)
- DESIGNER MUST CURB DESIRE FOR COMPLEX MODEL
- DIFFICULT TO REVISE SOLID MODEL (NOT SIGNIFICANT IF GENERIC SOLID PARTS CAN BE CREATED)

CONCEPT DESIGN AND ALTERNATIVE SELECTION- This section describes the method of alternative selection after each concept is modeled as a solid. The analytical results of each solid model (surface area, volume, and part weight) are used to generate various manufacturing data, from material cost to individual part cost. Figure 2 shows this in detail for four concept alternatives for a generic bumper beam. Each model generates its own individual analytical values.

The same solid model can be used as the starting point to generate the finite element model, with a great saving in time. All of the values from the FEA work can be used to assist in the concept alternative selection.

CAE METHODOLOGY

This technique uses a solid model of the part as a starting point for process driven design of concept plastic parts. A bumper beam has been used as the example, but any plastic part can be used. As described in previous section, solid models will be created of the various bumper beam alternatives. This could be done in any applicable solid modeling system. This method is general and independent of the solid modeling system used. Any of the top five software packages could be used to accomplish these tasks. These are shown in Table 1. Figure 3 shows the flow of the CAE methodology for a bumper beam. Figure 4 shows the large picture of the model.

All of the primary variables, obtainable from the CAE process, would be used to obtain the derived variables. These variables, listed in Table 2, would be used to determine the secondary or derived variables. The process requires manual calculations at the present time, but a longer term goal is to have these done by an expert systems type program.

Table 3 shows additional uses for the solid in process simulation.

Table 1 - Models Can Be Completed Via Solid Modeling Software Programs

- Aries
- Automation Technology Products - Cimplex
- CATIA
- Geomod
- Others

CREATE BUMPER CONCEPT
↓
CREATE CROSS SECTIONS
↓
EXPAND TO A SOLID MODEL ON ANY SPECIFIED SYSTEM
↓
OBTAIN MASS PROPERTY DATA
↓
PROCESSING DATA (DERIVED) SURFACE AREA - PROJECTED AREA
↓
FINITE ELEMENT ANALYSIS (STRESS, DEFLECTION, NATURAL FREQUENCY)
↓
COMPARE VALUES TO
  o GOAL
  o DESIGN CRITERIA
  o ACCEPTABLE DESIGN
↓
REVISE CONCEPT
↓
DECISION SELECTION CRITERIA
↓
DECIDE ON ACCEPTABLE CONCEPT (S)
↓
PROCEED WITH CONCEPT (S)
↓
HAND OFF - DOWNSTREAM & OTHER ACTIVITIES

FIGURE 1

CAE ASPECTS OF PROCESS-DRIVEN

PLASTIC PRODUCT DESIGN

[BUMPER SYSTEMS AS AN EXAMPLE]

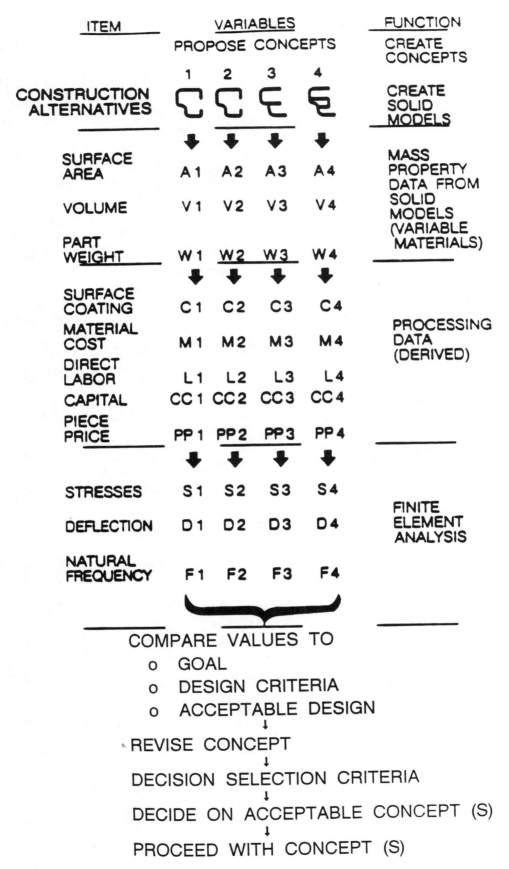

Fig. 2 BUMPER SYSTEM DESIGN

Table 2 - CAE Values of Part Stored in the Speadsheet

Primary Values

- Volume
- Surface Area
- Center of Gravity
- Moments of Inertia
- Products of Inertia
- Direction Cosines of Maximum Inertia Axes
- Material Density
- Projected Area
- Part Stress
- Part Deflection

Derived or Secondary Variables

- Weight (Using Part Material Density)
- Surface Coating Volume
- Manufacturing Press Force (From Projected Area and Molding Pressure)

Other Value

- Press Area
- Press Cost
- Number of Presses
- Area Around Press
- Total Factory Square Footage
- Total Cost - Equipment
- Total Cost - Plant
- Production Rate (Cycle Time)
- EAU Type
- Stroke of EAU
- Total Deflection of System (EAU Plus Beam)
- Comments

Table 3 - Use of Solids in Process Simulation

- NC TOOL PATH VERIFICATION
- ROBOTIC WORKCELL SIMULATION
- CLEARANCE CHECKING
  FOR LOADING AND UNLOADING
  IN PRESSES, ETC
- CREATION OF ALL FIXTURES, INSPECTION, ETC
- NESTING FOR PART SHIPMENT (IF NEEDED)
- VERIFICATION OF Z-AXIS LOADING

CREATION OF A BUMPER

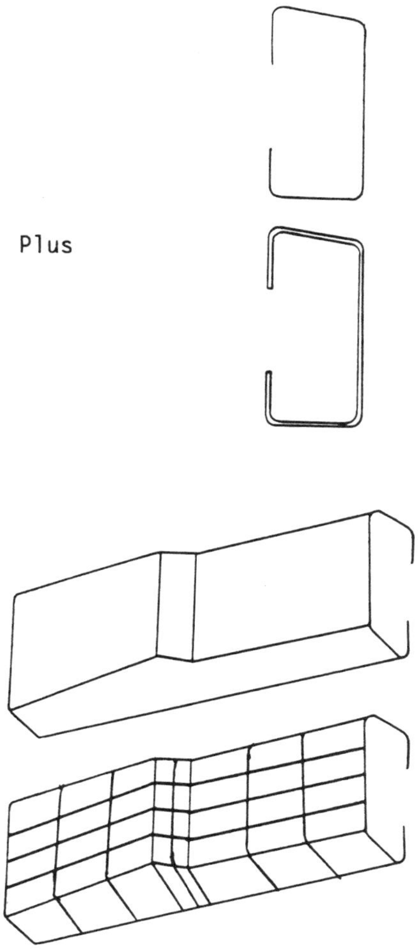

Enter CrossSection
Centerline Data
↓
Offset c Curve
for Total Thickness
↓
Construct Face &
Analyze Cross Section
↓
Decide on
FEA Analysis
↓
Extend Cross Section
for 3-D Bumper
↓
Created Midplane
Surface of Bumper
↓
Create Nodes &
Elements on Surface
↓
Analyze Model
↓
Display Results
Stress & Deflections

FIGURE 3   C A E ASPECTS OF BUMPER SYSTEM DESIGN

DATA USE AND STORAGE

The results obtained from these studies must be used to make product and process decisions. Since the objective is to look at a greater number of concept designs, this will mean more data is required. A systematic procedure to store this data is needed. This is stored in a spreadsheet.

Spreadsheet storage of the data has advantages attributable to these programs. These include sorting, ordering, and preforming repetitive calculations. Alternative design concepts could be listed as rows, with the described variables as columns. Figure 5 shows a portion of this type of arrangement. Now for example, the spreadsheet can be used in the decision making process. Design alternatives could be arranged by increase weight or sorted by the lowest stress design. Reports could be prepared for management showing the lowest capital investment or the lowest square footage requirement.

NEW METHOD OF USING CAE METHODOLOGY IN "ODD" AND "BLACK BOX" DESIGN

This proposed method has potential for various departments, groups, and outside sources to provide design concepts. By using these techniques to generate the numerical values and storing them on a spreadsheet, one central group could acquire and utilize design data from many sources. Each group could be told which row of the spreadsheet to enter their data on, and all data shipped to a central source and compiled into a single master design data spreadsheet. While this concept has been discussed, it has not been implemented. To date, eight different designs have been studies by this method.

SUMMARY AND FUTURE WORK

A CAE methodology has been proposed based on solid modeling of the part. It has been successfully applied to eight bumper beams of ferrous and plastic materials. Storage of the data has been completed on computer spreadsheets. This method has allowed the completion of one design and analysis in two-thirds of a working day.

Future work is planned in the area of improved spreadsheet storage, and incorporation of plastic materials property data. All equations for press capacity determination and capitol investment will be incorporated into an expert system program. These will be incorporated into a vertical integrated procedure for plastic part design which will design products for the year 2000 and beyond.

ACKNOWLEDGEMENT

The author wishes to express his gratitude to Chrysler Motors Corp. for making this publication possible. Special thanks are extended to Lisa Batzloff.

REFERENCES

1) Miller, B., "Plastics World to Show Ford's All-Plastic Car", Plastics World, May, 1988, p. 25.
2) Toensmeier, P. A., "Urethane Tire: Next Automotive Option?", Modern Plastics, December, 1987, p. 34.
3) Miller, B., "Plastics Take Center Stage at S.A.E.", Plastics World, April, 1988, p. 24.
4) Smoluk, G. R., "Large-parts Blow Molding Takes on New Measure of Cost-Efficiency", Modern Plastics, October, 1987, p. 61.
5) Anonymous, "A Bumper with Looks that Last", Dupont Magazine, July/August, 1988, p. 14.
6) Klein, A. J., "Plastics Lead the Drive Toward Part Consolidation", Plastics Design Forum, July/August, 1988, p. 33.
7) Martino, R. J., Editor, "Modern Plastics Encyclopedia", October, 1986, Volume 63, No. 10 A.
8) Rowland, R., "50% Rise in Plastics Seen in '98 Vehicles", Automotive News, June 20, 1988, p. 16.
9) Morley, B. C., "The MCAE Revolution - And Why", MS 86-1089, Society of Manufacturing Engineers, Dearborn, MI, 1986.

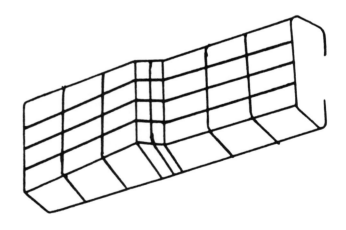

Fig. 4 FEA Analysis

MANAGING PLASTIC COMPONENT TECHNOLOGY

REV. 12/14/88

PROJECT DESC.	MATERIAL	PART VOLUME	WEIGHT	PROJECTED AREA	FORCE CAP. TONS	$/PRESS
1 CONCEPT	METAL					
2 CONCEPT B	METAL B					
3 CONCEPT C	METAL C					
4 CONCEPT D	PLASTIC D					
5 CONCEPT E	PLASTIC E					

TOTAL AREA	$/SQ. FT.	TOTAL COST	MAT'L SP. GR.	POISSON'S RATIO	E	G

TEST LOAD RESTRAINTS	MAXIMUM STRESS	MAX. DEFLECTION	TOTAL DEFLECTION	COMMENTS

FIGURE 5 TYPICAL SPREAD SHEET FOR C A E METHODOLOGY

NEED FOR A CAE BASED PLASTICS TECHNOLOGY PROGRAM

Robert G. Dubensky
Chrysler Motors
Highland Park, MI USA

Donald E. Jay
Chrysler Motors
New Mack Avenue Process
Detroit, MI USA

Ralph K. Salansky
Creative Industries Group
Auburn Hills, MI USA

ABSTRACT

A comprehensive educational plan for training plastic technologists is greatly needed. This paper presents the quantitative data supporting these projections. Needs statement to show how this work must be accomplished (by computer-aided-engineering techniques), discussion on the types of training required, and proposed methods to complete this training are discussed.

KEY WORDS

Plastics, composites, computer-aided-engineering, CAE, education, continuing education, continual skills enhancement.

PLASTIC PRODUCTS and the replacement of existing materials with plastics has grown at a tremendous rate (for items ranging from bicycle wheels to automobile panels). (Reference 1 - 7).* Since 1983, plastic compounders have developed approximately 1,000 new materials each year. Trained individuals will be needed to adapt these materials to product design and manufacturing. The automobile industry alone is projected to increase its use of plastics by 50% in the next decade. (Reference 8)

This projection is favored by sales and marketing personnel; but may be viewed with dismay by manufacturing and design personnel. This raises the following questions: "Where will the extra personnel needed to do all this work come from in the next five years?", "How will they be trained?", "Who will train them?". The companies producing plastic parts are already strained to the limit on personnel and resources. This brings up the question, "Where will they go for this training?".

* Numbers in parenthesis refer to references.

OBJECTIVE

The objective of this paper is to present a comprehensive educational plan for training plastic technologists, the background data supporting these projections, and a statement showing how this work must be accomplished by computer-aided-engineering techniques. In addition, a training plan is presented along with a method to complete this training.

DETERMINATION OF NEEDS

To determine these needs, the current literature was surveyed. See a few of the headlines copied in Figure 1. Additional techniques included contacts and conversation with suppliers, see Figure 2 for a few headlines. A survey was conducted of educational institutions.

EDUCATIONAL INSTITUTION SURVEY- Initially, the Society of Plastics Engineers was contacted for their publication, "1985 Listing of Institutions which offer Graduate or Undergraduate Plastics/Polymer Programs in the United States and Canada." The institutions listed in Table 1 were contacted for detailed information about their programs. An 85% response was returned to these contacts.

The program of study of each of these institutions was examined for the type of program and type of courses. Very few courses introduce the new computer techniques. Based on the projected need and current trends in all industries (Reference 8), as well as the plastics industry, a program is structured based on computer-aided techniques.

This is not to say that computer techniques are not taught today in these institutions, but the process of integration of all of these techniques must be completed into the plastic curriculum.

PROJECTED NEEDS - Based on discussions with industry contacts, conversations with suppliers, and a detailed study and synthesis of published facts, the following needs can be determined. In 1984, there were 1.09 million plastics industry employees (SPI data). Assume that there are approximately 1.3 million employees in the domestic plastics industry today. Reference 8 predicts a growth rate of 50%; therefore, it is assumed that a 33% increase in personnel would occur. This would necessitate 430,000 persons required in the ten year period. This is 43,000 persons per year (assuming a linear demand rate - highly unlikely with weight saving trends in the auto industry today). Now assume one out of eight of these will be in the engineering technology area, which gives 5,000 persons per year as a requirement.

Naturally, the demand will be intense initially and there will be a lack of trained individuals in the initial early years. Key industries will have to train their own interested personnel in these Computer-Aided-Engineering (CAE) techniques. These corporations will have the lead over their competitors. A few can be hired from other industries, but they will have to be trained in the ways of plastic product design (not as easy a task as it seems - see headlines in Figure 2). The remainder of this paper discusses a program to train individuals to meet the projected demands with new CAE techniques.

CAE TERMINOLOGY

COMPUTER-AIDED-ENGINEERING (CAE) - This term has been used for all aspects of the engineering design and processing procedure that have been computerized. This includes the items listed below.

(Design Area)

CAD -- Computer-Aided-Drafting
FEM -- Finite-Element-Modeling
FEA -- Finite-Element-Analysis
Solids -- Solid Modeling
CAI -- Computer-Aided-Inspection
CAT -- Computer-Aided-Testing
CAPP -- Computer-Aided-Process-Planning

(Manufacturing Area)

CAM -- Computer-Aided-Machining
CIM -- Computer-Integrated-Manufacturing
MOLDFLOW -- A computerized process used to analyze the flow of plastic compounds in molds.
PLASPEC -- A computerized plastic material property data base.
AUTOSIM -- A complete factory material flow simulation program.

SMC molders unite
New SMC Automotive Centre will propagate use of plastic

Niche-picker
Ashland stays strong by supplying material for special vehicles

How will we get rid of all that plastic?
As automakers embrace the lightweight materials, environmental concerns arise

Composites cooperative
The Big Three link up to be leaders in composite materials

COVER STORY

PLASTICS POTENTIAL
As the Big Three team up to tackle composites, suppliers face technological hurdles

Making a pitch for polyurethane
Mixing materials in cars calls for special coatings, Mobay says

Figure 1 - Typical Headlines Relating To Plastic Industry Developments

(From Automotive News, April 24, 1989, 32 page Plastic Supplement)

CAE BASED PLASTICS TECHNOLOGY PROGRAM

The program proposed should have course work based in three areas, including (Reference 10):

1) Plastic Product Design
2) Plastic Product Processing
3) Computer Integrated Manufacturing of Plastic Products

Courses in plastic and composite properties would be the foundation for all other courses. These would include courses in:

- Mechanical and Physical Behavior
- Effects of Processing on Behavior
- Strength of Plastic Materials
- Methods of Testing - Properties and Products

COURSES IN PLASTIC PRODUCT DESIGN - This would be composed of courses in:

- Mechanical Design Analysis
- Kinematics
- CAD Techniques
- Solid Modeling
- Product Design Application

COURSES IN PLASTIC PRODUCT PROCESSING - This area would include courses in:

- Conventional Manufacturing Techniques (Including Mold and Die Preparation)
- Plastic Processing Techniques (Introductory and Advanced)
- Plastic Process Simulation
- Mold and Die Performance Simulation

COURSES IN COMPUTER INTEGRATED MANUFACTURING - This area would include studies in:

- Principles of Business Operation
- Program Management Techniques
- Decision Making and Product Selection
- Plastic Product Process Plant Simulation
- Statistical Process Control (SPC) Techniques

All courses would be offered as options to a series of basic core courses (required by the student). Figure 3 shows this arrangement. (Reference 11). Detailed course descriptions are available in Reference 12. Students could also elect courses from the other options.

Figure 2 - Typical Headlines Relating To Plastic Material Potentials

(From Automotive News, April 24, 1989, 32 page Plastic Supplement, ** Plastic News, May 29, 1989, p. 5)

TABLE 1 - EDUCATIONAL INSTITUTIONS CONTACTED

Ball State University
Central Michigan University
Cerritos College
Cincinnati Technical College
College of DuPage
University of Detroit
Eastern Michigan University
Ferris State College
Grand Rapids Junior College
General Motors Institute
Hennepin Technical Center
Indiana Vocational Technical College
Indiana Vocational Technical College (Ind.)
Isothermal Community College
Kent State University
Lakeshore Technical Institute
Laney College
University of Lowell
Macomb Community College
Michigan State University
Michigan Technological University
Milwaukee Area Technical College
Morehead State University
Morrisville State College
New Jersey Institute of Technology
North Carolina A & T State University
North Georgia Tech and Vocational School
Northeast Wisconsin Technical Institute
Northern Alberta Institute of Technology
Northwestern Michigan College
Oakland University
Shawnee State Community College
Southern Illinois University
Syracuse University
Virginia Highland Community College
Western Michigan University
Western Wisconsin Technical Institute

TABLE 2 - TYPICAL MCAE SYSTEMS AND COMPUTER CODES TO BE USED IN THE PROGRAM

CAE SYSTEMS

- Aries MCAE System
- PC CAD
- PC FEA
- PC Solid Modeling

PC COMPUTER PROGRAMS

- Kinematics
- Solid Modeler
- FEA Model
- FEA Solver
- FEA Post Processor
- Moldflow
- Moldcool
- Metflow
- ASI Factory Simulator
- Database Programs
- Expert System Programs

NEW PROTOTYPE CREATION

- 3-D Systems
- Cubital Division of Scitex
- Light Sculpturing

IMPLEMENTATION

The planned program covers the main aspects of the proposed curriculum and is the easiest part of the entire task. The key to the program will be the successful implementation of these separate courses into a solid coherent program. This will require a number of features including:

1) A dedicated and trained staff (with industrial experience and a desire to teach in a project type program).

2) Educational leaders who support the staff and program.

3) Proper equipment to facilitate these goals and objectives including:

a) Laboratory equipment to support the basic sciences and plastic properties and processing courses. (Nothing reinforces a student's knowledge as much as the ability to conduct a fracture test on a plastic material and examine it, or study the solidification pattern of plastic parts as they actually occur.

b) Computer facilities (in today's terminology this would be hardware and software, but this could be called "firmware", "chipware", "designware", etc., in the near future). This includes all of the facilities to conceive, design, and electronically analyze a "software model" of the plastic part and the equipment and facilities needed to make sample prototype parts. This would include the new 3-D stereolithography equipment that prepares prototype parts directly from a solid model of the part on a CAE system. This is done by cross linking ultraviolet, light-sensitive, polyester molecules with an ultraviolet laser light source.

Table 2 lists some of the PC based systems and computer codes for CAE applications that can be used in a Plastics Technology Program.

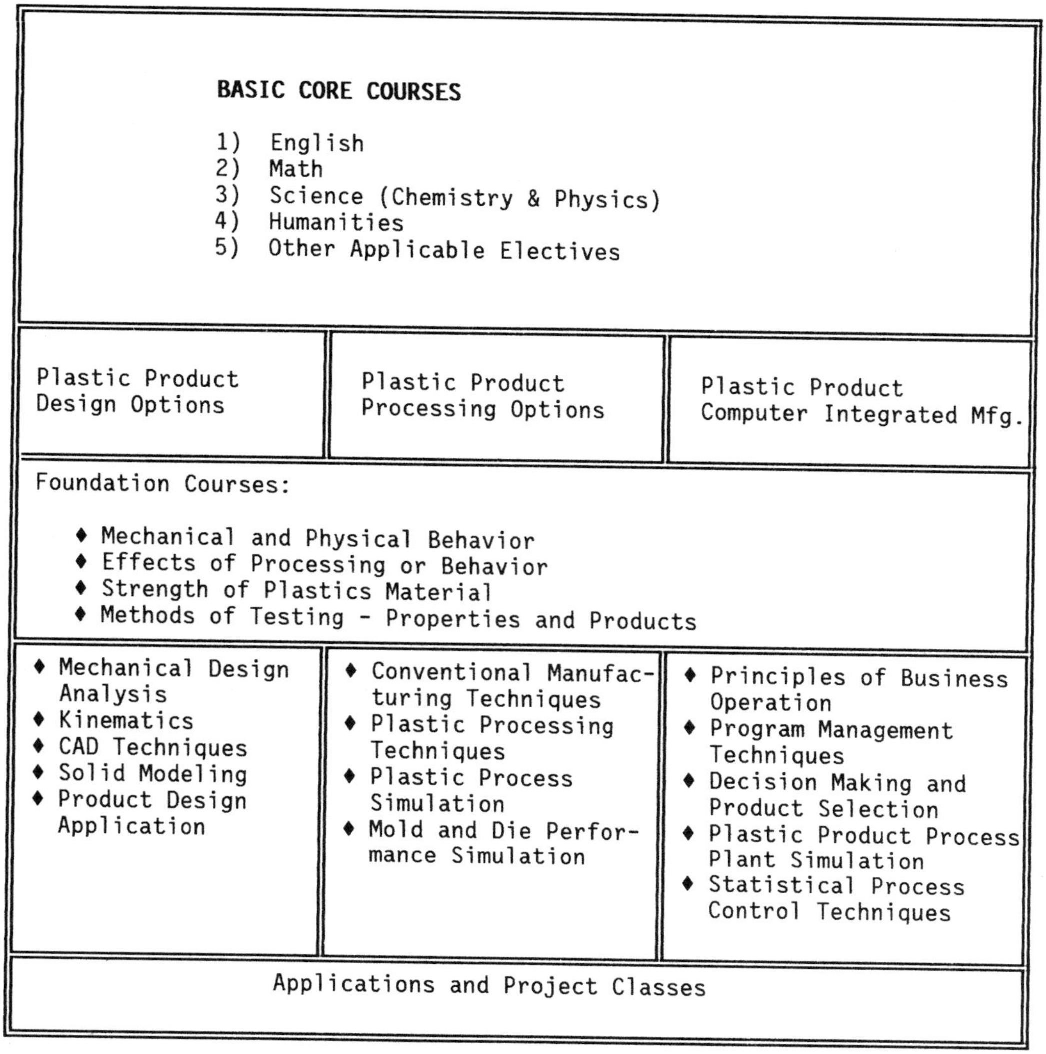

```
┌─────────────────────────────────────────────────────────────────────┐
│                     BASIC CORE COURSES                              │
│                                                                     │
│         1)  English                                                 │
│         2)  Math                                                    │
│         3)  Science (Chemistry & Physics)                           │
│         4)  Humanities                                              │
│         5)  Other Applicable Electives                              │
│                                                                     │
├───────────────────────┬───────────────────────┬─────────────────────┤
│ Plastic Product       │ Plastic Product       │ Plastic Product     │
│ Design Options        │ Processing Options    │ Computer Integrated │
│                       │                       │ Mfg.                │
├───────────────────────┴───────────────────────┴─────────────────────┤
│ Foundation Courses:                                                 │
│                                                                     │
│   ♦ Mechanical and Physical Behavior                                │
│   ♦ Effects of Processing or Behavior                               │
│   ♦ Strength of Plastics Material                                   │
│   ♦ Methods of Testing - Properties and Products                    │
├───────────────────────┬───────────────────────┬─────────────────────┤
│ ♦ Mechanical Design   │ ♦ Conventional Manu-  │ ♦ Principles of     │
│   Analysis            │   facturing Techniques│   Business          │
│ ♦ Kinematics          │ ♦ Plastic Processing  │   Operation         │
│ ♦ CAD Techniques      │   Techniques          │ ♦ Program Management│
│ ♦ Solid Modeling      │ ♦ Plastic Process     │   Techniques        │
│ ♦ Product Design      │   Simulation          │ ♦ Decision Making   │
│   Application         │ ♦ Mold and Die Perfor-│   and Product       │
│                       │   mance Simulation    │   Selection         │
│                       │                       │ ♦ Plastic Product   │
│                       │                       │   Process Plant     │
│                       │                       │   Simulation        │
│                       │                       │ ♦ Statistical       │
│                       │                       │   Process Control   │
│                       │                       │   Techniques        │
├───────────────────────┴───────────────────────┴─────────────────────┤
│                 Applications and Project Classes                    │
└─────────────────────────────────────────────────────────────────────┘
```

FIGURE 3   PLASTIC TECHNOLOGY PROGRAM OPTIONS

4) Specific application or project classes - During the final year of the program, a student in each of the areas must complete a project in the area of specialization. This will solidify all of the concepts the student has learned. Educational leaders must realize this is facilities and instructor intensive, but is critical to the success of the program. All of the techniques of proper case study implementation and creation must be used.

PREFERRED IMPLEMENTATION TECHNIQUE

There are at least four methods of implementing the educational programs and techniques described in this work. They include:

1. Mainframe Computers
2. Mini Computers
3. Stand Alone Workstations
4. Personal Computers

The CAE procedures, shown in Figures 4 and 5, must be vertically integrated. While these CAE procedures may appear to be easy to accomplish on the mainframe, this may be very difficult. This requires the interfacing of codes by different, large suppliers (a formidable task). (Reference 11)

It is recommended that a vertically integrated CAE package be used, or the software acquired and loaded on a P.C. or workstation. These hardware and software areas are changing so rapidly, that before this type of program can be implemented, workstation capabilities will exceed the needs of the student. (Reference 9 and "Engr. Workstation; A Technical Guide," by Ken Anderson, Computer Graphics World, April 1989, p. 81-86.)

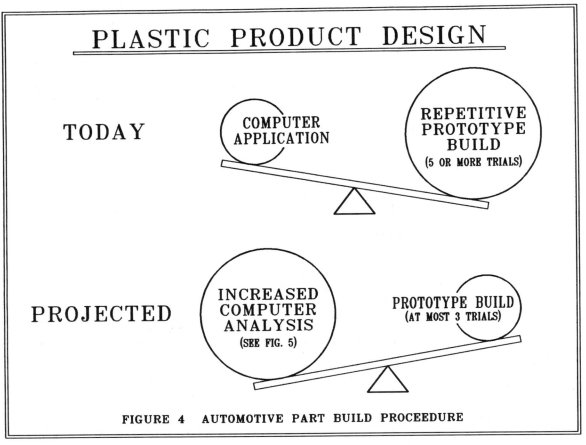

FIGURE 4 AUTOMOTIVE PART BUILD PROCEEDURE

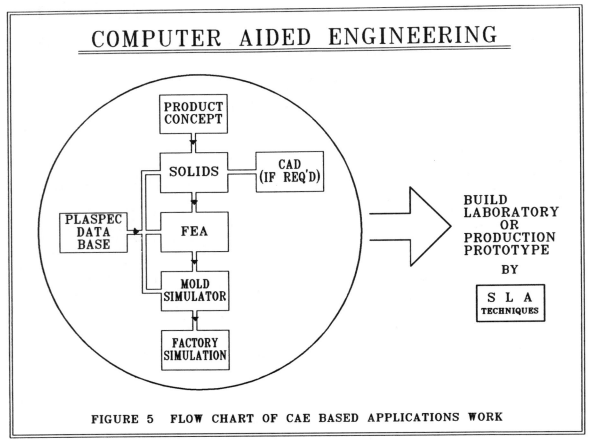

FIGURE 5 FLOW CHART OF CAE BASED APPLICATIONS WORK

SUMMARY AND NEEDS

This paper has provided the description of a CAE based plastics technology program. Such a program is required to provide the trained personnel to support the planned growth in the plastics industry.

What is needed, are forward-thinking, educational institutions to implement such a program. Alternately, local groups of industries, trade associations, and individual companies must undertake such training to provide the personnel needed three years from now. Plastic societies could take the lead in such programs. To miss this opportunity is to lose out on a coming trend toward providing trained CAE personnel in the plastic industry.

ACKNOWLEDGEMENTS

The authors wish to express their gratitude to Chrysler Motors Corporation for making this publication possible. Special thanks are extended to Lisa Batzloff.

REFERENCES

1) Miller, B., "Plastics World to Show Ford's All-Plastic Car", Plastics World, May, 1988, p.25
2) Toensmeier, P.A., "Urethane Tire: Next Automotive Option?", Modern Plastics, December, 1987, p.34
3) Miller, B., "Plastics Take Center Stage at S.A.E.", Plastics World, April, 1988, p.24
4) Smoluk, G. R., "Large-parts Blow Molding Takes on New Measure of Cost-Efficiency", Modern Plastics, October, 1987, p.61
5) Anonymous, "A Bumper with Looks that Last", Dupont Magazine, July/August, 1988, p.14
6) Klein, A. J., "Plastics Lead the Drive Toward Part Consolidation", Plastics Design Forum, July/August, 1988, p.33
7) Martino, R. J., Editor, "Modern Plastics Encyclopedia", October, 1986, Volume 63, No. 10 A
8) Rowland, R., "50% Rise in Plastics Seen in '98 Vehicles", Automotive News, June 20, 1988, p.16
9) Hamilton, C. Hayden, "MCAE Enters the 1990's", Manufacturing Engineering, June, 1989, pp. 80-82
10) Dubensky, R. G., Personal Communication, January 30, 1989
11) Dubensky, R. G., Personal Communication, April 5, 1989
12) Dubensky, R. G., Personal Communications on Course Descriptions, April 14, 1989.
13) Anderson, K., "Engineering Workstation: A Technical Guide", Computer Graphics World, April, 1989, pp. 81-86

CASE HISTORIES OF AN ADHESIVE INTERLEAF TO REDUCE STRESS CONCENTRATIONS BETWEEN PLYS OF STRUCTURAL COMPOSITES

Raymond B. Krieger, Jr
American Cyanamid Company
Havre De Grace, MD 21078 USA

Abstract

This paper describes a special high-strain, low-flow adhesive for use between strategic plys of structural composites. The purpose is to reduce inherent shear stress peaks, or concentrations, by providing a high-strain adhesive interleaf between plys. This interleaf has high strain capability because of its formulation and because it is much thicker than the usual matrix layer between plys. Shear stress-strain data is presented for the interleaf and for typical matrix resins. A linear stress analysis is presented for the plane between plys. This shows that stress peaks can be reduced by as much as ten times, in the linear range, and far more in the ultimate strength range. This interleaf technique can greatly improve the strength and fatigue durability of structural composites.

THE ADVENT OF COMPOSITE LAMINATES on the airframe scene has not always led to success comensurate with the high promise of material properties. In many secondary structures, the designs have saved weight and money. In primary structure, glass and Kevlar have, in some measure, given way to the superior properties of carbon fiber composites. The simpler designs in carbon fiber have generally met their structural goals. In more complex designs, premature failures have been unexpected and disconcerting. The reasons for these failures are not to be found in the fundamental properties of the laminate, i.e., tension, compression, flexure, and interlaminar shear.

The author suggests that a primary cause is the Achilles heel of transfer of shear between plies of structural composites.

These shears have not been wholly overlooked by stress analysis. Plies of adhesive film are often introduced at strategic locations where high shear flows are predicted. Nevertheless, the author suggests there are three questionable facets in the contemporary approach, namely:
1. While shear flows are readily calculated, the shear stress concentrations are not, because they depend on resin stiffness in carrying the shear flow transfer.
2. The adhesive ply will mix with the matrix resin. This increases the adhesive stiffness, and so raises the shear stress.
3. The hot-wet compression strength is reduced because the modulus of the matrix resin is lowered by mixing with the adhesive.

This paper addresses these problems by A) introducing a new adhesive film which will not miss with the matrix, and B) offering stress analysis techniques to more accurately calculate the shear stress concentrations in order to better predict the strength of the design.

Three case histories are presented which show quantitative improvement in structure by using interleaf.

SHEAR TRANSFER IN LAMINATES

In structure made from laminated plies of composite there are often serious shear forces which must be transferred between plies. Since this plane is not directly crossed by fibers of the composite, the shear loads must be carried by the matrix resin, acting as a thin glue line. This resin plane, or glue line, can become overloaded because of shear stress concentrations. Shear failure on this plane can preclude the structural test or even trigger the collapse of the entire beam, well below design strength.

To understand the situation more clearly, we can begin with Fig. 1. This shows a square increment of laminate loaded with a shear flow around its edges. The shear flow is equal and opposite on parallel sides, and causes a shear deformation from a square to a diamond shape. Moving to Fig. 2 the thickness of the laminate has been abruptly increased over half of the increment. The shear flow on the thinner half must now redistribute over the extra laminate thickness. This results in a sudden concentration of shear force at the edge of the thicker half. This can readily overload the matrix and cause splitting between the plies, as shown.

To visualize the shear concentration problem, it may help to add a layer of adhesive at the critical plane. Fig. 3 shows this glue line as far thicker than the matrix resin layer between plies. It is seen that the adhesive is deformed (loaded) at a maximum at the edge and fades to zero at some point when the laminate shear strain is equal through its entire thickness.

QUANTITATIVE CALCULATION OF SHEAR STRESS CONCENTRATIONS

In Fig. 3 it can be seen that the shear load transfer in the adhesive is stiffness driven in the same sense as in the classic metal to metal bonded skin doubler specimen, (1). Figure 4 shows this specimen deformed when the skin (or laminate) is under tension. The adhesive shear stress distribution is shown. The peak stress is given by the equation

$$\frac{KP}{\sqrt{\frac{t\, t_a\, E}{G}}} \qquad \text{Eq. (1).}$$

where $K = .7$, P = load per inch, t = laminate thickness, t_a = glue line thickness, E = laminate tensile modulus and G = adhesive shear modulus (1).

Fig. 3 shows the same principle for adhesive shear stress distribution. Equation (1) will apply since all the stiffness parameters are the same, except of course that E (laminate tensile modulus) must now become G_L (laminate shear modulus).

REDUCTION OF SHEAR PEAK BY USE OF AN ADHESIVE INTERLEAF

The addition of the glue line, shown in Fig. 3, is a great improvement in reducing shear stress concentrations. It is readily seen by inspection of equation (1) that the glue line stiffness is a function of the glue line thickness (t_a) and the shear modulus of the adhesive (G). This makes it possible to closely estimate the improvement percentage, or ratio, obtainable by adding the adhesive interleaf. For Fig. 2, no adhesive, the matrix shear modulus, G, is 200,000 psi and the "glue line thickness" of the single layer of matrix between plies is .00004 inches.

$$\text{Then } \frac{1}{\sqrt{\frac{t_a}{G}}} = 70,900 = \text{stiffness factor} \qquad \text{Eq. (2).}$$

Next, for Fig. 3, adhesive interleaf, $t_a = .005$ inches and $G = 100,000$

$$\text{Then } \frac{1}{\sqrt{\frac{t_a}{G}}} = 4460 = \text{stiffness factor} \qquad \text{Eq. (3).}$$

The shear stress reduction is proportional to these stiffness factors, and so the reduction is 70,900/4460=16 to 1.

Equations (2), (3) are for the linear range of shear modulus. This means that the stress reductions apply to the lower load range of fatigue and creep. For ultimate load, the matrix modulus will not change substantially, but the adhesive exhibits very high ultimate shear strain. The "effective modulus" could be 1/3 of the linear modulus and the stress reductions as high as 27 to 1.

The magnitude of these stress reductions is so great that success appears assured in spite of errors (tolerances) in measuring actual glue line thickness and modulus.

THREE CASE HISTORIES OF INTERLEAF APPLICATIONS

The first example involves a successful elimination of micro-cracking in graphite composites. Fig. 5 illustrates a definition of micro-cracking, which occurs when three or more tape plies exist with fibres all in one direction. The cracks appear during the cool-down portion of the cure cycle. They seem caused by shrinkage of matrix resin which is resisted by ±45° plies whose graphite does not shrink. It was seasoned that if high strain interleaf was used, the shrinkage would not be restrained. Three configurations were tested with 3,4,& 5 plies as shown in Fig. 5. Interleaf was added as shown so that no more than two plies were allowed without an interleaf. No microcracking occurred during cool down and after some 200 cycles of temperature change between -67°F and 200°F.

The second example involves an access door in a wing cover skin. This door has to perform as part of the wing cover skin, i.e., tension loads in the skin must be carried across the access opening by the door itself. This is done by bolts around the perimeter of the door. Fig. 6 shows a cross section of the design. The initial failure mode was a crack between the filler and wing skin at an unacceptably low load. The ultimate strength was not in question. An interleaf layer was added between the filler and the skin. The load carrying ability was increased by 16%, more than enough to make the design a success.

The third example involves "blade" stiffeners on a graphite skin subject to compression as shown in Fig. 7. These stiffeners were also subjected to secondary and tertiary loads. These loads caused failures in the stiffener as shown in Fig. 6. These failures precluded the structure from carrying its primary load successfully. Interleaf was added at those planes where peak shear and tension stress concentrations occurred. Fig. 7 shows these locations and the spectacular improvement in load carrying capacity.

CONCLUSIONS

1. It is possible to drastically reduce shear stress concentrations at critical locations in composite structure. This is done by introducing a ply, or interleaf, of adhesive film, with los stiffness, at the strategic location.
2. It is possible to predict the shear stress concentrations because the requisite adhesive shear stiffness can be known and the stress formulae have been developed.
3. It is possible to obtain maximum reduction of shear concentration and minimum loss of hot-wet compression strength by use of an adhesive formulated so as not to mix with the matrix resin.

REFERENCES

1. Evaluating Structural Adhesives Under Sustained Load in Hostile Environment. R. B. Krieger, SAMPE Tech. Conf. 5,634 (1973)

BIOGRAPHY

Raymond B. Krieger, Jr. graduated M.I.T. with a B.S. in Aeronautical Engineering in 1941, and from then until 1953 was with the Glenn L. Martin Co. Until 1956 he was with Luria-Cournand as Chief Engineer. From 1956 to the present he has been with American Cyanamid Company, Bloomingdale Department, first as Chief Engineer, then Sales Manager, now as Technical Manager. His duties encompass criteria development and testing of structural adhesives, technical service to the airframe industry including design, processing and quality control aspects of structural bonding metal to metal, sandwich and structural plastics. His is the author of numerous papers on structural bonding.

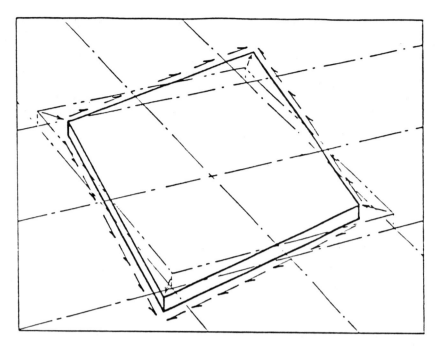

Fig. 1 Shear Flow and Deformation on Increment of Composite Laminate

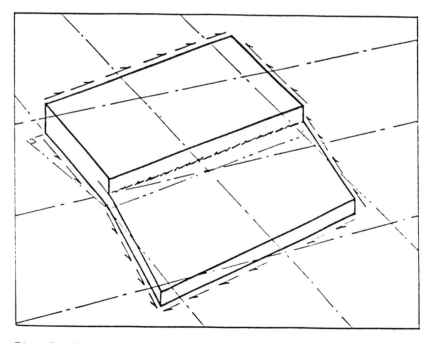

Fig. 2 Shear Flow and Deformation at Thickness Increase in Laminate

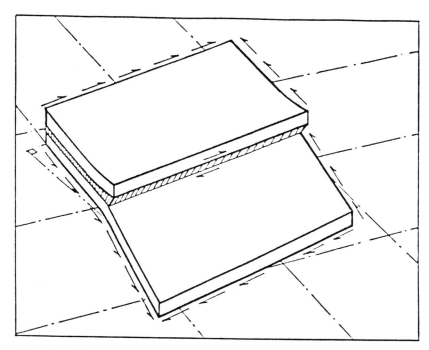

Fig. 3 Adhesive Shear Distribution When Laminate Increment is in Shear

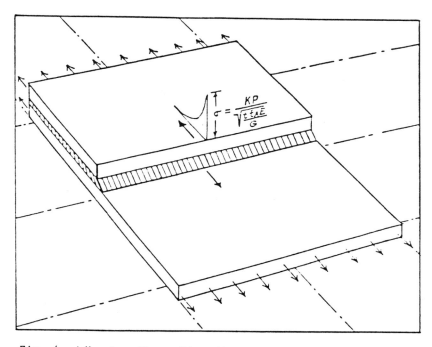

Fig. 4 Adhesive Shear Distribution when Laminate Increment is in Tension

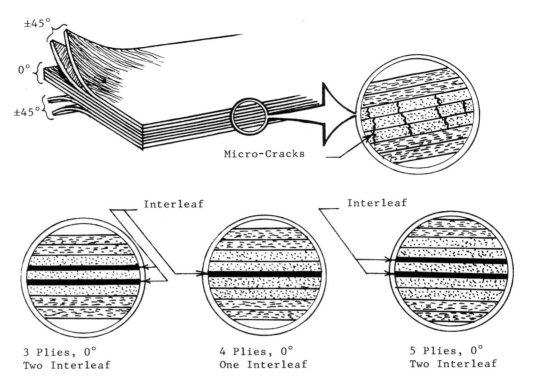

Fig. 5 Interleaf Used to Prevent Micro-Cracking

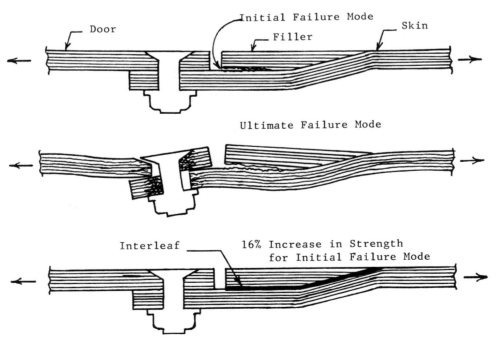

Fig. 6 Interleaf Used to Strengthen Access Door

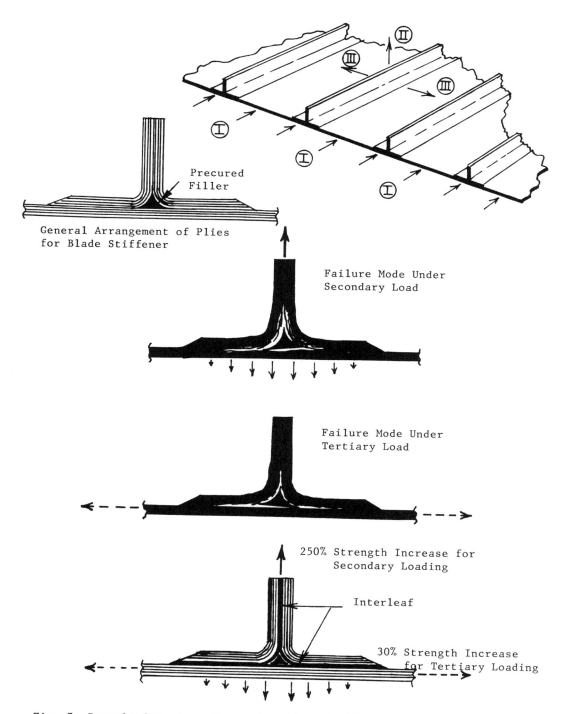

Fig. 7 Interleaf Used to Strengthen Blade Stiffener

IMPROVING THE PROCESSING CHARACTERISTICS OF STRUCTURAL RIM SYSTEMS

T. B. HOWELL, R. E. CAMARGO, D. A. BITYK
ICI Polyurethanes
Formulated Products Division
Sterling Heights, MI 48077

ABSTRACT

Applications for Structural RIM (SRIM) have continued to grow as molders and end users become more familiar with the process potential of this novel molding technology. A demand for increased design flexibility, larger parts and more weight savings means further challenges for chemists and engineers designing SRIM systems. Following the introduction of its first two SRIM systems, ICI Polyurethanes* has responded to the user's demands by developing systems featuring improved flow characteristics, and/or lower molded densities. A new family of systems has been designed for high fiber content composite moldings where, because of part size, longer injection times are needed. These new systems, which allow for injection times of 10 seconds or more, maintain the excellent properties that have characterized earlier SRIM formulations. In the areas of low density composites, a new family of materials will be described which, when used in constructions having 10% to 30% glass by weight, display excellent processability. The materials can be formed into composites having an overall specific gravity of 0.5 or less, thus offering an excellent strength to weight ratio.

STRUCTURAL REACTION INJECTION MOLDING (SRIM) is the process by which high performance composites can be produced using the fast cycle times and equipment associated with RIM. A low viscosity reactive mixture is injected into a closed mold containing a preplaced fiber reinforcement. After injection, the components react relatively fast, and are demolded in times comparable to those utilized for conventional RIM. Demold times as low as 30 seconds possible in many instances. Cycle times, on the other hand, vary with equipment, chemical systems, and the method used to place the reinforcement in the mold cavity. Under normal circumstances, part to part cycle times are three minutes or less. SRIM applications are growing at a very fast pace. The automotive industry has been the leader in the introduction of SRIM materials and processes for such applications as spare tire covers, bumper beams, floor pans, and certain under the hood components, etc. Non-automotive applications are also emerging, thus giving SRIM the potential to become a common method for the production of a variety of components in the 1990's.

The purpose of this paper is to discuss some of the different SRIM materials available to designers and product engineers today. Although SRIM is mainly associated with highly reactive materials capable of forming a high strength, temperature resistant composite, new materials have emerged that further expand the SRIM capabilities. High performance foams specifically designed for SRIM used with medium to low glass content are now available for applications requiring high strength to weight ratios, and for parts which incorporate design features such as bosses, ribs, etc. Another extension of the SRIM technology is the ability to encapsulate a high compressive strength, low density polyurethane core within a high strength composite. This type of structure provides a very stiff, low weight construction useful for a variety of applications. Examples of these materials will be provided throughout this paper.

HIGH STRENGTH COMPOSITE MATERIALS: As mentioned earlier, one of the most attractive concepts associated with the SRIM process is the manufacture of high strength engineered

---

*ICI Polyurethanes is a business unit of ICI Americas, Inc.

structural composites utilizing the cost efficiencies associated with the RIM process. Materials, such as ICI's RIMline* GMR-5000, have been specifically developed for molding these products. Most applications use a continuous strand glass mat reinforcement in levels varying from 20% to 50% by weight. Low material viscosities allow the use of a much wider range of reinforcements including chopped and woven glass mats, aramid and graphite fibers. Preforms made either from sprayed chopped fibers or a thermoformable continuous strand glass mat can also be used.

The matrix material in RIMline GMR-5000 itself is a high modulus, thermally stable thermoset polymer. Even without any reinforcement, the flexural modulus of the material is over 680 MPa (100,000 psi) up to temperatures in excess of 200°C (or 400°F). Decomposition temperatures exceed 300°C (600°F).

Figure 1 shows the elastic modulus of the RIMline GMR-5000 base resin measured in a dynamic mechanical mode. Softening temperature as indicated by a large increase in the loss tangent is approximately 250°C. The thermal and mechanical properties of the base resin translate into excellent thermal and mechanical properties of the composite. As seen from Table 1, the physical properties of composites produced from a continuous strand glass mat in levels up to 50% by weight improve with glass loading. Figures 2 and 3 show respectively the flexural modulus and impact strength as a function of glass loading. As shown in Table 2, the properties can be extended further by combining different types of reinforcement. For combinations of random mat with woven unidirectional or bidirectional materials, for instance, flexural moduli of the order of 15 GPa (2 million psi) or higher are possible.

In many of the new applications for which SRIM materials are being considered, thermal and chemical resistance are critical. As observed from Table 3, RIMline GMR-5000 has outstanding property retention when aged at 175°C (ca. 350°F) for approximately 200 hours. Key properties such as impact, flexural modulus and tensile strength are only slightly affected by this kind of heat exposure. Similarly, the material retains its original properties to a large extent when subjected to a 200 hour immersion test in engine oil at 120°C (250°F).

In addition to displaying outstanding thermal and mechanical properties, SRIM materials must also display good processing characteristics. The viscosity of SRIM components must remain low during the filling stage to allow the system to flow at moderate injection rates (typically less than 1.5 kg/sec. or 3.0 lbs/sec.) for different levels of reinforcement. Internal pressures during filling are a function of flow rate, type and level of reinforcement and system viscosity, and a reactive profile. A thorough review of the fluid mechanics and heat transfer governing the filling stage can be found elsewhere (Ref. 1) The viscosity of the reactive mixture for RIMline GMR-5000 at an average liquid temperature of 40°C-50°C is approximately 50 cp, low enough to produce good filling at high glass loading without the undesirable channeling effects that could result from very low viscosities.

System reactivity also plays an important role in part processability. Ideal reactivity profiles are such that very little reaction takes place during the filling step to avoid viscosity and subsequent pressure increases. Very short reaction times lead to premature gellation and incomplete filling. After filling is complete, the reaction must be fast enough to keep demold times economically competitive. Proprietary catalyst technology used in RIMline GMR-5000 displays what is known as a "snap cure". This is the ability of the material to change from a low viscosity liquid to a solid polymer in a very short time interval. As observed in Figure 4, it is possible to control the reactivity of the resin to extend adiabatic gel times to over a minute. In spite of these longer gel times, these slower systems still display the sharp temperature rise (i.e. snap cure) and achieve the same final reaction temperature as the faster systems. Achieving this maximum temperature is critical to develop optimum thermal and mechanical properties for the SRIM matrix material. Actual gel times under molding conditions are somewhat lower due to heat transfer effects. (cf. Ref. 1). Understanding these reaction profiles is critical when considering SRIM for the production of very large parts such as floor pans or truck beds, and when optimizing production of smaller parts.

LOW DENSITY STRUCTURAL RIM MATERIALS: RIMline GMR-5000 type resins, as discussed above, are designed to be molded at full polymer density. It is sometimes desirable to design a product that contains features such as bosses, thickness gradients or reinforcing ribs without having to use complex preforms that could drive the cost of the part to unacceptable levels. For these applications, a new family of foamed SRIM systems has been designed having excellent flow characteristics at reinforcement levels of 30% by weight or less. The durable SRIM foams used for these applications are capable of performing well even in the absence of any reinforcement. Thus, design features such as

---

*RIMline is a registered trademark of ICI Americas, Inc.

TABLE 1

PHYSICAL PROPERTIES VERSUS GLASS LOADINGS FOR AN SRIM SYSTEM

RIMLine® GMR-5000

Glass, %	0	20	30	40	50
Flexural Modulus, MPa (psi)	2,100 (300,000)	3,600 (520,000)	6,200 (900,000)	8,900 (1,300,000)	10,300 (1,500,000)
Notched Izod, J/m (ft-lb/inch)	37 (0.70)	428 (8.0)	588 (11.0)	748 (14.0)	962 (18.0)
Tensile Strength, MPa (psi)	38 (5,500)	63 (9,200)	110 (16,000)	172 (25,000)	220 (32,000)
HDT at 1.8 MPa (264 psi), °C (°F)	150 (300)	250 (480)	>250 (>480)	>250 (>480)	>250 (>480)
Specific Gravity	1.20	1.33	1.46	1.53	1.61

TABLE 2

EFFECT OF REINFORCEMENT CONFIGURATION ON PROPERTIES

RIMline® GMR-5000

Glass %	40	45	50
Flexural Modulus, MPa (psi)			
-Parallel	13,200 (1,920,000)	10,700 (1,560,000)	13,700 (2,000,000)
-Perpendicular	5,400 (780,000)	8,200 (1,200,000)	--- ---
Notched Izod, J/m (ft-lb/in)	2,670 (50.0)	2,460 (46.0)	1,760 (33.0)
Tensile Strength, MPa (psi)	380 (55,000)	330 (48,000)	235 (34,000)
Construction	UURUU	UBRRBU	BBRRBB

U = Unidirectional (3.6 Kg/m² or 12 oz/ft²)
B = Bidirectional (5.4 Kg/m² or 18 oz/ft²)
R = Random Mat (0.5 Kg/m² or 1.5 oz/ft²)

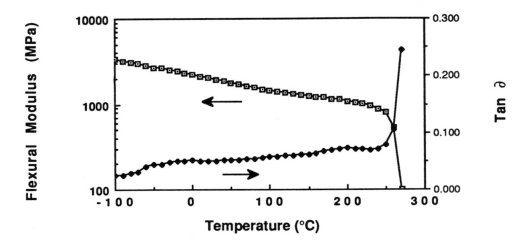

Figure 1: DMA Temperature Sweep for RIMline GMR-5000 base resin

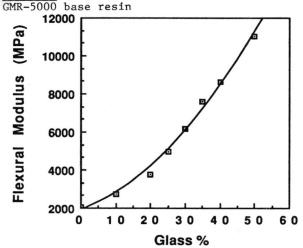

Figure 2: Flexural Modulus versus glass content for typical SRIM components.

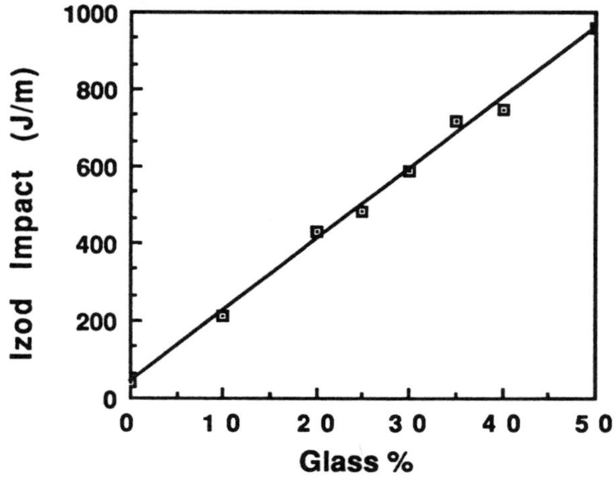

Figure 3: Izod impact (notched) versus glass glass content for typical RIMline GMR-5000 SRIM components

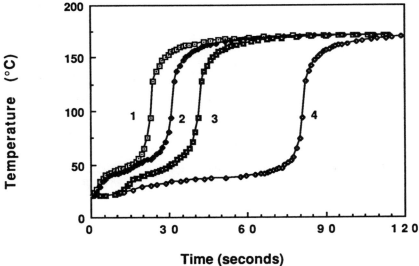

Figure 4: Adiabatic reactivity data for SRIM systems (1) Conventional SRIM system, (2) Early development SRIM system, (3) RIMline GMR-5000 system, (4) Extended flow experimental system.

those mentioned above can easily be incorporated. SRIM foams are not classified in the same level of performance as RIMline GMR-5000 composites. These materials do, however, provide well established cost benefits and production efficiencies typical of RIM. They compete effectively with other technologies, such as polyester FRP, in terms of performance, design flexibility and part cost.

The first commercial system of this type introduced by ICI Polyurethanes was RIMline GMR-200. Designed for composite constructions of 10% to 30% glass content by weight, the system is formulated with a blowing package to allow the user weight adjustments and maximum flow through the reinforcement during the filling operation. Properties of the system as a function of glass level are given in Table 4. Density of the structural foam can be reduced to as low as 0.5 g/cm^3 (32 pcf). The overall density of the composite depends on the level of reinforcement used. For a typical thickness of 6.3 mm (0.25 inches) the shrinkage of RIMline GMR-200 at 20% glass is less than 0.2%. With the proper sprue design, this material offers easy demolding in approximately 2 minutes.

A series of new automotive trim and weight saving applications, such as sunroof covers, interior door panels, etc. has led to the development of a newer family of very low density resins which combine excellent processability with the mechanical performance of SRIM foams. Molded at similar glass levels as RIMline GMR-200, these materials produce composites with overall densities of 0.5 g/cm^3 (32 pcf) or less. Typical physical properties are given in Table 5.

Although the foaming action of SRIM foams contributes to the improved surface appearance of the composite, the presence of a reinforcement prevents these products from achieving Class A surfaces. The surface quality obtained, however, is adequate for many applications. When the correct cleaning procedures are used most structural foam SRIM systems display excellent paintability. The availability of internal mold release and in-mold coating technologies further enhances the benefits of these materials.

POLYURETHANE FOAM CORES: As indicated earlier, a recent variation of the SRIM technology development deals with the encapsulation of a low density polyurethane foams in a continuous layer of a high strength composite. The density of the core foam may be as low as 0.1 g/cm^3 (6 pcf) molded to the required shape. The outer skins of the composite normally carry the tensile loads while the foam core transfers the shear loads across the part. This construction allows for a very high degree of stiffness at reduced weight.

An example of this new kind of materials is provided by the RIMline RS-100 series. This is a family of materials specially designed for foam core applications. The primary limiting factor to the integrity of a composite construction utilizing an inner core is the shear strength of the core. The new foam core materials specifically combine weight savings with good shear strength, particularly at molded densities of 0.2 g/cm^3 (ca. 15 pcf) or higher. Processability of the core material is very important as the core must be molded without defects in a uniform reproducible density. The adhesion to the external composite shell is also a major factor in determining the integrity of the overall composite. The RIMline RS-100 product line is specifically designed to satisfy these criteria. Typical properties of a representative system are in Table 6.

SUMMARY

Structural RIM applications have enjoyed a high level of growth in the last few years. Applications in the next few years are expected to expand and diversify. A major factor in this growth has been the utilization of RIM processing for the production of composite structures, thus providing the user and the designer with reduced tooling costs and faster cycle times than other composite manufacturing alternatives. Clearly, the availability of materials capable of delivering the processing and performance necessary for SRIM will play an important role in the expansion of these applications. Reviewed here are three kinds of materials that contribute to the increasing acceptance of SRIM in composite manufacturing. High strength, thermally stable composites are needed for demanding applications where metal structures are to be replaced. Applications requiring a combination of strength, design flexibility and weight savings can now make use of structural foams specifically designed to work in the SRIM process where excellent flow through the reinforcing medium is an important requirement. Finally, a complimentary family of systems has been developed for composite constructions that demand exceptional stiffness with very low weight.

References:

1. V. M. Gonzalez, Ph.D. Disertation, University of Minnesota, 1985.

TABLE 3

HEAT/CHEMICAL DURABILITY OF RIMline® GMR-5000

Property	Initial	100 hrs. at 175°C	200 hrs. at 120°C in Oil
Glass, %	36	36	36
Flexural Modulus, MPa (psi)	7500 (1,100,000)	<10% Change	No change*
Tensile Strength, MPa (psi)	166 (24,000)	<10% Change	No change
Tensile Modulus, MPa (psi)	7,100 (1,000,000)	<10% Change	No change
Notched Izod Impact, J/m (ft-lb/inch)	590 (11.0)	ca. +20%	No change

*Within experimental variation at 95% confidence limits.

TABLE 4

PROPERTIES VERSUS LOADING FOR LOW DENSITY SRIM MATERIALS

RIMline® GMR-200

Glass %	0	15	25
Density, g/cm^3 (pcf)	0.70 (44)	0.95 (59)	1.05 (65)
Flexural Modulus, MPa (psi)	1380 (200,000)	3240 (470,000)	4230 (615,000)
Tensile Strength, MPa (psi)	24.0 (3,500)	27.5 (4,000)	34.5 (5,000)
Unnotched Charpy Impact, J/m^2 (ft-lb/inch^2)	1.4 (9.0)	1.8 (12.0)	2.4 (16.0)
Heat Distortion Temperature 0.45 MPa (66 psi), °C (°F)	77 (170)	100 (212)	104 (220)
1.80 MPa (264 psi), °C (°F)	57 (135)	90 (195)	94 (202)

TABLE 5

PROPERTIES OF A LOW DENSITY SRIM SYSTEM

Glass %	16	20
Density, g/cm^3 (pcf)	0.25 (16)	0.40 (25)
Flexural Modulus, MPa (psi)	620 (90,000)	1030 (150,000)
Tensile Strength, psi	15 (2,100)	21 (3,100)
Elongation, %	2	2
Unnotched Charpy Impact, J/m^2 (ft-lb/inch^2)	0.5 (3.5)	1.0 (6.6)
Heat Distortion Temperature at 1.8 MPa (264 psi), °C (°F)	68 (155)	94 (200)

TABLE 6

PHYSICAL PROPERTIES OF A LOW DENSITY CORE FOAM

RIMline® RS-100-10

Molded Density, Kg/m^3 (pcf)	130 (8)	275 (17)
Tensile Strength, MPa (psi)	2.3 (340)	4.6 (670)
Elongation, %	11	9
Shore D Hardness	10-14	25-28
Flexural Modulus, MPa (psi)	118 (17,000)	250 (37,000)
Compressive Strength, Kg (lb)	12 (26)	36 (80)
HDT, at 1.8 MPa, °C (°F)	46 (115)	46 (115)
CLTE, mm/mm/°C x 10^6 (in/in/°F x 10^6)	5.7 (3.5)	5.9 (3.3)

DIMENSIONAL STABILITY, % change (measured at RT)

120°C/1 hour (250°F/1 hour)	l + -0.3 w = -0.2 t = +0.6	l = -0.02 w = 0.0 t = +0.2
175°C/1 hour (350°F/1 hour)	l = -1.0 w = -1.0 t = -0.8	l = -0.9 w = -0.8 t = 0.2

MOLD FILLING ANALYSIS OF STRUCTURAL REACTION INJECTION MOLDING (SRIM) AND RESIN TRANSFER MOLDING (RTM)

M. J. Liou, W. B. Young, K. Rupel
K. Han, L. J. Lee
Engineering Research Center for Net Shape Manufacturing
The Ohio State University
Columbus, OH 43120 USA

Abstract

This study is focused on characterizing the mechanism associated with mold filling in the SRIM and RTM processes by combining results obtained from flow visualization studies and from monitoring the pressure drop of a fluid as it flows through a mold containing pre-located glass fiber reinforcements. A numerical simulation model has been developed to predict the flow front progression and pressure field during mold filling. The simulation results are compared with those from experiments.

Introduction

In the reaction injection molding (RIM) process, two streams of reactive monomers are pumped and metered from separate tanks under high pressure (usually 1000 to 2000 psi) into a mix head (mix chamber), where they are directed at one another to cause impingement mixing. The monomer mixture then flows into the mold where it cures. Even though mixing occurs at a high pressure the monomer mixture is injected into the mold at a relatively low pressure (about 50 psi). The advantages of RIM over traditional thermoplastic injection molding include a much lower clamping pressure and lower energy requirements during processing. These advantages are due mostly to the low viscosity of the monomer mixture and the fast exothermic chemical reaction of the monomers.

Researchers have been trying to increase the strength of the parts produced by RIM process. One approach to strengthen RIM parts is to add fillers or fibrous reinforcements to the resin and, then, cure the resin so that the fillers and fibers trapped in the cured resin. Fillers, such as glass microspheres, calcium carbonate particles, and fibrous reinforcements, can be added to the RIM part during manufacturing in one of two ways. In reinforced reaction injection molding (RRIM), milled or chopped fibers are added to the monomer streams either before or immediately after the streams are mixed and before they enter the mold. Fillers can be added to the monomer streams in a similar way. This method of reinforcement is similar to thermoplastic injection molding with short fiber reinforcement and suffers from some of the same problems. The reinforcement is usually very abrasive and results in high wear rates for the injection equipment. In addition, once the fibrous reinforcements are in the mold, they tend to align with the flow direction creating directional properties in the final part. If the flow direction and the loading direction do not coincide, these directional properties may be detrimental to the material strength in the direction needed.

The second method of adding reinforcement material to a RIM part is called structural RIM or SRIM. In a SRIM process, the reinforcement fibers are preplaced in the mold. These fibers add extra resistance to the flow field so the pressures needed to fill the mold are not as low as those required for RIM. In fact, the filling pressure may reach 400 psi, which is still much lower than the pressures encountered in many other plastic and fiber reinforced plastic processes, such as thermoplastic injection molding and compression molding [Eckler, 1987].

Another similar process used to create continuous fiber reinforced thermoset polymeric composites is resin transfer molding (RTM). RTM follows the same process as SRIM except that the streams are mixed by mechanical means rather than impingement. The resin mixture is then pumped into the mold and through preplaced fiber mat. Different types of mixing allow the using of thermoset resins which can not be mixed via impingement. The different types of resins used for RTM and SRIM have very different viscosities and curing time, which lead to differences in filling time and injection rates. In a SRIM process, the monomers start to react very quickly meaning that the part must be filled within about 6 to 20 seconds [Eckler, 1987]. In a RTM process, however, cure time ranges from 20 minutes to several hours, therefore lowering the filling rate will not significantly add to the cycle time. Since the SRIM monomer mixture has a very low viscosity (less than 100 cP), high flow rates through the fibrous reinforcement can be achieved with reasonable pressures. The SRIM resins are designed to snap cure, meaning that very little or no reaction occurs until the mold is filled, after which the reaction proceeds very quickly [Molnar, 1988]. The viscosity of SRIM resins are usually in the range of 150 - 250 cP at room temperature, but at the molding temperature, they can drop to 100 cP or less. RTM resins, on the other hand, are typically in the range of 700 - 1000 cP [Eckler, 1987]. RTM resins are pumped into the mold at a much slower rate than the SRIM resins due to their higher viscosities. Since the RTM resins do not cure as quickly as the SRIM resins, RTM filling time can be greater than SRIM filling time without creating a short shot (resin setting up before the mold is filled). Slower filling rates mean lower filling pressures are required, and the force needed to keep the mold closed is also reduced. These low clamping forces often allow fiberglass-epoxy molds to be used, which are cheaper than the conventional steel molds.

The focus of this study is to characterize the mechanism associated with mold filling in the SRIM and RTM processes by combining results obtained from flow visualization studies and from monitoring the pressure drop of a fluid as it flows through a mold containing pre-located glass fiber reinforcements. To facilitate this study, an acrylic mold was constructed and the flow patterns of non-reactive fluid flowing through various layers, types, and combinations of preplaced glass fiber reinforcement mats were photographed. A computer filling simulation model that can handle permeability variations was also developed [Liou et al., 1989]. This model can predict flow front position and filling pressure as a function of time with a constant flow rate. The predictions of this model are compared with data obtained from experiments with spatial variations in permeability to test the validity of the model.

Experiments

The experimental work was conducted with the mold filling apparatus developed by Molnar [1988] and diagramed in figure 1. A constant speed pump was created by mounting a 3 1/4 inch diameter hydraulic cylinder in the test section of a Instron Universal Testing Instrument (model 1137). The cylinder was filled by a hand operated rotary feed pump. After the cylinder was filled and the valve to the transfer pump closed, a valve to the mold was opened and the Instron's crosshead was set to descend at a constant speed. The descending crosshead pushed the cylinder ram back into the body of the cylinder, thereby forcing the nonreactive fluid (DOP oil) into the mold. The fluid flow rates were varied by varying the speed of the crosshead.

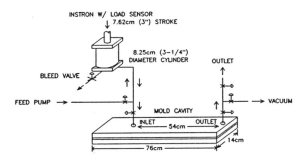

Figure 1 Schematic diagram of the SRIM/RIM mold filling apparatus

The 8 inch by 31.5 inch plattens of this mold were constructed out of 1/2 inch clear acrylic sheet so progress of the flow front could be observed during the filling process. The mold cavity thickness was created and varied by inserting different rectangular spacers with a large rectangular hole cut in them between the acrylic platens. The thickness of the spacer and the area of the hole in the spacer created the volume of the mold. The mold was sealed by glueing 1/32 inch rubber sheet to the spacer so that when the mold was shut the rubber sheet would seal the seams between the mold halves and the spacer. Super glue was found to hold rubber to the aluminum spacers particularly well. The assembly was clamped together with steel angle irons and C-clamps. Steel angle iron or flat plate was used between the C-clamps and the platens to spread the force applied by the C-clamps and prevent damage to the acrylic platens from the concentrated force of the clamps. When visual inspection of the flow front progress was not required from one or both sides of the acrylic platen, the platen on that side was reinforced with 1/4 inch steel plates to reduce the bending of the platen from the mat compression force. The only modification to the mold filling apparatus was the method in which filling pressure was measured. Molnar [1988] computed the pressure from force measurements taken by a load cell mounted on the Instron's crosshead. We measured the pressure with a pressure transducer tapped into the filling line just before it entered the mold.

Table 1 shows three types of fiber mats used in this study. Trevino [1989] conducted permeability tests for these three reinforcements. Two devices with well-defined flow field were constructed to measure the permeabilities in the flow direction (K_x) and the transverse direction (K_z). The permeabilities of the different types and combinations of glass fiber mats were obtained as functions of flow rate and mat porosity. The results were input to the database in the computer filling simulations.

Table 1 Reinforcement mats tested.

Mat Type	Mat Part Number
continuous random glass fiber	OCF-M8610
stitched bidirectional glass fiber	CoFab A1118B
stitched unidirectional glass fiber	CoFab A0108

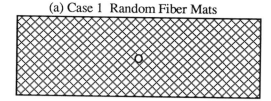

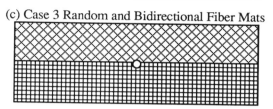

Figure 2 Fiber mat arrangements for three cases (a) 6 layers of random (b) 4 layers of random with 6 layers of random patches (c) 4 layers of random and 6 layers of bidirectional

Results and Discussion

Three different arrangements of fiber mats (figure 2) were used to investigate the filling characteristics. In case 1, the whole cavity is filled with six layers of random fiber mats. In case 2, four layers of random fiber mats are filled the cavity and two more layers of random fiber mats are added to two small rectangular areas as shown in the figure. Therefore, porosity and permeability variations are created in the plane. In case 3, random and bidirectional fiber mats are placed side-by-side in the mold cavity and the porosity as well as permeability is different in these two regions. Permeability variations in the part may occur for several different reasons. If a part needed to be stronger near an attachment point without changing the thickness of that part, extra reinforcement could be added to increase the fiber content and strength in that area. On the other hand, if sufficient strength couldn't be obtained by using all random reinforcement mat in the stacking sequence, some of the random mats could be replaced by directional reinforcement mats.

Figure 3 is a plot of the predicted position of the flow front with time from simulation and the actual position of the flow front from experiment at three different times. There is a good agreement between the simulation and the experiment for the flow front position as the nonreactive fluid fills this mold cavity. The shape of the flow front at 7 seconds, however, does not agree well with the simulation results. The leading edge of the flow front is at the wall due to flow channelling between the reinforcements and the edge of the mold. This often occurs even if the reinforcements were completely against the wall as they were in this case. Figure 4 is a plot of the transient pressure prediction from two dimensional simulation and pressure measurements. The flow contacted the side wall of the mold after approximately 3 seconds and it can be seen that the relationship between pressure and time becomes linear at approximately the same time. This is because the filling is a radial flow before the fluid contacts the side walls and it is a flow through constant cross section after contacting the side walls.

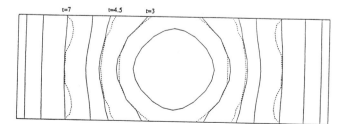

Figure 3 Flow front contour of case 2 (porosity = 0.66, center gate), ------experimental,——— numerical

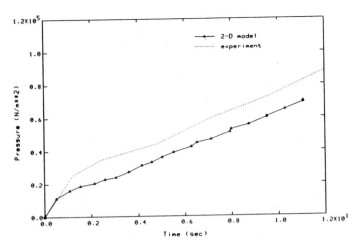

Figure 4 Mold filling pressure of case 1 (random fiber mats)

Two different variations of the experiment were run. For one experiment, a hole was cut through the random mats under the gate so that there would be no flow resistance in the thickness direction at that entrance. The experiment was also run without the hole through the mats under the gate. The two dimensional filling simulation ignored any flow resistance in the thickness direction. The difference in pressure between the experiment with the hole under the gate and the experiment without the hole was small, indicating that the flow resistance through the random mat in the thickness direction is relatively small and the out of plane flow near the gate does not add significantly to the filling pressure when this random reinforcement mat is used.

In the second case, extra reinforcement mats were added in specific areas of the mold to study the effects of local permeability variations. In industrial practice, these variations could be either intentional or due to variations in the porosities of the mats. The permeability variations were created by adding two more layers of mats. The two patches were an equal distance away from the flow entrance. The position of the patches, along with the predicted and actual flow front positions, can be seen in figure 5. As in the previous set of experiments a hole was again cut through the mats under the gate to eliminate flow resistance in the thickness direction. From figure 5 it can be seen that even by increasing the fiber content by 50% in the area of the patches the permeability in that area was not altered enough to create a dry (resin free) area in the part. In fact, the porosity decrease only slowed the advancement of the flow front a small amount. The pressure traces of the computer simulation and two experiments are presented in figure 6. Here, several different phases of the pressure trace can be distinguished. The flow front met the side wall after about 3.5 seconds, and it could be seen that there was a transition in the pressure increase from logarithmic to linear at this time as there was in the uniform random mat experiments. At about 7 seconds the flow front encountered the patches of lower porosity mat and the slope of the pressure trace increased and then decreased again (at about 9 seconds) when it cleared the patch. At about 10 seconds the experimental pressures increases start to trail off because the flow front had already penetrated all the mat and was being vented from the mold. The two experimental pressure traces are always diverging indicating that the permeability of the reinforcement was different for the two runs. The number of mats, mold thickness, and position of the mats were the same for both experiments, so the pressure variation must be due to a variation in the material properties of the mats. The surface density variation of the random mat can be rather large.

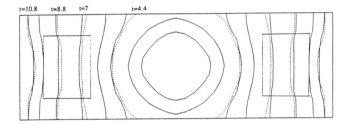

Figure 5 Flow front contour of case 2 (porosity 1 = 0.78, porosity 2 = 0.692, center gate), ------- experimental, ———numerical

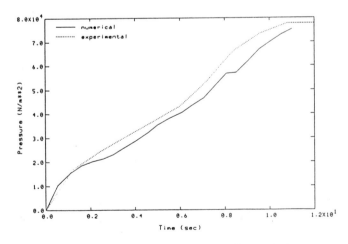

Figure 6 Mold filling pressure of case 2 (non-uniform random fiber mats)

Experiments were also run by using random and bidirectional mat side by side in the mold cavity. Since the flow front expands more slowly in the bidirectional mats than in the random mats as shown in figure 7, the random mat in this case has a higher permeability. Two experiments were run with the side by side, bidirectional and random, mat arrangement. One with a hole cut through the mats in the thickness direction under the gate and one without the hole. The pressure traces for both experiments and the two dimensional computer model are plotted in figure 8. The computer model did not consider flow resistances in the thickness direction and matched the experiment with the hole cut through the mats much better than the experiment without the hole.

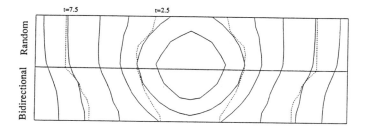

Figure 7 Flow front contour of case 3 (porosity R = 0.777, porosity B = 0.5235, center gate), ------- experimental, ———numerical

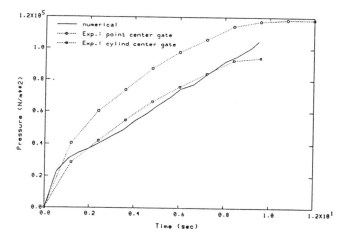

Figure 8 Mold filling pressure of case 3 (random and bidirectional fiber mats)

Conclusions

In a center-gated mold the flow enters from the top or bottom of the fiber stack. Therefore, in areas near the flow entrance the flow must pass through the thickness of the part. The permeability in the thickness direction has a large effect on the total flow resistance of the part. The two dimensional in-plane computer filling model was found to accurately predict the flow front position and filling pressure when the random mats, which have high permeability in the thickness direction, were used or when a hole was cut through the thickness of the mat stack under the flow entrance. The effects of in-plane permeability variations, which may be due to porosity variations in fiber mats or local strength reinforcement in composite parts were also characterized. Further research to characterize the flow in the thickness direction is under way.

Acknowledgement

The authors would like to thank the Engineering Research Center for Net Shape Manufacturing at the Ohio State University for the financial support. The participation and interests of the member companies, Ashland Chemical, Gencorp, General Motors, Owens Coring Fiberglass and Union Carbide, are gratefully appreciated.

References

Eckler, James H., Ashland Chemical Company, private communication, 1987.

Liou, M. J., W. B. Young, K. E. Rupel, K. Han and J. L Lee, "Filling Simulation and Experimental Verification in Resin Transfer Molding (RTM) and Structural Reaction Injection Molding (SRIM)" paper submitted to Polymer Composites.

Molnar, John A., "A Mixing and Molding Filling Study in Reaction Molding", M.S. Thesis, The Ohio State University, 1988.

Trevino, Lisandro, "Study of The Permeability of Fibrous Reinforcement Mats Used in Resin Transfer Molding and Structural Resin Injection Molding", M.S. Thesis, The Ohio State University, 1989.

THE PROBABILISTIC NATURE OF FRACTURE IN CARBON-CARBON COMPOSITES

H. Aglan, A. Moet
Case Western Reserve University
Cleveland, OH USA

Abstract

Three-directional (3D) carbon fiber reinforced carbon (C/C) composites are extremely heterogeneous materials. The fracture behavior of a 3D C/C composite is analyzed in view of tests conducted on a sample set of 14 identical specimens. Three-point bending tests reveal a twofold scatter in the maximum stress at which crack initiation occurs. Subsequertly, a fourfold scatter in the critical energy release rate G_{IC} is noted. The G_{IC} data fit a normal distribution with an average of 4.6 KJ/m^2 which ranks the composite tougher than other high temperature materials such as ceramics. It is also noted that this composite exhibits significant stable crack propagation before ultimate fracture occurs.

IN RECENT YEARS, the mechanical and thermal properties of carbonaceous materials have significantly improved due to the addition of reinforcing fibers to the bulk carbon as well as the arrangement (architecture) of these fibers in the matrix. Three-directional C/C composites represent one of the most important types of C/C composites. This is due to the fact that their preform structure can be tailored in three directions reflecting the final property required of the material.

These materials possess a coarse orthotropic texture due to the relatively large yarn bundles coupled with the anisotropy of the graphitic matrix. An important consequence of such anisotropy is the microcrack network induced by thermal stress during manufacturing [1,2]. These microcracks are formed both at the interface and within the yarn bundles (Figure 1). Obviously, microcracks are sites of stress concentration and some may propagate under load, but their network assembly would plausibly lead to the arrest of propagating microcracks. Hence the microcrack/void network

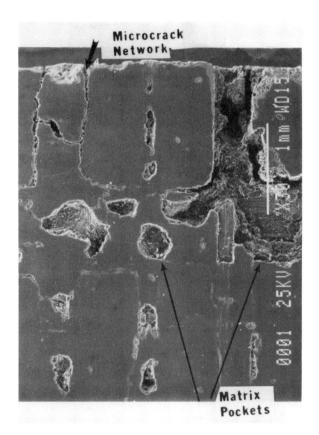

Fig. 1 - Photomicrograph of a polished cross-section of the 3D C/C composite showing some mini-mechanical features

could play two contrasting roles: embrittling and toughening.

The microstructural complexities in 3D C/C composites suggest that their fracture behavior is perhaps different from that of homogeneous isotropic materials. They can cause a wide scatter in their fracture behavior. The few reported studies which attempted to characterize the fracture toughness of other C/C composites [3-9] recognized these complexities.

In the present paper, the fracture behavior of a 3D C/C composite is explored in view of the results obtained from a sample set of macroscopically identical specimens fractured under the same test conditions. The study aims at exploring the influence of microstructural heterogeneities within the composite on the fracture toughness.

MATERIAL AND TESTING

The material considered in the present work is a 3D C/C composite ring segment (Fig. 2a) obtained from The Aerospace Corp., Los Angeles, CA. The average thickness of the ring is about 1/4".

Fracture tests were conducted on a three-point bend specimen under monotonic loading. The test specimens were machined into 56 x 14 mm beams from the billet section as shown in Figure 2b. A 3 mm notch was introduced at the middle of each specimen using a 0.38 mm thick milling cutter. The notch was cut in the radial direction of the ring, i.e., the crack was always initiated from the inside to the outside of the ring, as shown in Figure 2a.

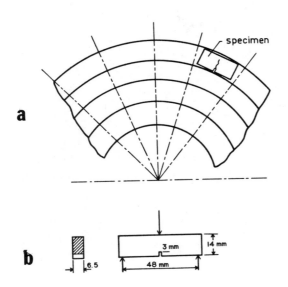

Fig. 2 - Specimen orientation and geometry

Fracture testing was carried out at room temperature using an MTS testing system equipped with a 1 kip load cell. The test was run using displacement control with a crosshead speed of 50 μm/min. A specially designed Instron stiff steel load fixture was used to insure minimum machine compliance

The load vs. the load-point displacement (LPD) curves were generated by continuously monitoring the load and crosshead displacement during tests. A clip gauge was also used to monitor the crack opening displacement (COD). Both load vs. LPD and load vs. COD curves were plotted instantaneously during the tests on X-Y plotters.

The tip of the notch was viewed using a traveling optical microscope. A video system was attached to the microscope in order to obtain interval records of crack propagation and to record any damage events which could be observed.

In the present study 14 macroscopically identical notched specimens were fractured under identical test conditions. Unnotched specimens were also fractured under the same conditions.

RESULTS AND DISCUSSION

Typical load versus LPD and load versus COD curves for one specimen of this composite are shown in Figures 3 and 4, respectively. Both curves exhibit the same trend. At the beginning there is a linear portion followed by another of increased nonlinearity. After a peak is reached, major damage events appear to have taken place causing successive load drops. Each load drop is preceded by a constant load plateau.

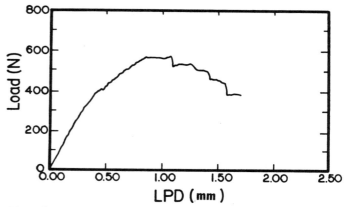

Fig. 3 - Load versus load-point displacement (LPD) for one specimen of the 3-D C/C composite

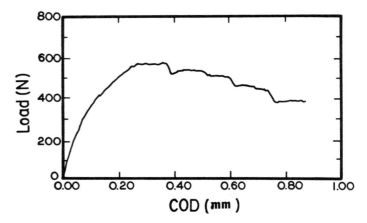

Fig. 4 - Load vs. crack opening displacement (COD) for the same specimen shown in Fig. 3

STRESS ANALYSIS

The maximum bending and shear stresses were evaluated in the middle of each specimen. The maximum bending stress is

$$\sigma_{max} = \frac{My}{I} \quad (1)$$

where M is the maximum bending moment sustained by the specimen, $y = W/2$ (W is the width of the specimen), and $I = BW^3/12$ is the moment of inertia of the section (B is the thickness of the specimen). The maximum shear stress is

$$\tau_{max} = \frac{3}{2}\frac{P_{max}}{A} \quad (2)$$

where P_{max} is the maximum load sustained by the specimen and A is its cross sectional area. The bending and shear stress fields are shown in Figure 5.

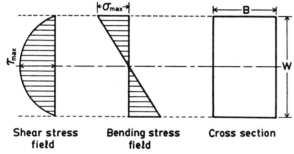

Fig. 5 - Bending stress and shear stress fields at the middle cross-section of the specimen

Young's modulus was calculated using linear elastic beam theory from the load vs. LPD curve of unnotched specimens. The expression for Young's modulus takes the form

$$E = \frac{Pl^3}{48I\Delta} \quad (3)$$

where P is the load, l is the span of the beam, Δ is the LPD at mid-span. Only the linear portion of the load vs. LPD curve was used in the Young's modulus evaluation. Any bimodular effects on the bending deflection were ignored. The value of Young's modulus for the 3D composite under investigation was found to be about 2.3 GPa.

A histogram and the associated normal probability distribution of the maximum bending and shear stresses for the 14 identical specimens are shown in Figs. 6 and 7, respectively. The wide scatter in both the values of σ_{max} and τ_{max} attests to the random nature of the strength field of this composite. A scatter in the values of σ_{max} from 25 to 45 MPa is observed for the "same" material (Fig. 6). Also, a twofold change in the values of the maximum shear stress is seen in Figure 7.

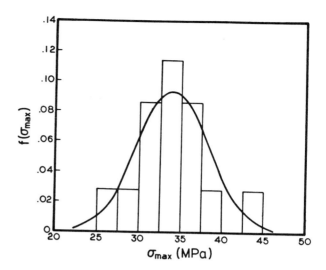

Fig. 6 - Histogram and the associated normal distribution of the maximum bending stress

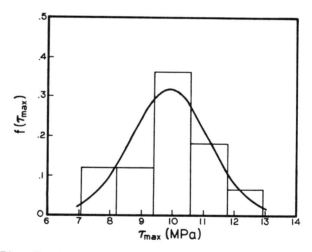

Fig. 7 - Histogram and the associated normal distribution of the maximum shear stress

FRACTURE TOUGHNESS

Fracture toughness is traditionally characterized by the critical stress intensity factor K_{Ic} [10] for brittle materials complying with linear elasticity. On the other hand the critical energy release rate J_{Ic} is employed for ductile material displaying elastic-plastic behavior. However, the following relationship generally holds under plane strain conditions.

$$J_{Ic} \equiv G_{Ic} = K_{Ic}^2(1-\nu^2)/E \quad (4)$$

where E is Young's modulus and ν is Poisson's ratio.

Neither the load-displacement behavior of this material (Figures 3, 4) nor the crack tip mechanisms indicate clear elasto-plastic bearing. Accordingly, G_{Ic} is used to characterize the fracture toughness of our sample set. Note, however, that stable crack growth did occur past the maximum load as indicated by the observed load drops in Figures 3 and 4. Thus G_{Ic} was computed as a measure of the energy released in association with the onset of stable crack growth.

For our material, Poisson's ratio is approximated as 0.1 [12]. Hence, it has been neglected and calculated as K_{Ic}^2/E. K_{Ic} for the bending specimen used is given by [13]

$$K_{Ic} = \sigma_o \sqrt{\pi a_o} \cdot f(a_o/W) \quad (5)$$

where σ_o is the maximum bending stress, a_o is the notch length and W is the specimen width. The geometric correction function $f(a_o/W)$ is given by [13]

$$f(a/W) = 1/\sqrt{\pi} \cdot \frac{1.99 - (a_o/W)(1-a_o/W)[2.15 - 3.93\, a_o/W + 2.7(a_o/W)^2]}{(1+2\, a_o/W)(1-a_o/W)^{3/2}} \quad (6)$$

A histogram and the associated normal probability distribution of G_{Ic} for the 14 specimens are shown in Figure 8. The scatter in the value of the energy release rate is evident. A fourfold change in the value of G_{Ic} is observed for the 14 identical specimens which reflects the random nature of the fracture toughness of the composite.

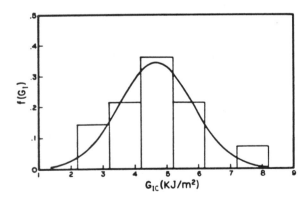

Fig. 8 - Histogram and associated normal distribution of the energy release rate at the notch tip corresponding to maximum load

The results discussed above indicate that fracture toughness of 3D C/C composite should not be considered as a deterministic parameter. In addition to the average value of G_{Ic} (G_{Ic} = 4.6 kJ/m^2), one ought to consider the standard deviation as well for accurate material selection. In comparison with other high temperature materials such as toughened ceramics [14], the examined composites appear to display higher fracture toughness.

SUBCRITICAL CRACK GROWTH

An important feature of the fracture behavior of 3D C/C composite is that it exhibits stable "crack" propagation past maximum load. Thus G_{Ic} reported here (Figure 8) is that associated with the onset of fracture. The material subsequently provides an increased resistance to further crack propagation. In addition, the crack trajectory diffuses in a complex path across and along the bundles, perhaps tracking microcracks and voids preexisting in the material.

A slow playback of the recorded videotape reveals that the crack initiated from the center of the notch. It then advanced by a series of steps involving matrix and cross bundle fracture processes. These steps correspond to distinct load drops. A stable crack in the specimen, Figure 9, advanced about 3.7 mm from the notch corresponding to about four yarn bundles; then unstable (critical) crack propagation occurred leading to ultimate failure.

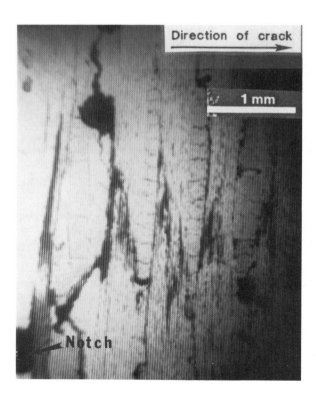

Fig. 9 - Photomicrograph of the propagating crack immediately before failure (taken from the video screen)

CONCLUSIONS

(1) Fracture of 3D C/C composites displays the main features of probabilistic brittle failure. This is manifested by a twofold scatter in the maximum bending stress, the maximum shear stress and in the deflection at maximum load. Accordingly, the scatter in the energy release rate is about fourfold.

(2) The average critical energy release rate of 3D C/C composite is about 4.6 kJ/m^2.

(3) The material displays significant stable crack propagation prior to its ultimate failure.

ACKNOWLEDGMENT

Financial support of this research by the Office of Naval Research is gratefully acknowledged. We are also thankful to Dr. G. Rellick of The Aerospace Corporation for providing the material, and to Drs. J. Jortner, L.H. Peebles, and A. Chudnovsky for their useful suggestions.

REFERENCES

1. Jortner, J., Carbon 24, 603 (1986)
2. Peebles, L., R. Mayer and J. Jortner, Proc. 2nd Int'l. Conf. on Compomposite Interface, Cleveland, OH, June 1988
3. Hettche, L. and T. Tucker, Crack and Fracture, ASTM STP, 602, 109, (1976)
4. Adam, D. and E. Odom, Composite 18, 5 (1987)
5. Guess, T. and W. Hoover, J. Comp. Materials 7, 2 (1973)
6. Kagawa, Y., S. Utsunomiya and M. Imazume, Proc. 30th Japan Congress of Material Research, 1987
7. Jenkins, M., A. Kobayashi, A. White and R. Brandt, Int. J. Fracture 34, 291 (1987)
8. Jenkins, M., J. Mikami, T. Chang and A. Ojura, SAMPE Journal (May 1988)
9. Phillips, D., Comp. Materials 8, 130 (1974)
10. ASTM Standard E399-81, Standard Test Methods for Plane Strain Fracture Toughness of Metallic Materials, 1981
11. ASTM Standard E813-81, Standard Test for J_{Ic} A Measure of Fracture Toughness, 1981
12. Jortner, J., private communications
13. H. Tada, "Stress Analysis of Cracks Handbook", 2nd Ed., p. 216, Del Research Corp., St. Louis (1985)
14. Faber, F. and A. Evans, in "Fracture in Ceramic Materials", p. 104, A.G. Evans, Editor, Noyes Publishing, Park Ridge, NJ (1984)

MECHANICAL RESPONSES FROM SCALE MODELS OF CARBON FIBER REINFORCED COMPOSITES

Yi Chen, Ajit K. Srivastava, Madhu S. Madhukar
Composite Materials and Structures Center
Michigan State University
East Lansing, MI 48824 USA

ABSTRACT

Scale models of beams made of unidirectional carbon fiber reinforced composites, CFRC (AS4/828/mPDA), were tested in a three-point loading system. The length scale factors were 1/3, 1/2, 2/3 and 1, corresponding to 12, 18, 24 and 36 plies of laminates, respectively. Scaled beams showed different moduli of elasticity in bending. The full scale beams showed the highest bending stiffness while the 1/3-size beams showed the lowest bending stiffness. Significant differences in rupture stresses and strains were also observed among these scaled beams. The smallest scale beams had the highest strength and strain at rupture, whereas the full scale model ruptured at the lowest stress and strain. Tensile tests showed similar tendencies that extensional stress and strain at rupture decreased as the number of plies in CFRC increased. These results indicated that traditional assumptions of common material properties may not be valid in scale modeling of CFRC structures. However, size had insignificant effects on Young's modulus and Poisson's ratio in tensile tests. Non-linear hardening behavior and failure without yield were commonly observed in the stress-strain curves of all beams and tensile specimens tested.

INTRODUCTION

THIS PAPER presents some experimental results of the effects of mechanical properties of carbon fiber reinforced composites (CFRC) on scale modeling under the conditions of flexural and tensile loading. The purpose of this on-going research is to understand and develop the scale laws that govern the prediction of mechanical response of CFRC material.

Despite the current tendencies toward computerization, researchers in engineering and physical sciences rely more than ever on experimentation. Scale models provide a means of studying physical systems or verifying mathematical models of real systems which may be too complex and too little explored. The scaled reproduction of a physical phenomenon or an engineering system can be very advantageous. Scale models permit transformation to manageable proportion of systems that, like a military bridge, may be too large and too dangerous for direct experimentation; or, like a flying space craft, inaccessible; or, like a fire storm, unmanageable; or, like seepage, too slow to work with. Scale models shorten experiments and promote a deeper understanding of the phenomenon under investigation. Scale models are especially cost-effective for failure analysis.

Scale models have been used efficiently and effectively to study response of large complex structures under various loading conditions. In literature, there are numerous publications on the application of similitude models in studying structural response to loads varying from quasi-static, dynamic, to impact in nature. Structural models of Saturn V rockets were used to study resonant frequencies, bending mode shapes, and damping [1]. Energy absorbing characteristics of automobile frames have been studied by using scale models [2]. Plastic models were used to develop frames for a grain harvester, and stress coating was applied to determine the points of stress concentration on the frame [3]. Scale modeling was used to study the deformation of roll-over protection structure (ROPS) for agricultural tractors [4], and rate effects on structures subjected to impact loading [5].

In space science, military and industry, there has been increasing interest in designing structures using modern composite materials due to their high performance and numerous desirable attributes including favorable strength to weight ratio. Because of insufficient information base for design, composite structures are extensively evaluated on the prototype scale, even through destructive testing. Hence, there is growing attention to apply model-scaling techniques in design of composite structures. Scale models were used in verifying buckling forms and explosion proof tests of naval ship hulls [6]. Recently, Morton [7] reported scaling of impact-loaded carbon fiber composites. Jackson and Fasanella [8 and 9] also studied scale effects in responses of graphite/epoxy composite beam-columns due to large deflection and impact loading. Significant size effects on strength were noticed [7, 8 and 9]. However, size effects on material properties have not yet been considered in literature. Classical assumption of common properties for a given material regardless of its size are still applied to composite materials. Scaling laws that govern model design of composite structures have not been fully developed. The anisotropic and non-homogeneous nature of composite materials requires deeper understanding and study of various behavior of composite materials with respect to size in order to establish appropriate model scaling laws.

In this study, static testing of carbon fiber reinforced composite beams, based upon dimensional analysis, were conducted, and some material properties were experimentally determined. The analysis and test results are discussed in following sections.

DIMENSIONAL ANALYSIS

Dimensional analysis is the basis of scale modeling. For engineering problems, physical quantities are all associated with certain dimensions. Relevant quantities and their dimensions are first identified. These quantities are then combined to formulate a set of dimensionless groups, called Pi-terms. Buckingham's Pi Theorem states that the number of Pi terms equals the total number of physical quantities minus the number of fundamental dimensions involved in the system [10]. For a static three-point flexural beam problem, pertinent quantities and their dimensions are listed in Table 1.

Table 1. Parameters and Dimensions for Three-Point Flexure Test

Parameter	Symbol	Dimension
Beam span	l	L
Beam width	b	L
Beam thickness	h	L
Central load	P	F
Deflection	d	L
Elastic modulus	E	FL^{-2}
Poisson's ratio	ν	1
Ultimate stress	σ_u	FL^{-2}

Generally, one is interested in predicting deformation of structures as a function of the external loading, the size of the structure, and the material properties. In the case of a simply supported beam, an equation of the following form may be written for the mid-span where deflection, bending moment, stress and strain are the largest:

$$d = f(P, l, b, h, E, \sigma_u, \nu) \quad (1)$$

The above equation may also be written in the following non-dimenional form using the Buckingham's Pi Theorem:

$$\left(\frac{d}{l}\right) = \phi\left(\frac{Pl^2}{EI}, \frac{l}{h}, \frac{b}{h}, \frac{\sigma_u}{E}, \nu\right) \quad (2)$$

or

$$\pi_1 = \phi(\pi_2, \pi_3, \pi_4, \pi_5, \pi_6) \quad (3)$$

whereas

$$\pi_1 = \frac{d}{l} \qquad \pi_4 = \frac{b}{h}$$

$$\pi_2 = \frac{Pl^2}{EI} \qquad \pi_5 = \frac{\sigma_u}{E}$$

$$\pi_3 = \frac{l}{h} \qquad \pi_6 = \nu$$

From the above Pi terms, design conditions for scale models may be developed as follows:

$$\frac{P_m l_m^2}{E_m I_m} = \frac{Pl^2}{EI} \quad (4)$$

$$\frac{l_m}{h_m} = \frac{l}{h} \quad (5)$$

$$\frac{b_m}{h_m} = \frac{b}{h} \quad (6)$$

$$\frac{\sigma_{um}}{E_m} = \frac{\sigma_u}{E} \quad (7)$$

$$\nu_m = \nu \quad (8)$$

whereas the subscript m denotes a variable in the model system.

If all of the above design conditions are satisfied by the model structure, then the following equation may be used to predict the deformation of the full-scale (prototype) structure:

$$d = \left(\frac{l}{l_m}\right) d_m \quad (9)$$

let $K_l = l_m/l$ be the length scale factor. If complete geometric similarity is maintained, the design conditions of (5) and (6) are satisfied. From (4), (7) and (8), we have,

$$K_P = \frac{P_m}{P} = K_E K_l K_l^{-2} \quad (10)$$

$$K_{\sigma_u} = \frac{\sigma_{um}}{\sigma_u} = K_E \quad (11)$$

$$K_\nu = \frac{\nu_m}{\nu} \quad (12)$$

Whereas K_P is the load scale factor; K_E is the scale factor for the elastic modulus; K_{σ_u} is the scale factor for the ultimate strength; and K_v is the scale factor for Poisson's ratio. It should be noted that the scale factor for area moment of inertia $K_I = K_L^4$. The relationships among the scale factors given in Equations (10), (11) and (12) should be used to design scale models.

In studies of scale modeling of structures made of conventional material, the model is generally fabricated out of the same material as the prototype. This would results in,

$$K_E = K_{\sigma_u} = K_v = 1$$

Thus, the design conditions of (11) and (12) would be inherently satisfied and from (10), $K_P = K_L^2$. For example, the load applied to a 1/4-scale model would be 1/16th that of the prototype load. In the study reported here, we were interested in investigating whether or not the design conditions of (10), (11), and (12) can be satisfied by model structures made of carbon fiber reinforced composites.

EXPERIMENTAL PROCEDURE

Material Preparation - The fibers used in this study were Hercules AS4 graphite fibers with a tensile modulus of 234 GPa (34,000 ksi) and a nominal tensile strength of 3.6 GPa (520 ksi). A diglycidyl ether of bisphenol-A (EPON 828, Shell Chemical Company) cured with 14.5 parts per hundred by weight of meta-phenylene diamine (m-PDA, Aldrich Chemical Company) was chosen as epoxy matrix.

Unidirectional composite prepregs were fabricated from the carbon fibers. The impregnation was done using a hot-melt prepregger with a slit die (Research Tool Corporation). Resin temperature was maintained at 52°C (125°F) while the fiber tow was drawn through a narrow outlet die which formed a thin impregnated layer of composite which was wound onto a cylindrical mandrel at a linear rate of about 3 m (9 feet) per minute. Approximately 1 kg (2 lbs) of tension was maintained on the tow during processing. Single layer tapes of fixed fiber volume content were prepared in this manner. Sections of dimensions 279 mm x 305 to 432 mm (11 in x 12 to 17 in) were cut from the tape and manually stacked in the desired sequence. A standard autoclave (United McGill Corporation) was used for consolidation. The curing cycle was 0.69 MPa (100 psi) for two hours at 75°C (167°F) followed by two hours at 125°C (275°F). The laminate was cooled to ambient temperature before removal from the autoclave. Approximately 12.7 mm (0.5 in) wide strips were cut from all edges of each plate and discarded. Specimens of dimensions recommended by ASTM specifications were cut from the remaining of each plate by use of a diamond coated wheel.

Tensile Tests - Specimens for 0° tensile tests were made from panels of four different lay-ups (12, 18, 24, and 36 plies). However, those tensile specimens had the same length (241.3 mm) and width (12.7 mm) rather than proportional to their thickness. Cross-ply glass/epoxy tabs were applied on tension specimens using epoxy glue. The tensile tests according to ASTM D-3039 were performed on a material testing system (MTS-880). The longitudinal and transverse strains were monitored by means of strain gages as well as a biaxial extensometer. Four to five specimens of each ply number were tested.

Flexural Tests - The dimensions and number of plies of those beam specimens for 3-point flexure are listed in Table 2. Fig. 1 shows sample specimens of four different sizes.

Table 2. Dimensions of Laminated Beams

S_l	n	l (mm)	b (mm)	h (mm)	l/h	b/h
1/3	12	121.9	8.5	2.04	59.9	4.1
1/2	18	182.8	12.7	3.08	59.4	4.1
2/3	24	243.8	16.9	4.00	61.0	4.2
1	36	365.7	25.4	5.52	66.2	4.6

It is seen that the ratio of thickness between those beam specimens deviated from the proportion of the number of plies, n. Thus, distortion existed in π_3, and π_4. The values of standard deviation of thickness were 0.001, 0.002, 0.004, and 0.008 for 12-, 18-, 24-, and 36-ply of CFRC, respectively. These showed the variation in thickness among four different lay-ups. Although all panels were cured in similar environment monitored by a pre-programmed controller, the uniformity of panel thickness became worse when the number of plies increased. This can be considered as an indication of the quality of cured panels, which might affect the mechanical properties of composites studied.

A three-point loading device was used according to ASTM D-790 specifications. Beam specimens were simply supported at two ends to manage a designated span. Vertical load was applied on the top face at mid-span. Strain gages were bonded on the bottom face at mid-span. Data of central load, central deflection, measured by MTS-880, and elongation strain were acquired and stored in a personal computer via a multi-channel analog signal interface. Four to five specimens of each lay-up were tested.

RESULTS

All results presented here are in normalized forms which are dimensionless. To normalize moduli and stresses, an E_o value of 138 GPa (20 Msi) was arbitrarily chosen.

Tensile Tests - There was good agreement among measured data obtained from both strain gage and extensometer techniques. Data of elastic modulus and Poisson's ratio are plotted against strain for four sample specimens in Figs. 2 and 3, respectively. Each curve in these plots extends to a point of failure, except for specimens of 24- and 36-ply specimens which could not be loaded to failure because end-tabs debonded from the specimens. Average values of the elastic moduli corresponding to 0.2% strain are plotted in Fig. 4 for four size categories. A similar plot for the Poisson's ratio is given in Fig. 5. Size influences on tensile strengths and strains are plotted in Figs. 6 and 7, respectively. These plots do not include data for 24 and 36 plies because of the tab-debonding problem aforementioned.

Flexural Test - Load-deflection curves for four sample beam specimens are shown in Fig. 8. All beams were loaded until failure. Fig. 9 shows the non-linear behavior of bending modulus for those four specimens. The bending modulus and strain were calculated by using the following equations:

$$E_b = \frac{Pl^3}{4dbh^3} \quad (13)$$

$$\varepsilon = \frac{6dh}{l^2} \quad (14)$$

It should be noted that Eqs. (13) and (14) ignore the effect of shear deformation in the flexural tests. However, the shear effect would be minimal since the span-to-thickness ratio, called "aspect ratio", was quite large (about 60).

The values of bending modulus corresponding to 0.2% strain are plotted in Fig. 10 for different beam sizes. Size effects on bending strength and bending rupture strain are plotted in Figs. 11 and 12, respectively.

DISCUSSION

Hardening and failure without yield - It was observed that all beam and tensile specimens failed in a brittle manner without yield. Data as presented in Fig. 2 show that the Young's modulus was not constant, and increased as the strain increased until fracture occurred. Similar phenomena for bending modulus were also observed from flexural tests (Fig. 9). This indicates hardening of CFRC $[0°]_n$ material as it deforms. This can be attributed to carbon fibers in CFRC, since the reinforcement exercised the majority of tensile and bending stresses. For tensile specimens, E increased rapidly before strain reached about 0.1%, and then turned to a slow and linear pattern. In beam tests, specimens showed rapid increase in the bending modulus, and then became constant after 0.1% strain. The behavior of CFRC $[0°]_n$ near zero strain in both tensile and beam tests was not yet clarified due to noise and zero shift of testing systems. In literature, hardening of carbon fibers have been reported [12 and 13].

If the rate of hardening is the same for all of four sizes, then it would not pose problems in similitude modeling. This would be indicated by parallel curves in both Figs. 2 and 9. For our calculations, representative values of tensile and bending moduli corresponding to a 0.2 % strain were selected. The values, as shown in Fig. 4, indicated that statistically there was no size effect on Young's modulus of CFRC $[0°]_n$. Thus, the scale factor for E, $K_E = 1$. However, results of beam tests, shown in Fig. 10, clearly indicated that there was a significant increase in E_b as the number of plies increased. This is in contradiction to the results obtained by the tensile tests. No explanation for this is offered at this time. However, if the nature of loading in the structures is primarily that of bending, one should not rely on data from tensile tests to build scale models.

Poisson's ratio - Fig. 4 shows Poisson's ratio increased as tensile strain increased up to a strain level about 0.1%, and then slightly decreased. Data as shown in Fig. 5 indicates that statistically there was no change in Poisson's' ratio due to the size of the specimen. This implies that the design condition (12) would be satisfied.

Rupture stress and strain - Rupture strains was less 1.4% for all specimens of 4 sizes in beam tests, and all specimens of 12 and 18 plies in tensile tests. Data of rupture stresses and strains as determined by tensile and bending tests (Figs. 6, 7, 11, and 12) indicate that these values decreased significantly as the number of plies increased in CFRC $[0°]_n$. Similar findings have been reported in [7, 8, and 9]

Design condition (11) requires that the scale factor for modulus of elasticity and rupture strength be equal. Based on the data of bending modulus, $K_E < 1$. However, data of the bending strength showed that $K_{\sigma_u} > 1$. This indicates that condition (11) can not be satisfied and poses a major difficulty in predicting failure of prototype CFRC $[0°]_n$ structures by sub-scale models. It should also be pointed out that a complete geometric similarity was not achieved for beam specimens of different plies as indicated by the data in Table 2, and discussed in that section.

The on-going research addresses the following issues: (1) identifying the effect of shear deformation on flexural response of CFRC; (2) determining the fiber distribution, fiber volume fraction, and material quality of various lay-ups of CFRC; and (3) studying possible effects of micro-failure modes on model scaling. Appropriate scale laws for CFRC can be developed after these issues are resolved.

CONCLUSIONS

Based on the data presented in this paper, it can be concluded that:

(1) CFRC $[0°]_n$ hardens as it deforms, and it is brittle upon fracture.

(2) There is a significant difference in the values of Young's modulus and bending modulus as measured in tensile and bending tests. The bending modulus increases as the number of plies increases in CFRC $[0°]_n$, whereas Young's modulus is not affected by size. Therefore, model scaling of structures subjected to bending loads should examine the scale factor for bending modulus rather than simply relying on that for Young's modulus.

(3) Poisson's ratio as measured in tensile test remains unchanged.

(4) The rupture stress and strain significantly decrease as the number of plies increases in CFRC $[0°]_n$.

(5) A sub-scale model of CFRC $[0°]_n$ structure fabricated under the conditions given in Eqs. (10), (11) and (12) would significantly under-predict the elastic stress and over-predict the elastic deformation and the failure strength of the prototype.

REFERENCES

1. Schuring, D.J., "Scale Models in Engineering", Pergamon Press, New York (1977)
2. Holmes, S.B. And L.D. Colton, 'Scale Model Experiments for Safety Car Development', Paper No. 730073, Society of Automotive Engineers, Inc. (1973)
3. Vandewiele, J., 'Combine Frame Development Using Plastic Model Situation', ASAE Paper No. 79-1577, American Society of Agricultural Engineers, St. Joseph, MI (1979)
4. Srivastava, A.K., G.E. Rehkugler and B.J. Masemore, 'Similitude Modeling Applied to ROPS Testing', Transactions of the ASAE, vol. 21, No. 4 (1978)
5. Srivastava, A.K. and G.E. Rehkugler, 'Strain Rate Effects in Similitude Modeling of Plastic Deformation of Structures Subject to Impact Loading', Transactions of the ASAE, vol. 19, No. 4 (1976)
6. Smith, C.S., M. Anderson and M.A. Clarke, 'Structural evaluation of GRP ship designs, In Conference Proceedings, Fiber Reinforced Materials: Design & Engineering Applications. Institution of Civil Engineers, London (1977)
7. Morton, John, 'Scaling of Impact-Loaded Carbon-Fiber Composites', AIAA Journal, vol. 26, No. 8, p. 989-994 (1988)
8. Jackson, K.E. And E.L. Fasanella, 'Scaling Effects in the Static Large Deflection Response of Graphite-Epoxy Composite Beam-Columns', Proceeding of the American Helicopter Society National Technical Specialists', Meeting on Advanced Rotorcraft Structures, Oct. 25-27, Williamsburg, VA (1988)
9. Jackson, K.E. and E.L. Fasanella, 'Scaling Effects in the Impact Response of Graphite-Epoxy Composite Beams', Paper No. 891014, 1989 General Aviation Aircraft Meeting & Exposition, Apr. 11-13, Wichita, Kansas (1989)
10. Langhaar, H.L., "Dimensional Analysis and Theory of Models", Robert E. Krieger Publishing Company, Huntington, NY (1980)
11. Zweben, C., W.S. Smith and M.W. Wardle, 'Test Methods for Fiber Tensile Strength, Composite Flexural Modulus, and Properties of Fiber-Reinforced Laminates', Composite Materials: Testing and Design (Fifth Conference), ASTM STP 674, p. 228-262, Tsai, S.W., Ed., American Society of Testing and Materials, (1979)
12. Hughes, J.D.H., 'Strength and Modulus of Current Carbon Fibers', Carbon, vol. 24, No. 5, p. 554-556 (1986)
13. Dresselhaus, M.S., G. Dresselhaus, K. Sugihara, I.L. Spain and H.A. Goldberg (Ed.), "Graphite Fibers and Filaments", p. 125-152, Springer-Verlag, Berlin (1988)

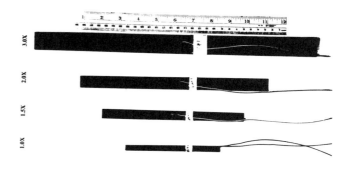

Figure 1. Sample CFRC $[0°]_n$ beams of 4 sizes

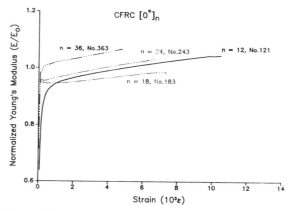

Figure 2. Typical non-linear hardening behavior of Young's modulus of CFRC $[0°]_n$ specimens

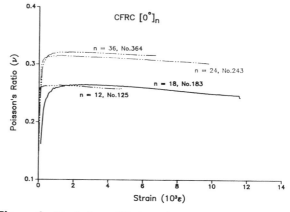

Figure 3. Variation of Poisson's ratio as strain changes

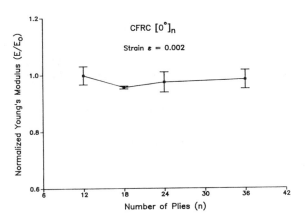

Figure 4. Size effects on Young's modulus at 0.002 strain

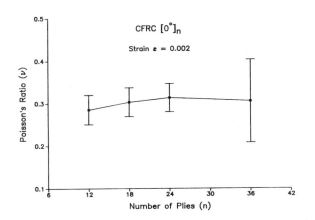

Figure 5. Size effect on Poisson's ratio at 0.002 strain

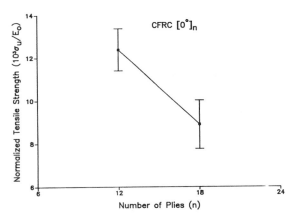

Figure 6. Size effect on tensile strength from 12- and 18-ply CFRC

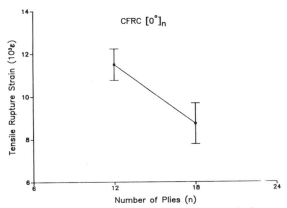

Figure 7. Size effect on tensile rupture strain between 12- and 18-ply CFRC

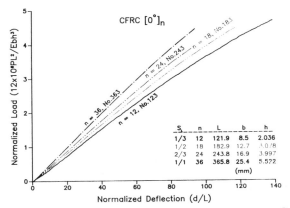

Figure 8. Typical load-deflection curves of CFRC $[0°]_n$ beams in flexural tests

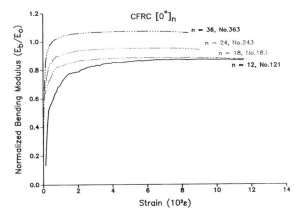

Figure 9. Typical non-linear hardening of bending modulus of CFRC $[0°]_n$ beams

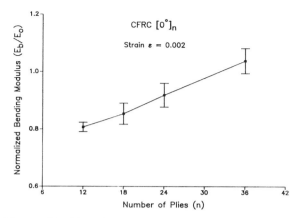

Figure 10. Size effect on bending modulus of CFRC $[0°]_n$ beams

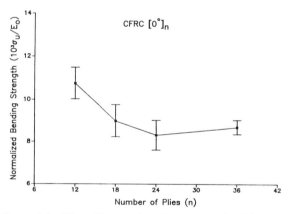

Figure 11. Size effect on bending strength of CFRC $[0°]_n$ beams

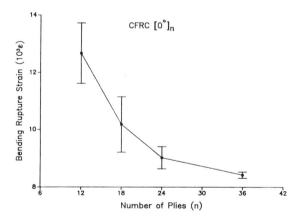

Figure 12. Size effect on bending rupture strain of CFRC $[0°]_n$ beams

DURABILITY IN ALUMINUM-SMC ADHESIVELY BONDED SYSTEMS

J. G. Dillard, I. Spinu
Chemistry Department
VPI & SU, Blacksburg, VA USA

J. W. Grant
ESM Department
VPI & SU, Blacksburg, VA USA

ABSTRACT

The durability of SMC - urethane - aluminum samples has been investigated. Lap shear test samples were exposed to fixed temperature and humidity conditions. During the test, samples were held at a stress of 0% and 50% of the ultimate failure force. The application of a constant load caused specimen failure during the 7 day exposure at 60°C (140°F) and 0% or 100 relative humidity. Failure under the conditions of (60°C) (140°F) 0% RH or 60°C (140°F) 100% RH occurred via adhesive failure such that a thin layer of adhesive was found on the metal surface. Surface analysis characterization measurements using scanning electron microscopy (SEM) and X-ray photoelectron spectroscopy (XPS) demonstrated that the locus of failure moved away from the adhesive and toward the adhesive - metal interface upon increasing the relative humidity from 0% to 100%.

INTRODUCTION

Structures prepared by adhesive bonding of composite materials are becoming more common in the automobile industry in order to save weight, to reduce costs, and to aid in corrosion prevention. These adhesively bonded structures must withstand the effects of environmental exposure and cyclic stresses. An important finding is that adhesive joints loose strength upon exposure to the service environment, particularly when high temperature and humidity are involved. Many factors affect the durability of adhesive bonds. The principal environmental factors that reduce durability are exposure to liquids and their vapors, salts, elevated temperature and stress. Usually combinations of these factors will operate to be potentially damaging.[1] Water is often the primary component that affects the environmental stability of adhesive bonds. Moderate temperatures by themselves do not have an adverse effect on structural bonds. However, the combination of high temperature and high humidity can have a serious effect on durability. [1] Joints may be subjected to intermittent or sustained stress. Minford [2] has shown that stressed joints tend to degrade faster when exposed to moisture than unstressed joints. Exposure to salt can have a very severe effect on bond durability.[1] Processes taking place in the adhesive during exposure to environmental conditions include creep, Hookian elasticity, retarded elasticity, moisture induced swelling, thermal degradation, stress induced chain scission, postcure, cracking, plasticization by water and hydrolysis.[3] For adhesives and fiber reinforced plastics there is a potential for viscoelastic failure of the product. Bonded SMC structures are a dual viscoelastic problem. Because the matrix of the composite is also polymeric, composites are subjected to time dependent behavior.[4] Another complication is the fact that the composites are not normally uniform (isotropic) in nature because most have either a gel coat at the surface or, if the later has not been deliberately created, the surface layer usually contains a higher proportion of resin than the interior. Both the gel coat and the resin rich surface layers are brittle and may display catastrophic failure when overloaded.[5]

A number of test methods have been developed over the years in the attempt to determine adhesive joint durability. Most attention has focused on the binding of metals; few studies have been concerned with other structurally bonded systems.[6] At present no reliable published data are available from durability studies involving SMC (sheet molded composite) adherends.

The objective of this research was to

carry out durability tests of SMC - bonded to aluminum using a urethane adhesive. Lap shear specimens were prepared and were tested at fixed temperature and humidity conditions with and without a load on the bonded joint. The failed sample surfaces were characterized via surface analysis in order to better understand the factors that affect durability. Surface chemistry was evaluated using X-ray photoelectron spectroscopy and topographical characterization was investigated by scanning electron microscopy (SEM).

EXPERIMENTAL

Preparation of samples

Lap shear samples were prepared using phase α-SMC and aluminum 6061. The SMC adherends (1"x4"x0.100") (Ashland Chemical Co.) were pretreated by priming with Pliogrip 6036 (Ashland Chemical Co.), a diisocyanate - containing primer. All aluminum adherends (1"x4"x0.125") were treated in a 5% (W/W) aqueous solution of NaOH at 70°C (158°F) for 2 minutes, washed with a 50% (W/W) aqueous solution of concentrated HNO_3 at room temperature for one minute, rinsed thoroughly with distilled water and dried 30 minutes in an oven at 110°C (230°F). Two types of lap shear specimens were prepared one with primed aluminum adherends and the other with unprimed aluminum adherends. Pliogrip 6600/6020, (Ashland Chemical Co), two part urethane adhesive was used. The thickness of the adhesive was 3.5 mm. Bonded lap shear specimens with one square inch bond area were cured at room temperature for 2 hours and then additionally cured for 30 min in an air oven maintained at 149°C (300°F).

The environmental cycles were followed for up to seven days. In some tests, specimens failed within 7 days. The specimens that did not fail during the exposure scheme were removed and the residual stress was measured at 82°C (180°F) using an Instron machine. Individual insulated containers with temperature and humidity controllers were used to house the samples during the exposure tests.

Failure force was determined using an Instron apparatus Model 1123 with a thermally controlled testing chamber, Model 3116. The cross-head speed was 1.27 cm/min (0.5 in/min). The lap shear samples were conditioned at 82°C (180°F) or 62°C (144°F) for 30 min before testing and were tested at 82°C (180°F) or 62°C (144°F). At least five measurements were made for each kind of sample.

Analytical Methods

In the failure experiments, two specimens were obtained: one specimen was principally adhesive and the other aluminum. From the failure experiments, samples were cut for XPS and SEM analysis. For these measurements the side showing bulk adhesive is designed the adhesive side and the other the aluminum side. The topographical analysis of failed samples was accomplished using an ISI-SX-40 scanning electron microscope. Samples were coated with a film of sputtered gold. Surface chemical analysis was accomplished with a Perkin-Elmer PHI 5300 X-ray photoelectron spectrometer.[7] Photoelectrons, generated using Mg K_α radiation (hν = 1253.6 eV), were analyzed in the hemispherical analyzer, and detected using a position sensitive detector. Although samples of approximately 15 x 17 mm were introduced into spectrometer, the area of the specimen sampled by the analyzer electron optics was approximately 10.0 x 2.0 mm. In the presentation of the elemental results, photoelectron peak areas were measured and subsequently scaled to account for ionization probability and an instrumental sensitivity factor to yield results which are indicative of surface concentration in atomic percent. The precision and accuracy for the concentration evaluations are about ±10% and ±15%, respectively.

RESULTS AND DISCUSSION

A summary of the experiments to explore the durability of SMC/urethane/aluminum bonded systems in summarized below.

1. Instron testing: SMC/urethane/aluminum samples at 82°C (180°F) and 62°C (144°F) to determine the failure force and failure mode (as prepared samples: base-line data).

2. Instron testing: bonded samples, no load, 60°C (140°F), dry air, 7 days.

3. Instron testing: bonded samples no load, 60°C (140°F), 100% RH, 7 days.

4. Instron testing: bonded samples, 50% load, 60°C (140°F), dry air, 7 days.

5. Instron testing: bonded sample, 50% load 60°C (140°F), 100% RH, 7 days.

A summary of the failure force and mode results is presented in Tables 1-4. The distribution of failure modes, _ie_ D, delamination; A, adhesive; C, cohesive, was determined by visual inspection of the failed samples. For all samples failure occurred adhesively and/or cohesively. No delamination of SMC was observed. Table 1 shows the Instron test results for as prepared samples (no load, room temperature and dry air). The failure force for samples prepared with unprimed and primed aluminum, tested at the same temperature are equivalent, within the experimental precision. Conducting the Instron tests at 62°C (144°F) and at 82°C (180°F) for samples prepared with unprimed aluminum resulted in a lower failure force at 82°C (180°F). More significantly, the failure mode for tests at 82°C (180°F) was mixed mode yielding 45% adhesive and 55% cohesive failure.

Lap shear results for specimens exposed at 60°C (140°F) with no load in dry air are given in Table 2. The samples in this experiment remained bonded for the 7 day test period. Examination of the data shows that

a) the average failure force is greater for the samples prepared using primed aluminum and

b) the failure force is reduced for samples exposed to 100% relative humidity, compared to materials tested in dry air.

It is also found that while the mode of failure is equivalent (66% A; 34% C) for lap shear samples exposed to dry air; exposure to moisture (100% RH) alters the mode of failure such that adhesive failure is dominant; unprimed aluminum/100% RH exhibited 98% adhesive failure and primed aluminum samples at 100% RH showed 87% adhesive failure.

Table 3 presents the Instron test results for samples stressed at 50%, exposed to 60°C (140°F) and dry air. Three unprimed Al/SMC samples failed during exposure in a very short time (<1.5 hrs). For these samples the failure mode was principally adhesive. For the samples which did not fail during exposure, the average failure force (172 lbs) was approximately the same as for as prepared samples 180 lbs (Table 2). For the samples assembled with primed aluminum, only one of the five failed at 5 days. The failure force and failure mode are similar to the corresponding samples exposed to dry air at 60°C (140°F) with no load (Table 2). Thus under the conditions of the test it is apparent that priming the aluminum adherend enhances durability.

The Instron test results for stressed samples (50%) maintained at 60°C (140°F) and 100% RH are presented in Table 4. All specimens failed in a very short time (~1hr). For both types of samples a substantial decrease in the proportion of cohesive failure and an increase in the fraction of adhesive failure mode occurred compared with unloaded samples exposed to dry air at room temperature. The specific results are 70% A vs 45% A for unprimed Al/SMC samples, and 74% A vs 28% A for primed Al/SMC specimens. Thus maintenance of load and high humidity are catastrophic conditions for the aluminum-composite bonded samples.

To investigate the failure mode in greater detail, surface analysis measurements were carried out for the adhesive failure regions on failed samples. The XPS results are collected in Tables 5, 6, and 7. In Table 5 are surface analysis data for a cast adhesive film, for sodium hydroxide cleaned aluminum, for primed aluminum, and for the respective failed surfaces of an as bonded Al/SMC sample (prepared using unprimed aluminum) that had been tested in the Instron instrument. The Al/SMC bonded specimen had not been subjected to any environmental conditioning nor was any load applied to the sample before the failure test. Carbon 1s XPS spectra for these materials are collected in Figure 1. The atomic concentration data and the XPS spectra are reference results that will be used in evaluating the chemistry at the failed surfaces. The important findings that will be used in the evaluation of the failure mode chemistry are that nitrogen at different concentrations is found on the adhesive film and on the primed aluminum surface, but nitrogen is not detected on the cleaned aluminum specimen. It should also be noted that the N/C atomic ratios are 0.042 for the adhesive film, and 0.085 for primed aluminum. It is also noteworthy that application of primer produces a sufficiently thick film on aluminum, that aluminum is below the detection level in the Al 2p region.

For the failed sample surfaces, nitrogen is noted on both the adhesive and aluminum surfaces. The nitrogen/carbon ratios are 0.053 and 0.047, respectively, for these adhesive and aluminum samples. The N/C ratio on the adhesive failure surface is equivalent to that for the pure adhesive film and the C 1s photoelectron spectrum exhibits a peak shape equal to that for the adhesive film (Fig. 1). The detection of nitrogen on the aluminum surface and the shoulder at 286.5 eV in the C 1s spectrum, attributable to urea functionality, is indicative of some adhesive on the aluminum failure surface. Examination of the SEM photomicrograph in Fig. 2 reveal depressions in the unprimed, primed aluminum, and the aluminum failed surface; and raised features on the adhesive failed surface. The presence of nitrogen on the failed aluminum surface and the absence of definitive features on the surface, that could be attributed to adhesive, lead to the suggestion that a film rather than fragments of adhesive may be present. The XPS and SEM results indicate that "adhesive failure" for the as prepared specimen occurs via a near surface "cohesive" failure in the adhesive.

The XPS surface analysis results for Al/SMC samples prepared using unprimed aluminum, and exposed to the selected conditions of temperature, humidity, and stress are presented in Table 6. Representative C 1s XPS spectra for these samples are given in Fig. 3. The C 1s spectra for the adhesive side failure are reasonably equivalent to the C 1s spectrum for the adhesive film. It is interesting to note that the urethane carbon functionality is greater for samples tested under load compared to specimens with no load. The detection of nitrogen on all aluminum failure surfaces is suggestive of adhesive on the surface. The N/C and N/Al ratios are affected by the conditions of the environmental treatment. The N/C ratio on Al failure surfaces decreases from 0.043 for the Al/SMC specimen exposed to 60°C/no load (NL)/Dry air (D) to 0.027 for the corresponding sample treated at 60°C/50% load (L)/Dry air (D). The N/Al ratio is equivalent for these two samples, 0.053 and 0.051. The decreased N/C ratio, but an invariant N/Al ratio could occur if other carbon-containing species are present on the failure surface. Viewing the C 1s spectrum for the 60°C/L/D sample, it is evident that the intensity in the binding energy region 286.5 eV is less - a result of an increase in the signal at 285.0 eV due to CH_n, hydrocarbon. The source of this additional hydrocarbon is unknown. The XPS results for aluminum failed surfaces for specimens treated at 100% RH (Table 6) show that the N/Al ratios are respectively less when compared to samples tested under dry conditions. Such results are consistent with the idea that less adhesive remains on the aluminum surface, and that the locus of failure is nearer the aluminum/adhesive interface and not deep within the adhesive.

In the SEM photomicrographs (Fig. 4) one notes that the surface texture of the adhesive failure side is a replica (inverse features) of the aluminum surface. No evidence for adhesive is apparent on the aluminum surface, yet the XPS data indicate the presence of nitrogen-containing constituents. Thus it is reasoned that a thin film of adhesive is present, a suggestion that was also offered above to account for the results found for as prepared Al/SMC specimens.

Surface analytical data for primed Al/SMC failed samples are given in Table 7 and representative XPS C 1s spectra are illustrated in Fig. 5. The data reveal nitrogen on the surfaces of failed-aluminum and -adhesive side samples. However, the C 1s XPS spectra on the adhesive side failure are not representative of adhesive. For the adhesive side failure samples the N/C atomic ratios are significantly greater than the ratio for an adhesive film. The carbon spectra and the N/C ratios are more representative of primer at the failed surface. Compare C 1s spectra in Fig. 1 with those in Fig. 5. Consideration of the N/C and N/Al ratios for samples exposed to dry air at 60°C with or without load, indicates that failure occurs within the primer. As load is applied under dry conditions or at 100% RH under no load, the extent of failure within the primer decreases, and the locus of failure moves more toward the primer-aluminum interface. This effect is supported by the decrease in the N/Al and N/C atomic ratios and the decrease in C 1s photopeak intensity in the region associate with primer nitrogen functionality, at BE=286.5 eV for aluminum side failure surfaces.

The N/C and N/Al ratios and the C 1s photoelectron spectra for aluminum failed sample surfaces resulting from treatments at 60°C, 50% Load, under dry or 100% RH conditions are similar. This finding indicates that the resulting surface chemistry at the "adhesive" failed surface is not significantly altered by moisture, ie. 100% RH. Such results suggest that even though failure occurs within the primer, the application of isocyanate or other primers may have potential to increase bond durability and resistance to moisture.

SUMMARY/CONCLUSIONS

Laboratory tests of aluminum/SMC lap shear specimens at a temperature above room temperature, under a fixed load, and in dry and moist atmospheres have been carried out. Increasing the severity of the test conditions, ie. increased load, or greater moisture content, results in accelerated failure of the specimens. In studies of primed and unprimed adherends primed aluminum materials exhibited greater durability for tests in dry air, but no improvement in 100% RH. Surface analysis results by XPS and SEM revealed that generally, "adhesive" failure occurred in the adhesive at the adhesive/aluminum interface for unprimed aluminum adherends, and in the primer at the primer/aluminum interface for primed aluminum.

ACKNOWLEDGMENTS

Thanks are expressed to Ashland Chemical Co., and General Motors who provided partial support for this study. Support was also provided through the NSF-STC program at VPI & SU - under grant DMR-8809714.

REFERENCES

1. D. M. Brewis in "Aluminium Adherends", Durability of Structural Adhesives, A. J. Kinloch, ed., Chapter 5, Applied Science Publishers, London, 1983.

2. J. D. Minford in "Adhesives", Durability of Structural Adhesives, A. J. Kinloch, ed., Chapter 4, Applied Science Publishers, London, 1983.

3. I. G. Zewi, F. Flashner, H. Dodiuk, and L. Drori, Int. J. Adhesion Adhesives, **4**, 137, 1984.

4. K. Liechti, W. S. Johnson, D. A. Dillard in "Experimentally Determined Strength of Adhesively Bonded Joints", F. L. Mattews, ed., Chapter 4, Elsevier Applied Science Publishers, London, 1987.

5. W. A. Lees, Int. J. Adhesion Adhesives, **6**, 171, 1986.

6. S. R. Hartshorn, "The Durability of Structural Adhesive Joints" in Structural Adhesives, S. R. Hartshorn, ed., Chapter 8, Plenum Press, New York, 1986.

7. J. G. Dillard, C. Burtoff, and T. Buhler, J. Adhesion, **25**, 203 (1988).

Table 1 - Instron Results for Samples Exposed to No Load, Room Temperature, Dry Air

(as prepared sample)

	Unprimed Al-urethane-SMC		Primed Al-urethane-SMC	
Test Temperature, °C	62	82	62	82
Failure Force*(psi)	299 ± 21	180 ± 18	283 ± 70	192 ± 18
(MPa)	2.06 ± .14	1.24 ± .12	1.95 ± .14	1.32 ± .12
Failure Mode*, %	14A86C	45A55C	25A75C	28A72C

*Results are average for 5 samples tested at each condition

Table 2 - Instron Results for Samples Exposed to No Load, 60°C (140°F), 7 days

	Unprimed Al-urethane-SMC		Primed Al-urethane-SMC	
Exposure	Dry Air	100% RH	Dry Air	100% RH
Failure Force*(psi)	190 ± 38	152 ± 56	226 ± 36	185 ± 27
(MPa)	1.31 ± .26	1.05 ± 0.39	1.56 ± .25	1.28 ± .18
Failure Mode*, %	66A34C	98A2C	66A34C	87A13C

*Results are average for 5 samples tested at 82°C (180°F), for each condition

Table 3 - Instron Results for Samples Exposed to 50% Load, 60°C (140°F), Dry Air, 7 Days

	Unprimed Al-urethane-SMC					Primed Al-urethane-SMC				
Time to Failure hrs	0.5			1.5	0.5		5 Days			
Failure Force psi		149*	195*			211*		237*	246*	225*
MPa		1.03	1.35			1.46		1.63	1.70	1.55
Failure Mode %	60A40C	35A65C	100A	75A25C	80A20C	35A65C	80A20C	25A75C	45A55C	80A20C

*Tested at 82°C (180°F)

Table 4 - Test Results for Samples Exposed to 50% Load, 60°C (140°F), 100% RH.

	Unprimed Al-urethane-SMC	Primed Al-urethane-SMC
Time to Failure, hrs	1.0	1.0
Failure Mode* %	70A30C	74A26C

*Results are the average for 5 samples at each condition.

Table 5 - XPS Results (Atomic Concentration) Failed Samples, No Load, No Environmental Exposure

Element	Adhesive Film	Non-Bonded Unprimed Aluminum	Non-Bonded Primed Aluminum	Unprimed Al/SMC	
				Adhesive Side	Aluminum Side
C	68.8	12.5	79.0	72.5	23.6
O	26.4	58.4	14.1	22.9	52.8
N	2.90	<0.1	6.72	3.88	1.12
Al	<0.1	28.9	<0.1	0.55	21.9
Mg	0.52	<0.1	<0.1	<0.1	<0.1
Others (Si,Cl)	1.35	<0.1	0.25	0.21	0.67
N/C	0.042	--	0.085	0.054	0.047
N/Al					0.051

Table 6 - XPS Results (Atomic Concentration) Failed Samples, Unprimed Al-Urethane-SMC

	60°C (140°F), Dry Air				60°C (140°F), 100% RH			
	No Load		50% Load		No Load		50% Load	
Element	Adhesive Side	Aluminum Side	Adhesive Side	Aluminum Side	Adhesive Side	Aluminum Side	Adhesive Side	Aluminum Side
C	72.8	24.0	73.7*	32.6*	71.6	17.1	73.2*	20.1*
O	21.5	52.9	22.6	48.3	23.2	55.7	22.9	55.5
N	5.19	1.04	3.56	0.89	4.43	0.88	3.94	0.64
Al	0.51	19.8	0.13	17.3	0.69	23.8	<0.1	21.6
Others (Si,Cl)	<0.1	2.19	<0.1	0.92	<0.1	2.42	<0.1	2.15
N/C	0.071	0.043	0.048	0.027	0.062	0.051	0.054	0.032
N/Al		0.053		0.051		0.037		0.030

*Failed during exposure test.

Table 7 - XPS Results (Atomic Concentration) Failed Samples, Primed Al-Urethane-SMC

	60°C (140°F), Dry Air				60°C (140°F), 100% RH			
	No Load		50% Load		No Load		50% Load	
Element	Adhesive Side	Aluminum Side	Adhesive Side	Aluminum Side	Adhesive Side	Aluminum Side	Adhesive Side	Aluminum Side
C	73.5	26.9	74.8	21.5	74.3	22.6	61.3	21.4
O	16.3	49.2	16.7	55.9	17.0	51.6	26.4	53.8
N	8.76	1.43	7.89	0.81	8.08	0.48	5.63	0.79
Al	0.77	20.2	<0.1	20.4	0.40	22.8	2.73	20.9
Others (Si,Cl)	0.69	2.19	0.55	1.51	0.22	2.50	3.98	3.12
N/C	0.119	0.053	0.105	0.038	0.109	0.021	0.092	0.037
N/Al		0.071		0.040		0.021		0.038

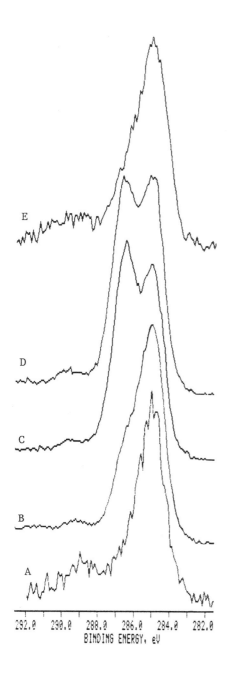

Fig. 1 C 1s Spectra

   A: Cleaned Aluminum
   B: Primed Aluminum
   C: Adhesive Film
   D: Adhesive Side for Failed As Prepared Unprimed Al-urethane-SMC Samples
   E: Aluminum Side for Failed As Prepared Unprimed Al-urethane-SMC Samples

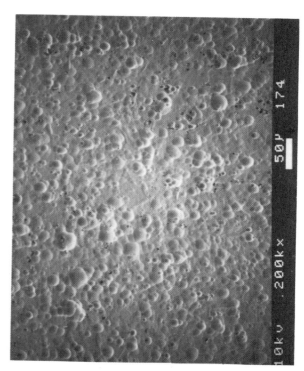

Unprimed Aluminum

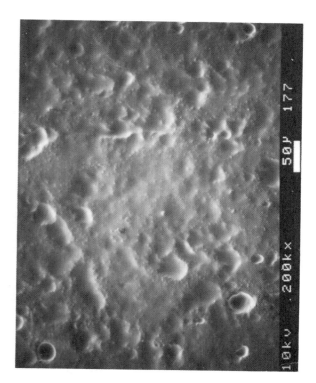

Primed Aluminum

Fig. 2a SEM Photos

Adhesive Side

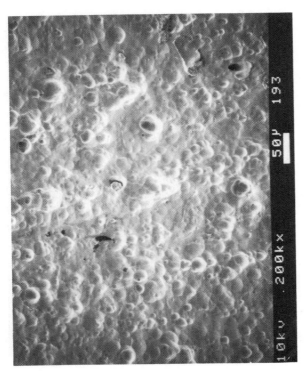

Aluminum Side

Fig. 2b SEM Photos for Failed As Prepared Unprimed Al-Urethane-SMC Samples.

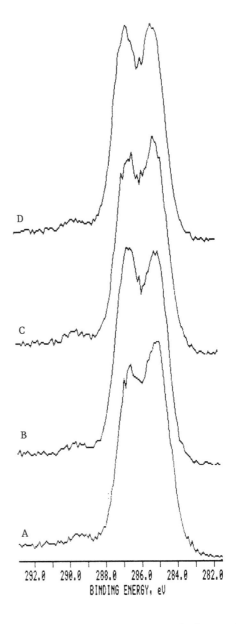

Fig. 3a C 1s Spectra for Failed Unprimed Al-Urethane-SMC Samples. Adhesive Side

A: No Load, 60°C (140°F), Dry Air
B: 50% Load, 60°C (140°F), Dry Air
C: No Load, 60°C (140°F), 100% RH
D: 50% Load, 60°C (140°F), 100% RH

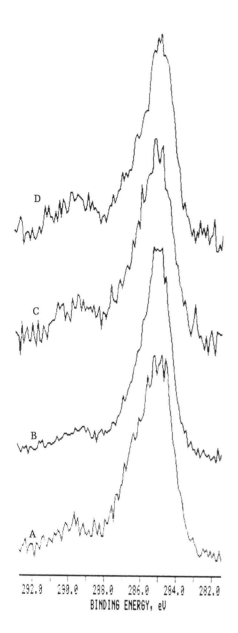

Fig. 3b C 1s Spectra for Failed Unprimed Al-Urethane-SMC Samples. Aluminum Side

A: No Load, 60°C (140°F), Dry Air
B: 50% Load, 60°C (140°F), Dry Air
C: No Load, 60°C (140°F), 100% RH
D: 50% Load, 60°C (140°F), 100% RH

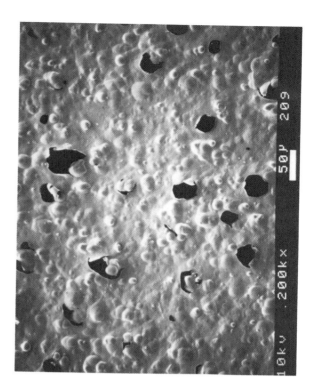

Adhesive Side

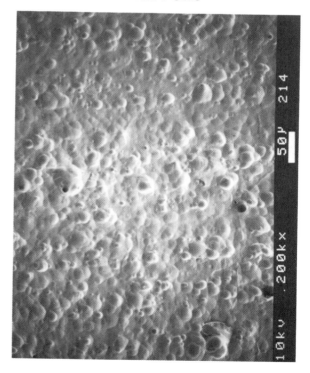

Aluminum Side

Fig. 4a SEM Photos for Failed Unprimed Al-Urethane-SMC Samples. No Load, 60°C (140°F), 100% RH

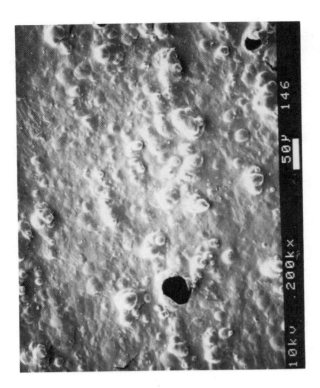

Adhesive Side

Aluminum Side

Fig. 4b SEM Photos for Failed Unprimed Al-Urethane-SMC Samples 50% Load, 60° C (140° F), 100% RH

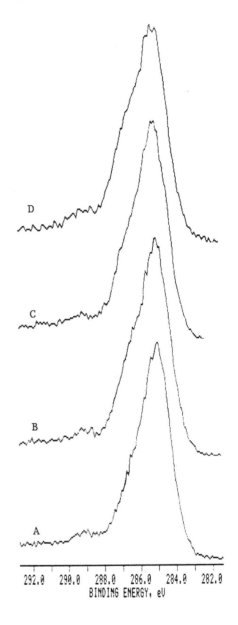

Fig. 5a C 1s Spectra for Failed Primed Al-Urethane-SMC Samples Adhesive Side

A: No Load, 60° C (140° F), Dry Air
B: 50% Load, 60° C (140° F), Dry Air
C: No Load, 60° C (140° F), 100% RH
D: 50% Load, 60° C (140° F), 100% RH

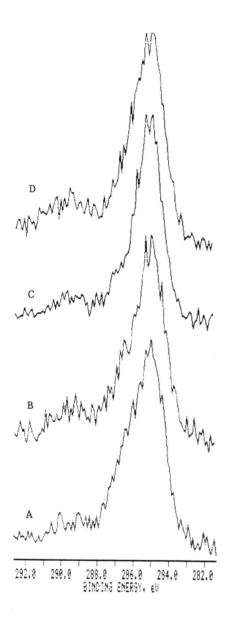

Fig. 5b. C 1s Spectra for Failed Primed Al-Urethane-SMC Samples. Aluminum Side

- A: No Load, 60° C (140° F), Dry Air
- B: 50% Load, 60° C (140° F), Dry Air
- C: No Load, 60° C (140° F), 100% RH
- D: 50% Load, 60°C (140° F), 100% RH

NON-DESTRUCTIVE DETERMINATION OF FIBER VOLUME AND RESIN CONTENT OF FIBER REINFORCED COMPOSITES

C. Salvadó
Applied Sciences Corporation
252½ So. Vista Way, Ste. 251
Carlsbad, CA 92008 USA

ABSTRACT

We regard fiber reinforced composites as a three phase heterogeneous mixture: fibers, matrix and pores. A nondestructive elastostatic method is used to determine their fractions by measuring two of those quantities and computing the third from the mixture relations for conservation of weight and volume. We compare the results to ones done using acid digestion.

The comparison of destructive and nondestructive results is done statistically by five statistical tests: comparison of their moments up to the fourth, linear least squares fit, and we attempt to disprove the null hypothesis by analysis of means and variances, and the Kolgomorov-Smirnov test.

The results of all five statistical tests is that there is no significant difference between the destructive and nondestructive results. Both the destructive and nondestructive values are drawn from the same distributions (i.e., the distributions of true fiber volume and resin content values). Therefore the nondestructive technique is an acceptable substitute for the destructive one.

1. INTRODUCTION

The advanced fiber reinforced composites industry expends much effort in testing the quality of their product. In particular, the testing for matrix content, a quantity that is important for the strength/weight quality of these composites, is currently being done destructively by a variety of techniques that includes acid digestion, burn-off, and, in the case of prepreg, solvent wash. These destructive tests, and the disposal of their chemical wastes, is costing the composite industry tens of millions of dollars per year (1). Since the use of composites is growing at an estimated rate of 15 to 20% per year (1), and the cost of lawful disposal of chemical wastes undoubtedly will increase considerably in the near future, the expense to the composite industry to perform these destructive tests will likely reach hundreds of millions of dollars per year in just a few years.

Recently, a technique to determine matrix content nondestructively has been introduced (2). In this paper we present further evidence that this technique is an acceptable substitute for the acid digestion method. In addition, we will show that with this technique it is also possible to determine the fiber volume of fiber reinforced composites, and therefore also porosity of these materials.

The comparison between the destructive and nondestructive tests is done statistically. We compare the moments of the distributions up to the fourth. We attempt to disprove the null hypothesis with three statistical tests: tests for significantly different means and variances, and the Kolgomorov-Smirnov test (3). Finally, because it seems to be the method of preference of the industry, with do a linear least squares fit to the data. All five statistical tests show that the nondestructive technique is an acceptable substitute to the acid digestion method, at least for the materials studied. This conclusion has also been reached independently by others (4).

2. DESCRIPTION OF THE NONDESTRUCTIVE TEST

The nondestructive test was done with an apparatus and method developed at Applied Sciences Corporation and we will refer to it herein as the ASC10 System, and for which patent applications are still being filed. In general terms, the method consists in making measurements of elastostatic displacement by the application of static stresses normal to the surface of the composite. The theory of the method is that the material responds to an imposed static stress in a unique way which depends on the materials fractions of matrix, fibers and voids, all of which can be discerned by performing previously calibration measurements. By making measurements at several stresses, the fractions of the phases can be determined.

The theory that interprets the elastostatic measurements is one cast in a material dependent way. This means that there are adjustable parameters in the theory which in general are different for every fiber/matrix (Fb/Mt) system, and must be phenomenologically determined (i.e., calibration is necessary for every Fb/Mt system).

Calibration proceeds as presented graphically in Figure (1). The operator prepares at least four specimens (although a good calibration has been achieved with two (5)). However it is preferable to have as many as possible to average out error due to destructive measurements. The specimens are then submitted to the ASC10 System for elastostatic measurements at two stresses, and the results are stored for reference after the destructive results are performed. In the destruction of the specimens, the operator must collect the fiber volume, and resin content by weight. The porosity fraction is computed from these using

$$f_v = 1 - \{1 + (d_f/d_m)(F_m/F_f)\} f_f , \qquad (1)$$

where d_m and d_f are the matrix and fiber densities, F_m and F_f are the matrix and fiber fractions by weight, and f_f and f_v are the fiber and void fraction by volume, and where it is assumed that the weight of the pores is small compared to the weight of the matrix or the fibers:

$$F_m + F_f = 1 . \qquad (2)$$

The destructively measured values for F_m and f_f, and the computed value f_v for every specimen are used to evaluate the adjustable parameters of the theory.

Some of the adjustable parameters of the theory must be positive for real materials. However it sometimes occurs in a calibration run that they are negative due to errors of the destructive results. For example, it frequently happens, as we are sure is the experience of most if not all chemists performing acid digestion measurements, that the porosity resulting for measured values of resin content and fiber volume, is negative. In such cases the calibration is rejected and the process is redone from the start as is indicated in Figure (1). If the parameters positivity constraint is passed, the calibration is accepted, and the Fb/Mt system is deemed completely calibrated as far as the ASC10 System is concerned, and the process of evaluating fiber volume and resin content of unknown specimens of that Fb/Mt system can commence as is shown in Figure (2).

This technique has been applied to several materials (2,4, and 5), and is currently being tested for several others including, Kevlar, fiber glass, thermoplastics, and prepreg. So far all tests have shown that the ASC10 System is an acceptable substitute for the corresponding destructive test, whether it be solvent wash, burn-off, or acid digestion. However, the same accuracy is not achieved in all cases. Since the ASC10 System measures fractions by volume, computes the density of the composite using

$$d_c = d_m f_m + d_f f_f , \qquad (3)$$

then converts to fractions by weight by the transformation

$$F_i = (d_i/d_c) f_i \quad (i = m,f) , \qquad (4)$$

if the density of the matrix is not well known, the accuracy of the ASC10 System may be less than in those cases where the density is well known and constant. The case of a varying matrix density is in thermoplastics, which have a range of states of crystalinity in which to settle, and which depends on the thermal history of the specimen.

3. DESCRIPTION OF THE TESTS AND STATISTICS OF THE RESULTS

In this test 15 specimens of a Gr/Ep were analyzed with the ASC10 System by performing 12 samples per specimen. Four samples had been used for calibration by using their average properties. The destructive tests had been performed already in the center of the specimen, and about which the nondestructive sampling was performed. It is important to realize an important difference between the destructive and nondestructive tests: the specimen of the destructive test was in this case a subset of the specimen of the nondestructive test.

Even so, the material tested was sufficiently uniform in these properties that it did not affect the comparison very much as can be seen in Tables 1 and 5. In these tables, DT and NDT stand respectively for destructive test and nondestructive test. In Tables 1 and 5 the greatest difference between the DT and NDT columns is 1.3%. Since the destructive test has an error certainly greater than +/- 1.3, the comparison of DT and NDT of fiber volume and resin content for each individual specimen in Tables 1 and 5 is considered to be good.

In Tables 2 and 6 we compare the moments of the DT and NDT results. Here we make the following abbreviations:

average	=	AVG
absolute deviation	=	ADEV
standard deviation	=	SDEV
variance	=	VAR
skewness	=	SKEW
kurtosis	=	KURT

In Tables 2 and 6, kurtosis has been defined so that the normal distribution has zero kurtosis, and those flatter than the normal distribution have negative kurtosis. As can be appreciated in Tables 2 and 6, the first and second moments of the DT and NDT distributions of respectively Tables 1 and 5, are not different by more than 0.1. This means that the analysis of means and variances shall show no significance difference between the distributions.

We also attempted to disprove the null hypothesis by three statistical tests: a t-test for significantly different means, a F-test for significantly different variances and the Kolgomorov-Smirnov test. The results are given in Tables 3 and 7. In these tables

$$0 \leq \text{Prob} \leq 1 . \qquad (6)$$

Typically the null hypothesis is deemed disproved (i.e., the observed difference is very significant), when Prob is equal to or less than 0.05 or 0.01. As can be seen in Tables 3 and 7, the distributions of DT and NDT values are not significantly different. We can further state that DT and NDT values are drawn from the same distribution which is the distribution of "true" values, since these values were derived from different test methods.

Finally, Tables 4 and 8 show the results of performing a linear least squares fit to the data: we attempt to plot a straight line of the form

$$(DT) = a + b \, (NDT). \qquad (7)$$

In these tables the following definitions hold:

$SIGMA_{(DT)}$ = uncertainty of DT ≥ 0

$SIGMA_a$ = uncertainty of parameter a ≥ 0

$SIGMA_b$ = uncertainty of parameter b ≥ 0

CHI^2 = a measure of the error of the fit ≥ 0

$0 \leq Q$ = goodness of fit or confidence in a and b ≤ 1

In these tables we have assumed the uncertainty of DT is no worse than 1.5 %. This is very small. Depending on who is doing the destructive tests, the uncertainty for laminates typically can be as high as 5.0 %. However, even by assuming a small uncertainty, the zero intercept and slope one line was included in the range of a and b by their resulting uncertainties. The fact that the estimate of a and b are not respectively near zero and unity, is in this case a function that the distribution of DT's and the NDT's have a small standard deviation (see Tables 2 and 6).

4. CONCLUSIONS

We have presented the nondestructive fiber volume and resin content values for a set of Gr/Ep specimens, and the corresponding destructive values for comparison. Both sets were compared by five statistical tests. All tests indicated that the nondestructive values are not significantly different from the corresponding destructive values. Therefore the nondestructive technique is an acceptable substitute for the acid digestion technique.

5. REFERENCES

(1) Martin Burg (1988), personal communication.

(2) Carlos A. Salvado, A. Birnie Hunter, and Martin Wai (1989). Proceedings of the SPE conference ANTEC '89, May 1-4, New York, New York.

(3) William H. Press, Brian P. Flannery, Saul A. Teukolsky, and William T. Vetterling (1988). Numerical Recipes In C, Cambridge University Press, Cambridge, England.

(4) A. B. Hunter, C. H. Sheppard, C. S. Schindler, and J. R. Linn (1988). Comparison Of Nondestructive And Acid Digestion Methods For Determining Resin Content, Quality Assurance Development Report No. 8112-02 Preliminary, Boeing Aerospace Company, Seattle, Washington.

(5) Carlos A. Salvado (1989), High Temple Workshop IX, Jet Propulsion Laboratory, Pasadena, California, Sponsored by Air Force Materials Laboratory, Wright-Paterson Air Force Base, Dayton, Ohio.

6. ACKNOWLEDGEMENTS

We wish to thank Walt Gill and Todd Huber of General Dynamics Corporation, Convair Division, San Diego, California, for providing the specimens that were tested at Applied Sciences Corporation, and performing the acid digestion tests of the same.

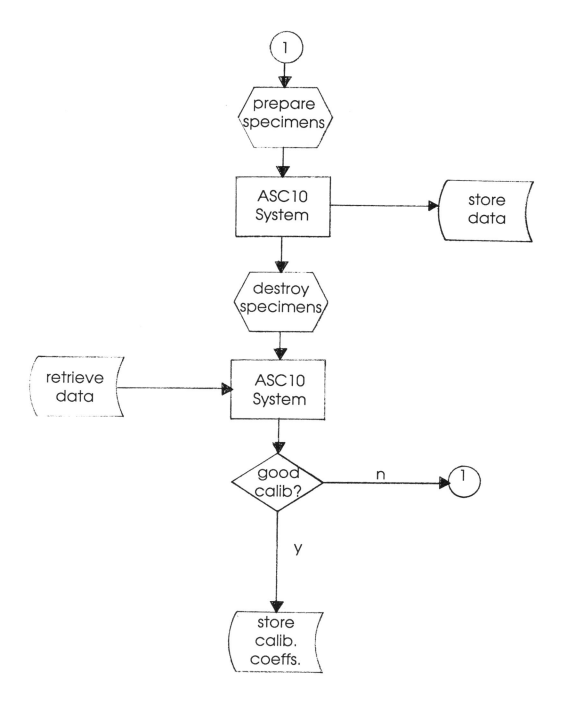

Fig. 1 - Calibration process for the ASC10

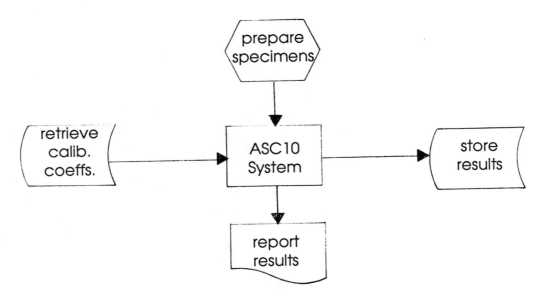

Fig. 2 - ASC10 System process for the determination of fiber volume and resin content.

Table 1

% FIBER BY VOLUME

SPECIMEN	DT	NDT
UP11	68.4	67.4
UP12	67.7	67.7
UP13	67.4	67.5
UP14	67.2	67.6
UP15	67.6	67.5
UP21	66.6	65.3
UP22	66.5	66.2
UP23	65.6	66.3
UP24	66.2	66.2
UP25	66.6	66.7
UP31	67.0	66.2
UP32	65.8	66.2
UP33	66.2	65.2
UP34	65.9	66.2
UP35	65.8	66.5

Table 2

MOMENTS OF THE DISTRIBUTIONS FOR % FIBER BY VOLUME

MOMENT	NDT	DT
AVG	66.7	66.6
ADEV	0.7	0.7
SDEV	0.8	0.8
VAR	0.7	0.7
SKEW	0.4	-0.1
KURT	-1.1	-1.2

Table 3

STATISTICAL ANALYSIS OF THE DISTRIBUTIONS FOR % FIBER BY VOLUME

Analysis of Means	$t = 0.40$	Prob = 0.69
Analysis of Variance	$F = 1.06$	Prob = 0.92
Kolgomorov-Smirnov	$D = 0.20$	Prob = 0.93

Table 4

LINEAR LEAST SQUARES FOR % FIBER BY VOLUME
(DT) = a + b (NDT)

$a = 18.8 \quad b = 0.72$

$SIGMA_{(DT)}$	$SIGMA_a$	$SIGMA_b$	CHI^2	Q
1.5	33.3	0.5	2.181	1.000
1.0	22.2	0.3	4.907	0.977
0.5	11.1	0.2	19.627	0.105

Table 5

% MATRIX BY WEIGHT

SPECIMEN	DT	NDT
UP11	24.2	25.1
UP12	25.4	25.0
UP13	25.9	25.3
UP14	25.9	25.3
UP15	25.4	25.6
UP21	25.8	27.0
UP22	26.2	26.3
UP23	26.1	26.2
UP24	26.3	26.3
UP25	26.0	26.0
UP31	25.9	26.4
UP32	27.0	26.4
UP33	26.9	27.1
UP34	27.0	26.4
UP35	26.5	25.8

Table 6

MOMENTS OF THE DISTRIBUTIONS FOR % MATRIX BY WEIGHT

MOMENT	DT	NDT
AVG	26.0	26.0
ADEV	0.5	0.5
SDEV	0.7	0.7
VAR	0.5	0.4
SKEW	-0.7	-0.0
KURT	0.5	-1.2

Table 7

STATISTICAL ANALYSIS OF THE DISTRIBUTIONS FOR % MATRIX BY WEIGHT

Analysis of Means	t	= 0.08	Prob = 0.94
Analysis of Variances	F	= 1.22	Prob = 0.71
Kolgomorov-Smirnov	D	= 0.20	Prob = 0.93

Table 8

LINEAR LEAST SQUARES FOR % MATRIX BY WEIGHT
(DT) = a + b (NDT)

$a = 7.4 \quad b = 0.72$

$SIGMA_{(DT)}$	$SIGMA_a$	$SIGMA_b$	CHI^2	Q
1.5	16.1	0.62	1.868	1.000
1.0	10.7	0.41	4.203	0.988
0.5	5.4	0.21	16.813	0.208

HEAT TRANSFER AND CURE ANALYSIS FOR PULTRUSION*

*presented at the SME Pultrusion Clinic, Los Angeles, California, April 4-6, 1989.

Gibson L. Batch, Christopher W. Macosko
University of Minnesota
Minneapolis, MN 55455 USA

Abstract

Computer-aided analysis of pultrusion is an area of active research. A successful model is important for technological advancement and penetration into new product markets. In this paper, heat transfer and cure inside a pultrusion die are analyzed by numerical modeling. First the reaction kinetics of unsaturated polyester and vinyl ester resins are modeled using equations derived from the chemical mechanism. Then, the pultrusion analysis is formulated to include the effects of RF preheating, heat conduction from the wall, and separation of the profile from the die after cure. The analysis is tested with process data with a 1-inch round die and vinyl ester resin matrix. Several die temperatures and initiator concentrations were used for the verification. The analysis is accurate for the experimental cases studied.

PULRUSION is one of the few continuous processes for composites. In pultrusion, continuous provings and mats are saturated with resin in a wet-out tank before being pulled through a heated die for curing (Figure 1). A basic understanding of the variables affecting cure is critical to fabrication. New resin formulations and evolving part geometries invoke difficult processing challenges, and no longer can manufacturers depend solely on experience when selecting process variables. The difficulties are compounded by high materials cost and low part counts, so the costs of startup and process debugging can be high. Processing rates have been increased by fast-curing resins and by process enhancements, but they only scratch the surface of processing problems.

Substantial improvements in the processing are only possible with process analysis which can predict the conditions during processing given the machine set points. A computer-aided process analysis is useful for:

1) *Process Optimization.* Processing temperature, pressure, and demold time can be optimized to provide both uniform cure without excessive heating of the resin.
2) *Characterization of New Resins.* The curing of new resins can be modeled without performing expensive and time consuming experimental process studies.
3) *Process Control.* New control schemes would be possible in which process set points are material intensive (i.e. centerline temperature and cure) as opposed to process intensive (i.e. die temperature, demold time, or line speed).
4) *Startup.* Proper machine settings can be predicted by without expensive trial and error debugging and quality checks.
5) *Design.* As a design tool, the analysis indicates how new profiles should be heated, what rates of production are possible, and how changes in part geometry might ease processibility.

One intangible advantage of a process model is that it demonstrates that a process is well understood. If pursuit of applications held by aluminum, steel, and wood is to continue, producers will need to demonstrate that adequate knowledge and control is at hand to insure product consistency, reliability, and production volume for new markets. Most metal-working processes have computer-aided design programs; composites processes will need to have the same.

An analysis of pultrusion will predict temperature, reaction rate, and extent of cure. Assuming that the gel point of the resin is known, the heat transfer analysis can predict where the resin gels, and perhaps more importantly, the extent of cure upon leaving the die. The analysis can be used to assist in the selection of line speed, die temperature, and resin formulation (initiator concentration).

Deriving a realistic model, however, is often difficult. The problem of describing "real world" manufacturing conditions was described by Price (1979):

"An examination of an actual process is difficult in many cases because the process[or] often must meet certain 'boundary conditions', such as the need to meet shipping

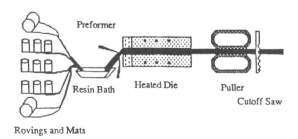

Figure 1. Pultrusion process schematic.

schedules, satisfy customers' demands, and deal with varying sources of supply. In many respects, an examination of a working process will not be as clean as might be desired. On the other hand, a working process does represent the real world, however inefficient it may be."

The purpose of this paper is to develop a heat transfer model from first principles which predicts temperature and extent of cure inside a pultrusion die. Heat transfer events outside the die, such as RF preheating and convective cooling, will be included in the analysis. Predicted temperatures will be compared with experimental data for pultrusion of a vinyl ester resin at several operating conditions.

PREVIOUS WORK

Reactive processing is used for a large number of rubbers and reinforced plastics: polyurethane reaction injection molding (RIM), silicone rubber molding, epoxies, and unsaturated polyester. Processing rate of these materials is limited by heat transfer and reaction. Thus, analysis of temperature during cure for a variety of chemical systems have one common issue: equations of reaction kinetics are nonlinearly coupled to heat transfer to and from the mold. Unfortunately, this coupling sometimes makes modelling computations difficult.

Because pultrusion is equivalent to a plug resin flow reactor, a mathematical model is deceptively simple. During the last few years, several workers have published curing models which describe the ideal manufacturing setting. Unfortunately, they used untested kinetic models and mentioned no experimental verification. Without experimental validation, it is unknown whether or not process fundamentals important during cure (i.e. reaction kinetics, changes in heat transfer coefficient, and preheating) are incorrect or overlooked.

Price (1979), the first to use a heat transfer model for pultrusion, examined two limiting cases: an isothermal case with a uniform die wall temperature and an adiabatic case where heat conduction was negligible. The former is appropriate for thin profiles, the latter for thick profiles. Neither, however, are realistic because die temperature varies along its length and heat conduction is usually significant. The model used first order kinetics for epoxy resins. No experimental results were presented. Aylward, et al. (1985), repeated the calculations by Price using a generalized finite element code.

Tulig (1985) used finite elements to model pultrusion cure of epoxy resins in round and irregular profiles. Special boundary conditions simulated both the heat input from the heater bands outside the die and heat losses due to convection with air. Tulig's work is the only published model to date which has been successfully verified with experimental data for epoxy resin. No published work has successfully verified a pultrusion analysis of unsaturated polyester resin.

Richard (1986) analyzed pultrusion of epoxies using first order kinetics. Unlike previous work, two-dimensional heat transfer in irregular geometries (i.e. I-beams) was modeled, and guidelines were presented on approximating low aspect ratio shapes as a rectangular die. Langan (1986) used a lumped parameter model intended for process control calculations. Though the model could not predict experimental data for unsaturated polyester pultrusion, goals for development of pultrusion process control were outlined. Han, et al. (1986) used an autocatalytic model for unsaturated polyesters and epoxies, and allowed density, thermal conductivity, and heat capacity to change with degree of cure. No experimental data were presented. Ma, et al. (1986) published a model similar to Han, but for the first time axial conduction was included in the calculations. No evidence was given, however, to show that axial conduction was significant.

Work by Batch and Macosko (1987a) was the first pultrusion curing analysis to use a mechanistic kinetic model for unsaturated polyester resin. They also presented a model for resin pressing and pulling force. No experimental data was presented. Later, Han, et al. (1987) used a similar kinetic model for analyzing heat transfer.

Because of the large heat of polymerization of many thermoset resins, the reaction kinetics are important in a heat transfer analysis for pultrusion. Hence, we will examine the reaction kinetics first.

REACTION KINETICS

To be most useful, a kinetic model should be able to predict the effect of temperature and resin formulation on the rate of cure. The predictive capability is necessary to optimize a resin formulation for pultrusion. However, an accurate kinetic model must be specific to the particular resin chemistry used in a composite. The same kinetic model, for example, may not be used in both the polymerization of unsaturated polyester and epoxy, because the reaction mechanisms are drastically different. Most polymer matricies used in thermosetting composites (e.g. unsaturated polyester and vinyl ester) react with the free radical mechanism. We will limit our scope to the kinetics of these resins.

Many commercial resins which react with the free radical mechanism are mixtures of polyester or vinyl ester oligomers and styrene as a reactive diluent. Crosslinking of the oligomers and styrene occurs by copolymerization at carbon-carbon double bond moieties located on each species. Formulations may have several inhibitors, initiators, and other additives which may affect the rate of cure.

Though reaction kinetics are affected by complex phenomena (such as microgel formation and morphology, unequal monomer reactivity, decreasing initiator efficiency, radical trapping, and termination), relatively simple models have been derived by Stevenson (1980), Lee (1981), and Batch and Macosko (1987b). In the kinetic model discussed below, the major assumptions are that radical termination is negligible, that initiator efficiency is constant, and that reactivity differences between double bonds on the comonomers is negligible.

The rate of polymerization is proportional to the concentration of free radicals R

$$\frac{dX}{dt} = k_p (1-X) R \qquad (1)$$

where X is the conversion of double bonds ($X \equiv (M_0-M)/M_0$, $0 \leq X \leq 1$) and k_p is the rate constant of chain propagation. Kinetic models differ on the treatment of k_p and R. In the model developed by Batch and Macosko (1987b), k_p is given by the lesser of either the intrinsic propagation rate constant k_0 or the rate from an empirically-derived model.

$$k_p = \text{MIN}(k_0, k_s(X_{max}-X)) \qquad (2)$$

k_p is a constant k_0 at low conversion (X<0.5), but at higher conversion it decreases linearly as X approaches X_{max}. The decrease in k_p is attributed to diffusion limitations in a crosslinking network, but also lumped together with diffusion are the effects of decreasing initiator efficiency, radical trapping, and unequal reactivity. Note that X_{max} is the conversion where polymerization would stop by vitrification at low cure temperature. An expression for X_{max} is derived from a model for vitrification by Hale (1988).

$$X_{max} = \frac{1}{a}\left(\frac{1}{T_o} - \frac{1}{T}\right) \quad (3)$$

At high cure temperature ($T > T_o/(1-aT_o)$), X_{max} is unity.

The radical concentration R can change dramatically during cure. Initially, free radicals are rapidly consumed by inhibitor Z present in commercial resin systems. The rate of consumption of inhibitor is equal to the rate of initiation of free radicals

$$\frac{dZ}{dt} = -2fIk_d \quad (4)$$

where $I(=(I_o+I_1)(1-\alpha_i))$ is the effective initiator concentration (which includes intrinsic initiator I_1 and an initiator depletion factor $1-\alpha_i$), f is initiator efficiency, and k_d is the rate constant of decomposition. R is approximately zero until nearly all the inhibitor is consumed.

$$R \sim 0 \text{ when } Z > 0 \quad (5)$$

After inhibitor is consumed, R increases with time according to the expression

$$\frac{dR}{dt} = 2fIk_d \quad (6)$$

The rate constants k_d and k_p have an Arrhenius temperature dependence. The rate of polymerization (Equation 1) is very rapid at high temperatures because of large values of dR/dt from Equation 6 and because of large k_p.

During isothermal polymerization, the rate of reaction can be segregated into three zones (Figure 2). In the inhibition zone, dX/dt is first zero because all radicals are consumed by the inhibitor. In the propagation zone, the rate of polymerization increases due to increasing R from Equation 6. Finally, at conversion above roughly 50 percent, the diffusion-limited propagation zone occurs and reaction rate decreases due to decreasing k_p from Equation 2.

This kinetic model has been verified with several temperatures and initiator concentrations in Batch (1989) for both a model chemical system (divinyl benzene with AIBN initiator) and a commercial vinyl ester resin with TBPO initiator. Further discussion of the kinetics is beyond the scope of this paper. We now apply these kinetics to a heat transfer analysis for pultrusion.

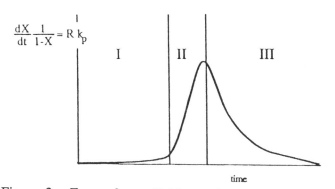

Figure 2. Zone of crosslinking polymerization during isothermal cure: I) inhibition zone, II) propagation zone, III) diffusion-limited propagation zone.

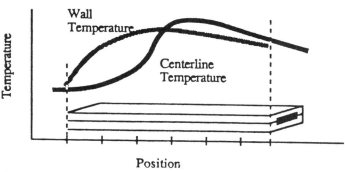

Figure 3. Temperature profiles at the wall and center of a pultruded laminate.

THEORETICAL

The ultimate goal of the heat transfer model is to apply to real world pultrusion processing conditions yet be computationally simple to run on small computers. A useful model must not only be accurate, realistic, and grasp all important fundamentals of pultrusion, but it must also be easy to use and portable (on microcomputers) to manufacturing facilities where it is needed. Most pultrusion companies are small and cannot afford large computers or the highly-trained personnel to operate them.

Temperature profiles in a pultrusion die are shown in Figure 3. After leaving the resin bath, the resin and fibers may pass through a preheater to assist the curing inside the die. At the die entrance there is a short taper section where excess resin is squeezed from the fibers. Once in the die, heat conducts from the die wall, so resin at the wall heats and gels before resin at the center. The reaction exotherm at the center can cause centerline temperature to exceed the temperature at the die wall after cure. The resin may shrink away from the die wall after cure due to volumetric contraction.

The assumptions necessary to simplify the heat transfer model are:

1) Neglect the effects of resin backflow at the die entrance on temperature. Because backflow is present only for the entrance taper of the die (less than an inch from the die entrance), heat transfer during backflow can be neglected.
2) Heating rate inside the radio frequency (RF) preheater is constant. Generally, heating rate is a function of temperature, but we will neglect this effect as in Jones (1974).
3) Die temperature is at steady state, and resin is in good thermal contact with the die until the profile shrinks away from the die wall after cure.
4) Axial heat conduction is negligible. Because the aspect ratio (length/thickness) of most dies is 40:1 to 200:1, heat conduction towards the wall is considered much greater than longitudinal conduction. It will be seen below, however, that local longitudinal temperature gradients may cause axial conduction to be significant at the reaction front.
5) The resin and fibers at any point inside the die have the same local temperature. Also thermal properties such as density, heat capacity, and thermal conductivity can be represented in the model as some average value of the resin and fibers.
6) Thermal properties are independent of temperature and conversion. During cure, density and thermal conductivity

are expected to increase and heat capacity will decrease (Pusatcioglu, et al., 1979, Mijovic and Wang, 1988) resulting in an increase in thermal diffusivity ($k/\rho C_p$) of approximately 50 percent. Until a model for these property changes has been formulated and verified, changes in thermal properties will be neglected.

7) Reaction rate is unaffected by the fibers. Little work has been done to observe the effects of the coupling agent on the reaction kinetics of unsaturated polyesters. The effect of glass on the kinetics is important for the processing of composites, and it needs more study in the future.

8) Reaction rate is unaffected by pressure of the resin inside the die. Pressures up to 500 psi can be achieved during pultrusion (Sumerak, 1985), which is far below the point where pressure is expected to alter polymerization rate constants (Weale, 1967).

These assumptions allows the composite to be modeled as a continuum during heat transfer, even though the fibers and resin are in two distinct phases. Also, reaction kinetics determined by analytical methods such as DSC will be used directly in the analysis without further parameter fitting or modification.

MODEL FORMULATION - Once the resin enters the die, heat conducts from the die wall to cure the resin. The entrance taper creates recycling of resin at the die entrance only, and the remaining distance through the die resin travels with the fibers as a plug flow reactor. Neglecting the effect of backflow on heat transfer (Assumption 1), the equation of energy for points within the profile in the Lagrangian (material) frame of reference is as follows:

$$\frac{d(\rho C_p T)}{dt} = \nabla \cdot (k \nabla T) + \Delta H_r \rho \frac{dX}{dt} \quad (7)$$

transient = conduction + generation

where ΔH_r is the heat of reaction per unit mass and dX/dt is the rate of conversion of the reactive resin matrix, as given in the kinetic model above. Conduction will occur in both longitudinal and transverse to the fiber orientation. Typically, neglecting longitudinal conduction (Assumption 4), and assuming constant density ρ, heat capacity C_p, and thermal conductivity k (Assumption 6), Equation 7 becomes

$$\frac{dT}{dt} = \alpha_{th} \frac{d^2T}{dz^2} + \alpha_{th} \frac{G}{z} \frac{dT}{dz} + \Delta T_{ad} \frac{dX}{dt} \quad (8)$$

where G=0 in rectangular coordinates and G=1 in cylindrical coordinates, α_{th} is thermal diffusivity (=$k/\rho C_p$), ΔT_{ad} is the adiabatic temperature rise (=$\Delta H_r/C_p$), z is the transverse spatial coordinate which is zero at the center and B at the wall (see Figure 4). Rectangular coordinates will apply to flat profiles where B is the half thickness; cylindrical coordinates will apply to round profiles where B is radius.

BOUNDARY CONDITIONS - Boundary conditions are important because they specify how the resin interacts with the die as it cures. The symmetry condition applies at the center of the profile.

$$dT/dz = 0 \text{ at } z = 0 \quad (9)$$

The boundary conditions at the die wall are either "isotracking" (e.g. temperature follows a prescribed die temperature T_{die} at a distance x from the die entrance)

$$T(B,x) = T_{die}(x) \quad (10)$$

or based on a convective heat transfer coefficient h_{die}.

$$dT/dz(B,x) = -h_{die}/k \, (T(B,x) - T_{die}(x)) \quad (11)$$

Initially, the resin is in good thermal contact with the die (h_{die} is large) so the isotracking condition is applicable. As resin shrinks away from the die, h_{die} decreases because of poor thermal contact, so the convective boundary conditions are used. The point of separation is determined when the thermal expansion of resin during heating is less than the shrinkage during cure, i.e.

$$\alpha_v (\overline{T} - T_{amb}) - \gamma \overline{X} < 0 \quad (12)$$

thermal expansion - shrinkage < 0

where $\overline{T}$ and $\overline{X}$ are average temperature and conversion across the profile, T_{amb} is ambient temperature. For the purpose of predicting the point of separation, the coefficient of volumetric thermal expansion α_v and coefficient of shrinkage during polymerization γ are assumed independent of temperature, conversion, or pressure.

Once the profile leaves the die, it begins to cool in air. Cooling below the glass transition temperature is important to resist cracking the pultrudate in the grips of the puller. Also at high line speeds, some reaction may occur after leaving the die. To predict the rate of cooling outside the die, the heat transfer coefficient in the convective boundary conditions is replaced with h_{air}

$$dT/dz(B,x) = -h_{air}/k \, (T(B,x) - T_{amb}) \quad (13)$$

h_{air} is larger if air is blown against the profile during cooling. Either heat transfer coefficient correlations (Perry

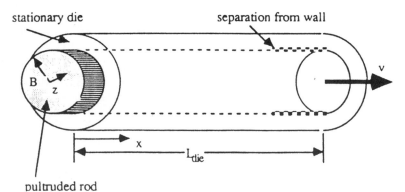

Figure 4. Geometry for heat transfer analysis in cylindrical coordinates (G=1).

and Chilton, 1973) or surface temperature measurements can be used to determine h_{air}.

PREHEATER MODEL - The equations above will model heat transfer both inside the die during cure and outside during cooling. Another important aspect of heat transfer is upstream from the die, where a preheater is sometimes used. Without a preheater, the initial temperature of Equation 8 is the ambient temperature ($T=T_{amb}$). With a preheater, the temperature at the die entrance is higher, and a preheater heat transfer model is necessary.

Geometry for the preheater model is in Figure 5. Inside the preheater, the rate of energy input, Q_{ph}, is much larger than the reaction exotherm. Outside the preheater, Q_{ph} will be zero. Also, the thermal diffusivity α_{th} before the die is decreased by an empirical insulating factor ϑ ($0<\vartheta<1$) due to trapped air and excess resin between the fibers. Because air is squeezed from the resin at the die entrance, ϑ is assigned a value of unity inside the die. Hence, Equation 8 is replaced by a more general expression which includes the preheater.

$$\frac{dT}{dt} = \vartheta \alpha_{th} \frac{d^2T}{dz^2} + \vartheta \alpha_{th} \frac{G}{z} \frac{dT}{dz} + \Delta T_{ad} \frac{dX}{dt} + P \quad (14)$$

transient = conduction + reaction + preheat

where $P (= Q_{ph}/(\rho C_p))$ is the rate of temperature increase inside the preheater, which for now is assumed to be constant (Assumption 2). The initial conditions of Equation 14 are ambient temperature T_{amb} and zero conversion. The boundary conditions at the center and at the surface of the profile before entering the die are given by Equations 9 and 13, respectively.

NUMERICAL METHOD - Because of the complexity of the kinetics and the boundary conditions, Equations 3-8 cannot be solved analytically. Instead, the problem is solved numerically by finite differences. The finite difference method (FDM) was selected because of its familiarity, wide-spread acceptance, and its simplicity. The goal of the pultrusion analysis is to be able to run on small computers, such as microcomputers, so that the analysis can be performed at the manufacturing facility.

The FDM uses equally spaced nodes in the solution domain, substituting difference formulae for derivatives in the heat equation. For the calculations below, 21 nodes are between the centerline and the wall. Calculations with Equation 14 are solved in time increments Δt by the Crank-Nicolson scheme,

$$\frac{T(i,j+1)-T(i,j)}{\Delta t} = \frac{1}{2}(F(i,j+1) + F(i,j)) + \Delta T_{ad}\frac{X(i,j+1) - X(i,j)}{\Delta t} + P \quad (15)$$

where

$$F(i,j) = \frac{\vartheta \alpha_{th}}{\Delta z^2}(T(i+1,j) - 2T(i,j) + T(i-1,j)) + \frac{\vartheta \alpha_{th} G}{2 z_i \Delta z}(T(i+1,j) - T(i-1,j)) \quad (16)$$

Note that the reaction term of Equation 15 is expressed in terms of conversion instead of rate of conversion, as seen in Equation 14. This partially decouples the equations of heat transfer and reaction kinetics, which improves computational speed but requires use of an iterative procedure between solving the heat transfer equations and the kinetics. Also, the time step Δt is varied in the analysis to keep local truncation errors small: Δt is large away from the reaction rate is small but decreases when reaction is rapid. Details of this solution technique and of the selection of Δt are presented in Batch (1989).

SUMMARY OF CURING MODEL - The kinetic model used for the pultrusion analysis is based on the mechanism of crosslinking free radical polymerization. Assumptions of the kinetic model are that termination is negligible, unequal reactivity is neglected, initiator efficiency is constant. The kinetic model uses an empirical expression for k_p at high conversion which lumps together effects of diffusion-limited propagation, decreasing initiator efficiency, and radical trapping.

The kinetic model was then applied to the analysis of heat transfer and cure during pultrusion. Included in the model are preheating before the die, changes in heat conduction due to shrinkage away from the die, and cooling after leaving the die. The equations of transient one-dimensional heat transfer and kinetics have been implemented by finite differences. The equations of heat transfer and reaction were decoupled by and iterative solution technique which reduces computational effort. Lagrangian (material) coordinates were used in the model, and the size of each time step in the calculations was based on criteria using the local truncation error.

Next, process temperature data will be compared to the predictions of the heat transfer analysis for verification.

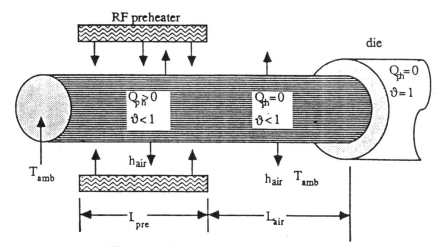

Figure 5. Geometry for preheater model.

EXPERIMENTAL

The experimental data will test the ability of the model to predict the effect of changing die temperature and resin formulation. Pultrusion experiments were performed on a 1-inch round die at the Dow Chemical Company (Freeport, Texas) on a commercial scale machine manufactured by Morrison Molded Fiberglass (Model TM570). The resin was Derakane 411-35LI vinyl ester resin (Dow Chemical) with 15 phr ASP400 clay filler and 1 phr Zeloc UN mold release. The initiator Trigonox 21 (Noury Chemical) (97 percent assay *tert*-butyl peroctoate (TBPO)) was at concentrations of both 1 and 2 phr. The glass rovings was at 62.4 percent by volume. Resin and fibers were preheated before entering the die by a radio-frequency (RF) preheater. The die was 48 inches (122 cm) long and had strip heaters mounted on the four lateral external surfaces near the die entrance (Figure 6). The remainder of the die was unheated and uninsulated.

The pultrusion experiments are summarized in Table 1. The experiments include several die temperature set points, T_{die} (at location CNTL in Figure 6), several preheater temperatures, T_{pre}, and two resin formulations (1 and 2 phr TBPO). All data was collected at a line speed of 12 in/min (0.505 cm/s). For each condition, die and resin temperatures were measured. Thermocouples inside the die block measured temperature at 10 locations. Profile temperatures were measured by feeding long welded Type J thermocouple wires (0.01 inch diameter, 0.01 seconds response time) between fibers at the die entrance. Axial position was determined by multiplying line speed by the time elapsed since the entry into the die. These profile temperatures were collected after steady state die temperature was achieved (approximately an hour after a change in set point). The estimated uncertainty in temperature is estimated from replicated thermocouples to be ±5 °C, and uncertainty in axial distance is ±5 cm.

Before comparing the temperature data to predictions, parameters are needed in the model for thermal properties (e.g. density, heat capacity, thermal diffusivity, heat of polymerization), reaction kinetics, preheater conditions, and boundary conditions (i.e. γ, α_v, and h_{die}). Details of finding these parameters and the parameter values used for the predictions below are in Batch (1989).

Table 1. Experimental variables in pultrusion experiments.

Run	initiator conc.	line speed v (cm/s)	die temp T_{die}(°C)	temp T_{pre}(°C)
1	1 phr	0.505	99	79
2	1 phr	0.505	124	84
3	1 phr	0.505	135	84
4	2 phr	0.505	125	73

RESULTS AND DISCUSSION

Verification of the analysis will be based on two criteria. First, predictions will be compared with data from one run to see if the model is accurate. Adjustments may be necessary to the model to improve the agreement. Then, using these adjustments, the analysis is compared with data for other operating conditions.

CHECKING FOR MODEL ACCURACY - The predicted temperatures near the center for Run 3 are compared to experimental data in Figure 7. Temperatures are in good agreement before and after polymerization (50-60 cm inside the die), but the predicted temperature increase is faster and greater than expected. Examination of the calculations using different number of nodes and time steps showed no significant change in the sharp peak. Thus, it may be concluded that the sharp peak is due inadequacies of the heat transfer model or reaction kinetics.

One likely cause for the sharp peak is axial heat conduction, which is ignored in the analysis, but would be significant for the steep longitudinal temperature gradients shown in Figure 7. Heat transfer models with axial conduction are very difficult to formulate, however, due to computational difficulties. Instead of rigorously modeling axial heat conduction, an approximation can be used to smooth the sharp temperature peak and hence account for the effects of axial conduction. One simple method is to subject temperatures to a first order time lag.

$$\frac{dT_s}{dt} = \frac{T_p - T_s}{\tau_{lag}} \quad (17)$$

where T_s is the smoothed temperature, T_p is the temperature predicted from the one-dimensional analysis. Figure 8 shows satisfactory agreement with the experimental data when τ_{lag} is 6 seconds.

As expected from axial conduction, the time lag distributes exothermal heat flow over a greater axial distance than predicted by the one-dimensional heat transfer model alone. Assuming the same τ_{lag} for other runs, comparisons will now be made after a change in die temperature and initiator concentration. The same time lag will be used to smooth the temperature predictions both at the center and near the wall.

CHANGING OPERATING CONDITIONS - Comparisons between the data and analysis will be based on three criteria: the location of the peak temperature near the center, the magnitude of peak temperature, and the qualitative agreement between theory and experiment over

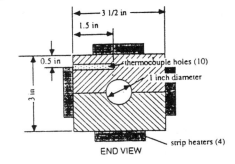
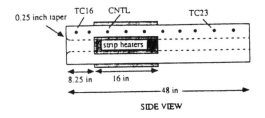

Figure 6. Geometry of die used for pultrusion trials.

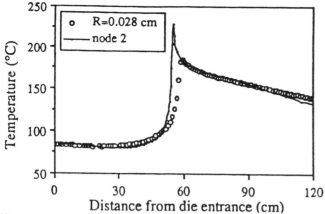

Figure 7. First comparison of model predicitons (lines) to experimental data (points) for Run 3.

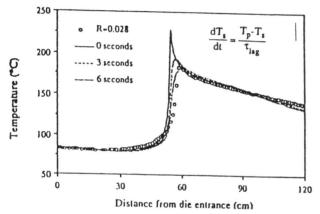

Figure 8. Effects of first order time lag τ_{lag} on predicted temperature to simulate axial heat conduction for Run 3.

the length of the die. Temperatures near the centerline after a change in die temperature (Runs 1, 2, and 3) are in Figures 9-11. The predictions are plotted for nodes closest to the experimental radial location of each thermocouple. In each of these cases, the analysis predicts the peak centerline temperature T_{max} within 3 °C and peak location x_{max} within 5 cm, which is below the estimated uncertainty of the data. The peak centerline temperatures and locations are compared in Table 2.

Also shown are data and predictions for temperatures close to the wall. Differences between measured and predicted temperatures near the wall are in part due to the time lag correction discussed above. Though the time lag correction made the analysis more accurate at the center, the predicted wall temperature now lags behind the experimental data.

When initiator concentration is doubled to 2 phr (Run 4, Figure 12), the analysis predicts maximum temperature within 8 °C, and peak location within 9 cm. Differences between predictions and experiment are slightly larger than experimental uncertainty, but the agreement is still adequate for verifying the analysis. Errors in the analysis after the peak exotherm are unexpected. The data indicate that shrinkage after cure increases at higher concentrations of initiator. If shrinkage with 2 phr initiator is faster than with 1 phr initiator, more shrinkage will occur in the die. Only a thin gap is between the resin and die wall after separation, so slight differences in shrinkage will have large effects on heat transfer.

From Figures 8-11, it may be concluded that the analysis qualitatively and quantitatively agrees with temperature data both at the centerline and wall for a change in die temperature or resin formulation.

CONCLUSIONS

In summary, the pultrusion process was analyzed with heat transfer and curing analysis. Reaction kinetics used in the analysis were derived specifically for unsaturated polyester resins. The heat transfer analysis included the effects of preheating, axial changes in die temperature, and resistance to heat conduction due to shrinkage after cure. An initial comparison between predicted and experimental temperatures showed that the analysis predicts the location of cure accurately, but the temperature has a sharp peak value much greater than experimentally observed. The sharp peak was removed by smoothing, which simulated the effects of axial heat conduction. Comparisons to data at different die temperatures and concentrations of initiator proved the analysis to be accurate.

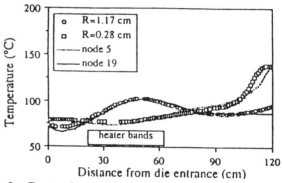

Figure 9. Comparison between predicted and experimental profile temperatures for Run 1 (99 °C die with 1 phr initiator).

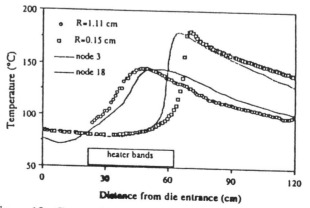

Figure 10. Comparison between predicted and experimental profile temperatures for Run 2 (124 °C die with 1 phr initiator).

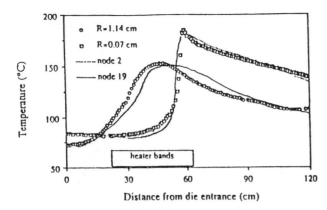

Figure 11. Comparison between predicted and experimental profile temperatures for Run 3 (135 °C die with 1 phr initiator).

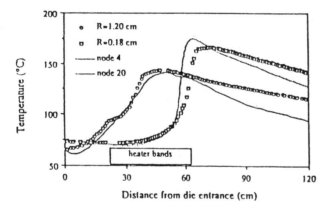

Figure 12. Comparison between predicted and experimental profile temperatures for Run 4 (125 °C die with 2 phr initiator).

This work is just one step towards developing a program for pultrusion die design and set point selection. Future efforts will 1) study irregular and hollow dies, 2) examine reaction the kinetics in the presence of fibers, 3) write a pre- and post-processor for the pultrusion analysis for convenient parameter input and display of results, and 4) model heat conduction through the die metal. One limitation of the current analysis is that it requires a temperature profile along the die length. To be most useful, however, the analysis must predict die temperature from the heater location and output, and the die geometry. These areas should be addressed as the pultrusion analysis becomes recognized as a feasible and useful design tool for pultrusion processors.

REFERENCES

Aylward, L., C. Douglas, D. Roylance, *Poly. Proc. Eng.*, 3, 247 (1985).
Batch, G.L., C.W. Macosko, 42nd Ann. Conf., Compos. Inst., Soc. Plast. Ind., 12-B (1987a).
Batch, G.L., C.W. Macosko, *Technical Papers*, Soc. Plast. Eng., 33, 974 (1987b).
Batch, G.L., Ph.D. Thesis, University of Minnesota (1989).
Hale, A., Ph.D. Thesis, University of Minnesota (1988).
Han, C.D., D.S. Lee, H.B. Chin, *Poly. Eng. Sci.*, 26, 393 (1986).
Han, C.D., H.B. Chin, *Technical Papers*, Soc. Plast. Eng., 33, 690 (1987).
Jones, B.H., 29th Ann. Conf., Reinf. Plast./Compos. Inst., Soc. Plast. Ind. (1974).
Langan, L.J., MS Thesis, MIT (1986).
Lee, L.J., *Poly. Eng. Sci.*, 21, 483 (1981).
Ma, C-C. M., K-Y. Lee, Y-D. Lee, J-S. Hwang, *SAMPE J.*, 22 (5), 42 (1986).
Mijovic, H., H.T. Wang, *SAMPE J.*, 24(2), 42 (1988).
Perry, R.H., C.H. Chilton, Eds., Chemical Engineers' Handbook, 5th ed., McGraw-Hill, New York (1973).
Price, H.L., Ph.D. Thesis, Old Dominion University (1979).
Pusatcioglu, S.Y. A.L. Fricke, J.C. Hassler, *J. Appl. Poly. Sci.*, 24, 937 (1979).
Richard, R.V., Ph.D. Thesis, Tufts University (1986).
Stevenson, J.F., *Technical Papers*, Soc. Plast. Eng., 26, 452 (1980).
Sumerak, J.E., Modern Plastics, March, p. 58 (1985).
Tulig, T.J., presented at AIChE Meeting, Chicago, Nov. (1985).
Weale, K.E., Chemical Reactions at High Pressures, Barnes and Noble, New York (1967).

Table 2. Comparison of peak location and temperature between experimental data and prediction of the analysis.

Run	Experimental		Analysis		Fit Quality
	T_{max}(°C)	x_{max}(cm)	T_{max}(°C)	x_{max}(cm)	
1	138.2	119.6	139.4	122.4	excellent
2	180.7	70.5	179.5	64.9	good
3	184.6	59.3	181.0	60.7	excellent
4	166.5	72.8	174.6	64.2	good

THE COMPOSITE INTENSIVE VEHICLE—THE 3RD GENERATION AUTOMOBILE!—A BIO-CYBERNETICAL ENGINEERING APPROACH

Jürgen Köster
MOBIK GmbH, Gerlingen, FRG

Abstract

The growing discrepancy between present 'linear' automotive engineering and its social and political environment calls for a new generation of automobiles. These automobiles result from a bio-cybernetical systems approach. They weigh much less than present cars and they are less expensive to manufacture, to run and to maintain. To be light, durable and elegant, they have to be 'Composite Intensive Vehicles' (CIV's), i.e.: the primary structure of the new car generation is based on advanced composite (AC) materials. An experts commission has recently valued the present (!) market potential at 1 million such CIV's per year in Europe alone. Since the necessary high-volume production techniques for AC components will soon be available, the CIV-engineering must now be pushed since still being in its infancy. What are the basic features of these CIV's? Will the traditional automotive industry be able to manage this change and scope with this challenge?

1. The 1st Generation: Yesteryear's Wood + Steel 'Composite' Automobiles

In 1886, when incorporating a gas-engine into a coach, Carl Benz opened a new dimension to individual mobility. Since then, these 'automobiles' have been developed to their present perfection, giving birth, at the same time, to one of the most powerful industrial complexes of our days and to one of the most sophisticated engineering sciences. The emerging concepts of composite intensive vehicles (CIV's) are but the latest phase in this evolution, though possibly the origin of a new generation of automobiles and of automotive engineering.

When analysing the history of automotive engineering, one may roughly distinguish three phases. Until world-war one, almost all mechanical features of today's automobiles were developed, until world-war II all electric appliances had been introduced, and during the 70th and the 80th, the electronics have conquered the automotive world. Nowadays, 100 years after its invention, the automobile is a mature product both mechanically and electronically.

Looking at this evolution from the materials point of view, one recognises a steadily progressing substitution of wood by steel in the vehicle's primary structure: The first automobiles were motorised coaches with all wooden bodies, composed of intelligently engineered wooden lathwork and thin wooden planking, covered by special coach lacquer, each car a masterpiece of coach craftmanship. By the way: these wall panels could be considered as early composites, since to prevent distortion, strong cotton cloth was firmly glued to the inside of the planks.

Almost from the beginning and to scope with the rapidly increasing engine power, steel frames were introduced (Daimler/Maybach) to carry the heavy dynamic loads. The then still open wooden bodies were bolted on those frames. The engine was mounted between their two beams and in the front to improve the road hold.

Since 1904, first the wooden planks were replaced by steel or aluminium sheets, then, since 1914, also the wooden primary structure was substituted by welded steel-profiles. This then revolutionary technique had been pioneered by the railcar engineers Edward Budd and Joe Ledwinka, who sold their first all-steel body to Dave Dunbar Buick in 1912. Thus the typical first generation steel vehicle was born, where two separate steel main assemblies were bolted together, the body and the drivable chassis, both intrinsincly stiff but structurally separated and not meant to complement each other.

2. The 2nd Generation: Today's All Steel Automobiles

In the 30th, this kind of hybrid structure, i.e. a vehicle consisting of two separate primary structures, dissolved so-to-say into what may be called the second generation type of vehicle, where the two structures, i.e. the body and the frame, were integrated into one homogeneous primary ('monocoque') steel structure. In Europe, GM's Opel was the first car company to introduce this type of steel-sheet-engineering into an automotive volume production in 1937 with Lancia's Lambda as its predecessor as early as 1921.

Today's automobiles are still being built in this way. A large number of deep drawn steel-sheet components has to be welded together to form a stiff and highly protective cage-type passenger compartment, to which the drive units and outside design-dictated panels are fixed. Take the latest Daimler-Benz roadster: its rather small and open 2+1 passenger body must be welded together from 356 major steel components. This type of engineering is called **differential engineering**, the resulting vehicle structure is a **discontinuous structure**.

The following parameters typically describe this type of '2nd generation-vehicle':

- the assembly cost accounts for some 30% of the total manufacturing cost (a typical result of the differential engineering approach),

- the dead-weight/payload ratio is well above 2:1 (due to the steel primary structure and limousine type of vehicle design),

- the engine-power/passenger ratio is mostly above 20:1 (for sufficient speed and acceleration in heavy traffic situations),

- the life-time of a car design is above 7 years (to allow for a reasonable amortisation of the expensive steel transformation equipment),

- the car design suggests power, weight and speed (since it still is prestige-oriented, narcissistic and 'eco-negligent').

The '2nd generation' automotive industry, too, shows very specific attributes:

- Innovation becomes more and more difficult; ever so often new 'innovative' features, which have been developed with large investments, turn out to be gadgets only with little or no marginal benefits.

- The marketing efficiency decreases; it requires more funds every year to maintain a given market position; it is image and niche oriented, which is typical for mature and 'mee-too' products.

- The product development is 'linear'; it more or less concentrates on face-lifts, engine-improvements and comfort electronics; it spreads and uptrades the current limousine oriented programs eventually even adding eco-hostile features.

- The manufacturing organisation follows the chain production principle once pioneered by Henry Ford and Andre Citroen, which is best suited for volume and assembly oriented productions.

- Stagnating markets squeeze the profits; efficiency programs such as JIT, CIM, CAD, CAM etc. render only temporary relief.

In short: the automotive industry, at least in the advanced countries, shows the typical attributes of a mature industry trying to survive with a mature product. Yet, since this product serves a basic need, it will survive and so will its industry. It will even escape its profit squeeze - under condition however, it succeeds to adapt its product, its image and its structure to its **already changed** technological, social and political environment.

3. The 3rd Generation: Tomorrow's Composite Intensive Vehicle ('CIV')

Tomorrow's automobile **must** be a low price low cost function oriented utility and must therefore differ from today's vehicles by quite a number of significant features. In fact, tomorrow's 3rd generation of automobiles must and therefore will:

- be **lighter**; the dead-weight/pay-load ratio will be below 2:1, may be even 1.5:1, since new light-weight structural materials will be available on the factory floor level;

- have **low power engines**; the power/passenger ratio could probably be 10:1 only, since less space and less dead-weight is to be moved and less top-speed is required for the 'new image car';

- require **substantially less assembly cost**; the assembly could cost less than 25%, probably even only 20% of total manufacturing cost, since less parts will have to be assembled (integration target 25:1?)

- **last longer**; 15 years for the primary structure with almost no maintenance seems within reach based on new structural materials (outside panels are being changed more often);

- require **less repair**, since the present 2,5 miles/h damage limit will be extended to 5 miles/h, eventually even to 7,5 miles/h without extra cost; the surface finish moreover will be more scratch resistant than today;

- be **compatible with public transport** systems for long distance hauls;

- be **safer** due to advanced electronic control devices, namely with respect to road hold (='active' chassis concept) and active driving safety: the 4-wheel drive will be generalised for this purpose amongst other (a must in light cars!);

- be **function oriented**: thus multi-purpose/2-passenger cars with trailors for example could become one of thee 'new' standard vehicles (for commuting + shopping); 4-wheel-steering would possibly be another 'normal' feature for easy parking; cabriolets will be less expensive than limousines since the new cars will be engineered from the floor-pan upwards;

- be **smaller** and of **'unusual' proportions**, i.e. they would normally be higher than large and higher than long;

- **more comfortable** due to generalised comfort electronics (which **are** no longer prestige attributes);

- **easy** to be **face-lifted**, even by the owner; niche-cars do not any longer require special factories.

This list could be extended, though already quite long. Most of the listed features are state-of-the-art or have at least left the laboratory. The material of these 3rd generation cars is mostly advanced composite, with either plastic or metal matrix. Steel will be reduced to a minimum. Both the structure and the philosophy of this car will be entirely different from today's car - and so may well be its manufacturer.

4. The CIV's: A Bio-Cybernetical Approach:

Looking at a car from a bio-cybernetical position, one recognizes an increasing conflict between today's automobiles and their biological and political environment. The energy balance is extremely poor: we neglect the waste during acceleration and braking or due to the poor aerodynamics of a clefted floor or of an open wheel house, we use energy when stopping at a traffic light; We moreover move our cars with much noise including the door slam; we accept much empty space in the car inspite little space on the road (the limousine type of vehicle still being the industry's standard product). But above all: we generally move more than 1 ton of material to move one person and we concentrate on energy sources, which destroy our vital environment.

Though this ecological incompatibility of present automobiles is well known since long, it only now has become a 'hot' political issue, at least in the densely populated areas. It has since started to accentuate the discrepancy between the automotive industry and its social environment. A 1989 Deluxe(!)4-seat(!) passenger 'concept' car is but one of the examples of this (yet to be recognised?) mismatch (2,2 to, V8, 8L, 568 HP!).

In fact: excessive noise, high speeds, immoderate use of natural ressources are being considered as even criminal acts by more and more automotive clients. Local electorates accept, that their town authorities simply ban automobiles alltogether from their cities and that scarce funds be attributed rather to noise protection than to additional traffic space.

There thus cannot be any doubt: tomorrow's cars may not only be different cars, they _must_ be different cars, since they have to reconcile our legitimate need for individual transportation with our liability to protect our vital ressources. Public conscience is increasingly sensitized almost everywhere around the globe, in Tokyo as well as in Florence or in Rio. Experts estimate that for the common market alone the immediate (!) volume of such a car would be some 1 million units/year.

5. The CIV: A Different Material and Technology:

Advanced composites (AC's) outdo steel in strength but weigh only 25% of it (glass). They furthermore

. damp noise,

. integrate parts (a prototype study has integrated 36 steel components into one composite component!),

. corrode less,

. need less tooling and equipment for transformation,

. render perfect ('class A') surfaces,

. can be recycled,

. need less maintenance

. may be engineered to more varied specifications,

. offer more design freedom

— to only enumerate the most important features, where AC's (and metal matrix composites) are superior to steel. Given the increasing political pressure to overcome the above mentioned environmental incompatibility of present automobiles, there is no doubt, that this material will be the material of choice for the primary structures of any 'bio-cybernetically integrated' car.

These automobiles will cost less to produce and therefore be cheaper to buy and they will cost less to run: They cost less to produce in as much as the higher material cost is offset by less assembly cost. They cost less to run, since they consume less energy, they last longer and they cost less to maintain and repair. Damage is limited in case of accident.

Yet three (if not more) obstacles have as yet to be overcome to produce and successfully sell such an automobile: (1) a high-speed manufacturing technology for advanced composites must be available, (2) the different material requires different engineering solutions. Namely present **structural discontinuities have to be overcome** by applying the rules of **integral engineering**. Also (3) the marketing position of an automobile must be changed away from power, speed and prestige to the function it serves, to its cost to run and maintain, to its low environmental load.

With respect to the high-speed manufacturing technology of advanced composite components, recent progress permits to anticipate, that such a production technology will be available as early as 1990. It must be recalled, that a first car with AC-components (though no 'CIV' yet) is already successfully (!) being marketed since 1988 (BMW Z1).

6. The CIV: To Be Built By A Different Manufacturer?

The question is provocative, yet justified given the present historically grown, consolidated structure of the automotive industry and its 'linear' approach to automotive development. Recent experiences seem to demonstrate the strong resistance to change; several attempts to adopt composite technologies on a larger scale so far have not resulted in accountable changes, may be for several reasons:

. Differential (metal-) engineering approaches are not compatible nor simply interchangeable with the (composite-) integral engineering approach. 'Metal' engineers experience great difficulties when trying to follow the reasoning of composite engineers. Isotropic matrix oriented engineering differs profoundly from anisotropic fiber oriented engineering. Composite applications have thus typically been limited to

- quasi-isotropic materials like SMC, GMT, BMC, RRIM and alike,

- secondary structural components, namely outside panels, where the composite component remains at any time interchangeable with a same metal component,

'Metallists' typically request, when considering an AC-component alternatively, that it compares favorably to steel or low-profile composite components; the specific structural and weight benefits of AC's must be ignored then, since they cannot be valued in a differentially engineered environment.

- The metal manufacturing equipment in the factories has to be amortised through the further production of metal components.

- The engine-and-power oriented corporate identity of the traditional car manufacturer renders a bio-cybernetical engineering/marketing difficult if not impossible

The few successful attempts to launch a 'CIV' have thus either been limited to a special series like the PORSCHE 959, or been planned and executed in an entirely different automotive entity with an entirely new (composite-)engineering team like the BMW Z1. This separation thus seems to be the basic condition for break-throughs also in the automotive industry namely when considering the change required to conceive, manufacture and sell a 3rd generation vehicle.

This condition is not new. There are several comparable situations in the history of automotive engineering: Edward Budd and Joe Ledwinka left the Hale & Kilburn Coach Company to create The Budd Company and build the first steel monocoque body for Dave Buick in 1912. The introduction of steel monocoque bodies on a larger scale happened in new factories after world war II. LAMBORGHINI has built the first all-composite monocoque road car after having been taken over by new owners. The BMW Z1 was designed and engineered by a newly created affiliate of the parent BMW AG, the BMW Technik GmbH with its **entirely autonomous** management and organisation.

7. Resumé: Beware The 'Metal Engineering' Pitfalls!

With the recent transfer of the Z1 production from the affiliate's (composite-) eggheads to the parent (Metal-) practiciens, AC's have finally arrived at the volume manufacturing level - at a moment however, where the volume manufacturing techniques are just only leaving the laboratory. AC-production is still low-volume and high cost manufacturing. Will the metal engineers return to 'their' lower cost metal components?

During the engineering phase CIV engineers are permanently confronted with composite/metal manufacturing cost comparisons. It must be kept in mind then, that AC-components can demonstrate their cost saving potential only, if the total car concept follows the rules of integral engineering, namely if the primary structure is what one may call 'AC-like': such structures are being developed from the bottom up to the waste line and roof.

The floor pan and the bumper energy system are the basic elements in this concept (both AC); vertical and horizontal panels are at least intrinsicly stiff (=AC sandwiches) if not load carrying; hoods, doors and lids may then be frictionally connected to the primary structure; metals, namely aluminium, and composites are technologically reconciled in this concept and no longer enemies.

AC-components, if properly engineered, cannot simply be replaced by metal components and if they are, the structure would probably not be an AC-engineering optimum.

Most important of all: a CIV project depends on its philosophy and on its support by the **entire** management. Early compromises in favor of steel prevent only too often parts integration. The AC-component's cost may then well kill such a project before it reaches the factory. Many a CIV thus was and may still in future be dead-born.

8. Addendum: Advanced Composites Are Profitably Recyclable!

The updated definition of 'simultaneous engineering' requires the team work of design, manufacturing <u>and</u> environmental engineering in a development project. New parts are only admitted if they may be easily dissembled and profitably recycled (=closed loop engineering).

Recent studies in Europe show both the problem and the opportunity: the average scrap car weighs some 880 kg and contains some 7% plastic materials (i.e. 53 kg). It takes 90 minutes and no special tools to dissemble this car. Plastic material is not recyclable, since it contains undeterminable grades (it is 'dirty') and recycled grades are not admitted for 'new' products. If recycled (TP only), market value is largely below recycling cost.

In contrast to this advanced composite thermoplastics are easily identified, dissembled, shredded into short fiber reinforced material and marketed as high quality engineering materials. Their market value is well above recycling cost.

Which demonstrates in addition to the outstanding engineering value of advanced composite thermplastics their equally outstanding advantage in the upcoming environmental issue. Plastics are not generally develish!

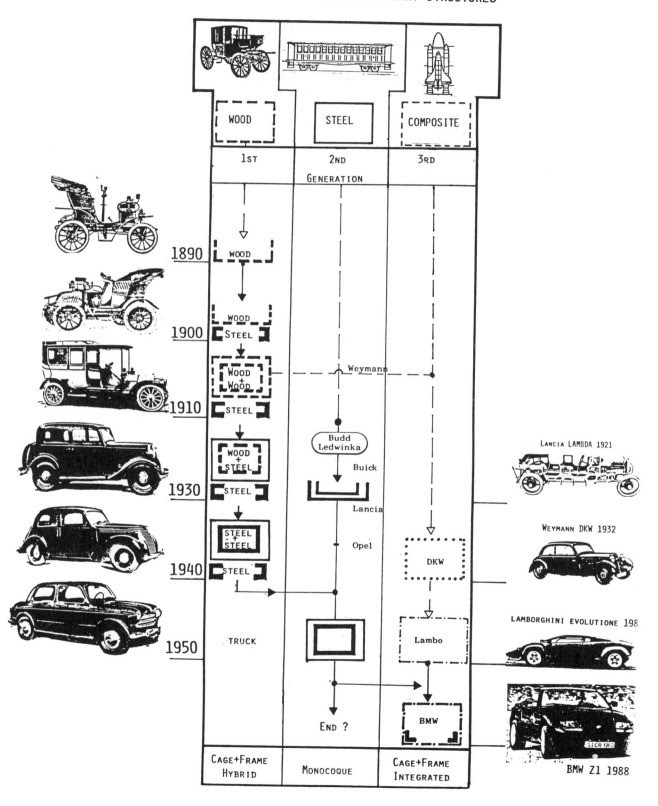

GENEALOGY of AUTOMOTIVE PRIMARY STRUCTURES

ULTRASONIC MOLDING OF PLASTIC POWDERS

Satinder K. Nayar, Avraham Benatar
Department of Welding Engineering
The Ohio State University
Columbus, OH 43210 USA

ABSTRACT

In ultrasonic molding, polymeric powders and pellets are compacted and fused together with the help of ultrasonic energy. The objectives of the work are to study the feasibility of ultrasonic molding and perform an experimental study of the effect of the process parameters on part quality, leading to the development of a prototype ultrasonic molder. Experiments were performed using a conventional plastic welder to determine the effect of ultrasonic compaction time, pressure, and the amplitude of vibration on tensile strength and percent elongation. Optimization of mechanical properties such as strength and ductility in comparison with compression molded samples was attempted. It was found that a minimum compaction time was required for each of the materials to fuse. The tensile strength and elongation increased with increasing compaction time to a maximum value, beyond which there was no significant change. For polystyrene the maximum tensile strength of ultrasonically molded samples was $85\pm8\%$ of the strength of compression molded samples and the maximum elongation was $95\pm10\%$ of compression molded samples. Similarly, for ABS the strength was $65\pm10\%$ and the elongation was $60\pm5\%$, for high density polyethylene the strength was $85\pm10\%$ and the elongation was $27\pm5\%$ and for polycarbonate the strength was $75\pm5\%$ and the elongation was $40\pm7\%$. This feasibility study showed that the polymers which are easy to ultrasonically weld are also easy to ultrasonically mold. It also shows that this method is very promising especially for compaction and fusion of materials which are difficult to mold by conventional methods.

INTRODUCTION

Ultrasonic molding of thermoplastic powders and pellets is a process where the ultrasonic energy is used to compact and fuse the powders and form a part. Recent research by Crawford, and Paul (1,2) has shown that it is possible to mold plastics by ultrasonic techniques. This has a number of potential advantages over conventional molding methods. As the compaction is performed at room temperature, heating, softening and melting occur without the application of external heat source. Materials having high viscosity and which are therefore difficult to injection mold can easily be processed by this technique. The ultrasonic molder is easy to handle and occupies less space. Filled plastics, which are difficult to mold by conventional methods, could also be ultrasonically molded.

Thermoplastics can be divided into two main groups: Amorphous and semicrystalline. For amorphous thermoplastics, flow and diffusion can not occur until the glass transition temperature is exceeded(3). For semicrystalline polymers, the melting temperature must be exceeded before flow and diffusion can occur(3). Therefore the ultrasonic compaction of amorphous and semicrystalline polymers is expected to be quite different.

In this work we studied the feasibility of ultrasonic compaction of polystyrene, polycarbonate and Acrylonitrile Butadiene Styrene (ABS) which are amorphous, and high density polyethylene which is semicrystalline.

Fairbanks[4] has previously established that the ultrasonic compaction of plastic powder provides energy of fusion for the powders, without the addition of external heat. He also found that the ultrasound not only fuses the powders but reduces the friction of plastic flow through extrusion orifice in ultrsonic extrusion. Paul and Crawford [1-2] studied the feasibility of molding polypropylene powder using ultrasonic vibration. They reported a tensile strength of 20MPa which was 85% of the value achieved by injection molding. They found the elongation to failure to be only 15% of the elongation of the injection molded parts. They also found that compaction time and particle size were directly proportional to the strength whereas pressure had little or no effect on the strength.

In ultrasonic welding of thermoplastics, heat is generated when the plastic is subjected to cyclical strain [3,5]. The power dissipated depends upon the loss modulus of the polymer and the cyclical strain amplitude [3]

$$Q = \omega \varepsilon_o^2 E'' / 2 \qquad (1)$$

where Q is the average power dissipated, ω is the frequency, and ε_o is the strain amplitude. Eq. 1 can also be used to estimate the power dissipated during ultrasonic molding of powders and pellets.

Determination of the dynamic moduli of the polymers is necessary for determination of the energy dissipated. By measuring the longitudinal and shear velocities one can determine the longitudinal and shear storage moduli [6]:

$$L' = \rho V_L^2 \qquad (2)$$
$$G' = \rho V_T^2 \qquad (3)$$

where ρ is the density, G' and L' are shear and longitudinal storage moduli respectively, and V_L and V_T are the longitudinal and shear velocities respectively. The tensile storage modulus is determined by [6]:

$$E' = G' (3L' - 4G') / (L' - G') \qquad (4)$$

Tensile loss tangent tan δ_E relates tensile loss modulus E'' and the storage modulus.

$$\tan \delta_E = E'' / E' \qquad (5)$$

The tensile loss tangent is given by [6]:

$$\tan \delta_E = \tan \delta_G - \frac{G'L' (\tan \delta_G - \tan \delta_L)}{(L' - G')(3L' - 4G')} \qquad (6)$$

where tan δ_G is the shear loss tangent and tan δ_L is the longitudinal loss tangent. Phase velocity and absorption coefficient determine the shear and longitudinal loss tangents. Longitudinal absorption coefficient can be calculated by the following [6]:

$$\alpha_L = \frac{1}{2(h_2-h_1)} \log_e \frac{A(h_1)}{A(h_2)} \qquad (7)$$

where h is the thickness and $A(h_i)$ is the attenuation for sample i.
Hence the loss tangents are determined from the following relations [6]:

$$\tan \delta_L = 2 V_L \alpha_L / \omega \qquad (8)$$

$$\tan \delta_G = 2 V_T \alpha_T / \omega \qquad (9)$$

Using measured shear and longitudinal wave speeds and attenuations in equations 2 through 8, the dynamic moduli of the polymers can be determined.

EXPERIMENTAL PROCEDURE

Ultrasonic Compaction

Branson ultrasonic 900M welder was used in all experiments. An aluminium mold and horn were designed to make parts that are 1" x 2" x t (variable thickness 't'). Under a constant amplitude of vibration (0.001"), hold time (10 sec) and pressure (50 psi), compaction time was varied (0 - 10 secs) for the following thermoplastics:

1. Cycolac ABS - T-1000 Borg Warner
2. High Impact polystyrene Grade 6800 (Chevron)
3. High Density polythylene Grade 6206 (Norchem)
4. Polycarbonate 141 Lexan (GE Plastics)

The pressure was also varied (20 - 100 psi) while keeping hold time (10 sec), amplitude of vibration (0.001") and compaction time 2 sec for polystyrene, 4 sec for polycarbonate, 4 sec ABS, and 8 sec for polyethylene constant. Tensile specimen were cut from each compacted samples using ASTM standard D638-86 with smaller sample size as shown in Fig 1. These specimen were then pulled at a rate of 0.05 cm/min on an Instron 4200 machine. Then their ultimate tensile strength and elongation were measured. These were compared with the strength and elongation of compression molded samples which were prepared as shown in Table 1. All compression molded samples were air cooled to room temperature.

Measurement of dynamic modulli

Figure 2 shows the experimental set up for measurement of dynamic modulli of thermoplastics. A Hewlett packard 1743A oscilloscope, a Panametric 5052PR pulse transmitter and receiver, longitudnal Panametric transducer of frequencies 0.5, 1.0, 2.0, 2.25 and 3.5 MHz and a shear Panametric broadband transducer of frequency 2.25 MHz were used. The wave velocity was determined from the time required for the wave to travel through the sample thickness and back. The attenuation was calculated using Equation 7. For the broadband shear transducer, the wideband signal was passed through a fast fourier transform providing information for the frequency range from 0.5 - 3.5 MHz. The dynamic modulli of the thermoplastics were determined from Equation 4, 5 and 6.

RESULTS AND DISCUSSION.

Dynamic Moduli

Fig 3 and Fig 4 shows the storage moduli and loss moduli for a range of frequencies. It can be seen that the storage modulus increases slightly with increasing frequency while there is a decrease in the loss modulus. This response is due to the inability of the polymer chains to respond to the high frequency loading [6,7]. Also the reduced motion of the chains increases the apparent stiffness of the polymer and reduces intermolecular friction. The moduli at 20 KHz were estimated by extrapolation of the high frequency data to low frequencies as shown in Figures 3 and 4.

Effect of compaction time

Figure 5 and 6 show a minimum time is required to compact and fuse the powders and pellets. As compaction time is increased there is an increase in %strength and %elongation, as compared to the compresion molded samples. Increasing compaction time increases the energy dissipation, which increases the flow and melting of the plastic and therefore results in the increase in strength and elongation.

It can also be seen that increasing the compaction time beyond a threshold value did not effect the strength and elongation, as seen in Figure 7 and 8. This is becaues eventually sufficient energy is dissipated to cause enough flow to achieve the maximum strength and elongation, and any further increase in the compaction time does not have any effect on strength.

During compaction the heat is generated within the plastic. The energy dissipated was calculated using Equation 1 Assuming: adiabatic heating, E" not a function of temperature, and all of the amplitude of vibration is taken up uniformly by the pellet or for powder. Figures 9 to 12 show an increase in energy dissipation with increase in compaction time. Also these figures show the theoretical prediction which are surprisingly close considering that E" can vary by more than an order of magnitude over the temperature range of interest.

For polystyrene the maximum tensile strength of 85±8% and maximum elongation of 95±10% of that of the compression molded sample was achieved. Similarly, for ABS the strength was 65±5% and elongation was 60±5%, for high density polyethylene the strength was 85±10% and elongation was 27±5% and for polycarbonate the strength was 75±5% and elongation was 40±7%.

Effect of Pressure

Application of pressure during ultrasonic compaction serves two major functions, (1) it provides good contact between the horn and the pellets for good energy transmission and (2) it causes the molten surface on the pellet to flow and fuse with other pellets. Figure 13 to 16 show the effect of pressure on strength and elongation. It can be seen that there is a slight increase in the strength and elongation with increase in pressure. This is similar to the results found by Paul and Crawford(1). This can be explained as low pressures results in poor horn to part contact and therefore poor transmission of the energy. Degradation of the material causes decrease of strength and elongation at higher pressures. This could be due to the presence of moisture in the plastics.

CONCLUSION

1. Both amorphous and semicrystaline polymers can be compacted by the application of ultrasound.
2. The strength and elongation is dependent upon the compaction time, and it increases as the compaction time increases.
3. The strength and elongation is relatively independent of the compaction pressure.
4. Values of E' and E" helped in predicting the energy dissipated which was in surprisingly good agreement with experiments considering the assumptions made.
5. ABS and polystyrene that perform well in ultrasonic welding also perform well in ultrasonic molding.
6. This work demonstrates that the ultrasonic molding parameters do have an effect on the strength and elongation of amorphous and semi-crystalline thermoplastics.

REFERENCES

1. Crawford, R.J., Paul D.W., Ultrasonics, 23-27 (1981)
2. Crawford, R.J., Paul D.W., Polymer Communications 21,138-139 (1980)
3. A. Benatar, " Weldability of Thermoplastics, " Proceedings of EWI N.America Welding Research Seminar, Columbus,OH 1987
4. Fairbanks, H.V., .Ultrasonics.12, 22-24 (1974)
5. Aloisio, C.J., Wahl, D.J.. and Whetsel, E.E., 1972. "A simplified Viscoelastic Analysis of Ultrasonic bonding". SPE Technical papers, Brookfield CT.
6. Read, B.E., Dean G.D.,"The determination of Dynamic Properties of Polymers and Composites". John Wiley and sons, Inc., New York, NY.1978
7. Ferry, John . D., "Viscoelastic properties of Polymers ". John Wiley and sons, Inc., New York, 1970

TABLE 1 Conditions For Compression Molding

Material	Peak Temp. oC	Time at Peak Temp.	Pressure (psi)	Tensile strength (psi)	% Elongation
ABS	125	45 min	2000	5712 ± 8%	41 ± 15%
Polystyrene	125	60 min	2200	2680 ± 10%	32 ± 10%
Polycarbonate	130	45 min	3000	8500 ± 9%	100 ± 12%
High Density Polyethylene	135	60 min	3500	4110 ± 7%	72.6 ± 11%

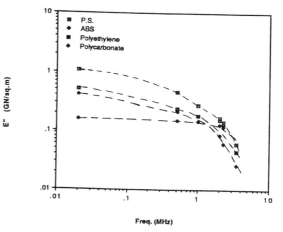

Figure 4 E" as a function of frequency.

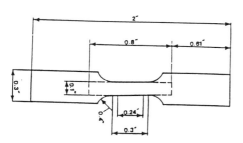

Figure 1 Typical Dimensions of Tensile Specimen.

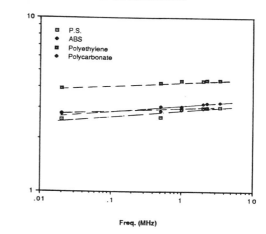

Figure 3 E' as a function of frequency.

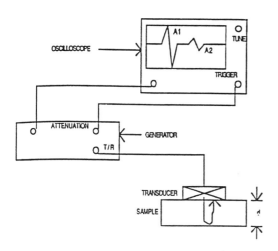

Figure 2 Measurement of sound velocity and Attenuation.

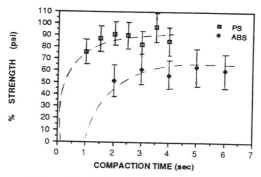

Figure 5 Effect of compaction time on strength for ABS and polystyrene

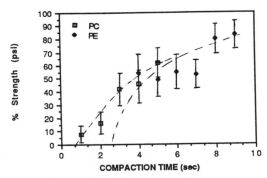

Figure 6 Effect of compaction time on strength for polycarbonate and polyethylene.

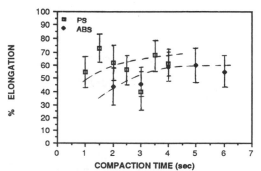

Figure 7 Effect of compaction time on elongation for ABS and Polystyrene.

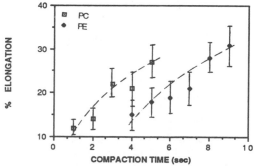

Figure 8 Effect of compaction time on Elongation for polycarbonate and polyethylene

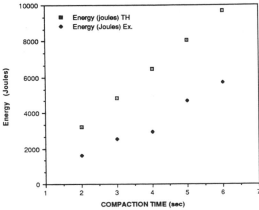

Figure 9 Effect of compaction time on energy (ABS)

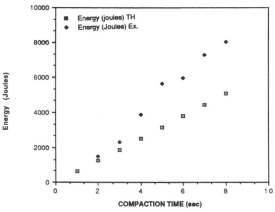

Figure 10 Effect of compaction time on energy (PC)

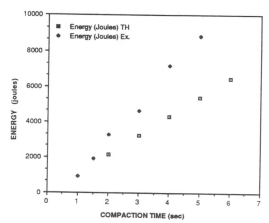

Figure 11 Effect of compaction time on energy (PS)

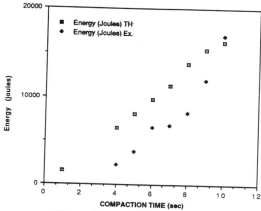

Figure 12 Effect of compaction time on energy (HDPE)

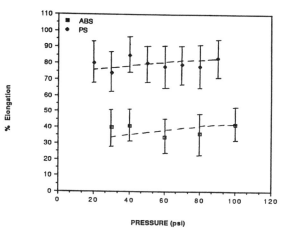

Figure 15 Effect of pressure on elongation for ABS and polystyrene.

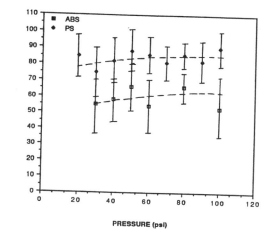

Figure 13 Effect of pressure on strength for ABS and polystyrene.

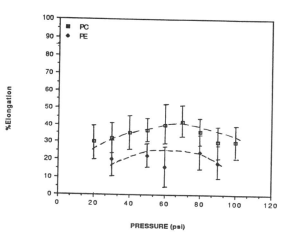

Figure 16 Effect of pressure on elongation for polycarbonate and polyethylene.

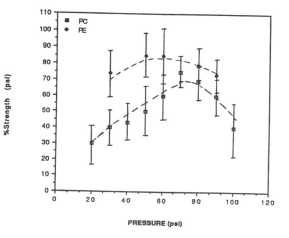

Figure 14 Effect of pressure on strength for polycarbonate and polyethylene.

CORRELATION OF DIELECTRIC CURE INDEX TO DEGREE OF CURE FOR 3501-6 GRAPHITE EPOXY

David R. Day, David D. Shepard
Micromet Instruments, Inc.
Cambridge, MA USA

INTRODUCTION

Dielectric sensing is currently the most promising technique for monitoring and controlling composite curing. Over the past several years several sensors have been developed to withstand the rugged and abusive environments found in production presses and autoclaves (1). Much work has been done interpreting the dielectric signals and using them to control various processes in real time (2-7). Measurements are typically taken at several frequencies over several orders of magnitude (0.1 Hz to 100 KHz) and the ionic conductivity is extracted from the response. During the course of a cure the ionic conductivity is a strong function of cure state and the temperature (8). In this work we have measured the temperature dependence of the uncured and cured resin. The temperature dependence of the dielectric response during cure is then removed from the data (9). Finally the dielectric cure index (the change in ionic conductivity with temperature dependence removed), is compared to degree of cure data from both the Springer cure model (10) and the Landuyt (11) cure model.

EXPERIMENTAL

All data were obtained from Hercules 3501-6 graphite epoxy (unidirectional). Dielectric data were measured and logged using a Micromet Instruments' System III Microdielectrometer. The Micromet Filtered IDEX sensor, which contains a built in filter layer to screen out graphite, was used in the Micromet Programmable Oven (Figure 1 A and B). For these studies the controller thermocouple was placed directly in the sample adjacent to the sensor so that exact sample temperature control could be obtained. Ionic conductivity, the temperature dependence, and cure index were determined using the Micromet Control Software package. The time and temperature data for two cures were used as input to the Springer model which was used to predict degree of cure. In a third experiment, the cure temperature profile was matched to a published 3501-6 cure profile from the Landuyt model. The cure index was compared the the published degree of cure from the Landuyt Model.

RESULTS AND DISCUSSION

The temperature and resulting ionic conductivity profiles for the three cures are shown in Figure 2A, 2B, and 2C. Although the ionic conductivity curves change with cure, due to the additional temperature dependence, they do not directly relate to degree of cure. In order to determine the temperature dependence of the ionic conductivity, a plot must be constructed of ionic conductivity vs. temperature (Figure 3, data from cure in Figure 2A). Although the conductivity dependence on temperature follows a WLF relation (8) and is slightly curved, it can be represented over a small temperature range as being linear. The temperature dependence of the uncured material (0%) can be estimated from the data during the initial heating stage of the resin. At the end of cure, the fully cured temperature dependence (100%) can be estimated from the behavior during cool down (see dotted lines in Figure 3). Between the initial heating and final cooling significant reaction is occurring and so intermediate temperature dependencies could only be determined by quenching the reaction at various points. This type of experiment has been previously carried out (9) and results showed that it is sufficient to measure only the 0% and 100% temperature dependencies and assume a linear change between the two during the cure. The following 0% and 100% temperature dependencies were estimated:

	A	B	Assumed T Range
0% (uncured)	.036	-9.50	80 - 180
100% (cured)	.045	-18.14	80 - 180

where:

log conductivity = A * Temp. (C) + B

The above A and B values were assumed constant and were used for determining cure index for all three cures. Cure index was determined by calculating the percent change in log conductivity from the 0% log conductivity value to the 100% log conductivity value for the particular temperature of the sample at any given point in time (9):

$$\text{Cure Index} = \frac{LC - (.036T - 9.50)}{((.045T - 18.14) - (.036T - 9.50))} * 100\%$$

where LC is log conductivity and T is temperature at a given instant.

It is important to note that with the temperature dependence removal, the cure index is expected to increase monotonically with cure state, although it is not necessarily expected to be EQUAL to the degree of cure. This is why the term INDEX is used instead of degree of cure. However, the cure index is expected to be a reproducible indicator of changes in cure state. Nonreproducibility could arise from large variation in ionic concentrations from batch to batch. A goal of this work and future work is to determine exactly how much ionic concentration can influence dielectric data.

Figures 4 and 5 show the cure index as well as the degree of cure (as determined from the Springer model) which correspond to Figures 2A and 2B respectively. The shapes of the cure index and modeled degree of cure (alpha) curves are similar, however, in both cases the cure index falls below the alpha curves. The cure index near the end of both cures exhibits a slow increase indicating slow but continuing reaction while the model alpha curves are nearly level. This is perhaps not unreasonable since the model was based on kinetic data derived from differential scanning calorimetry, a technique which is known to have low sensitivity near end of cure. Figure 6 shows the cure index and degree of cure determined from the Landuyt Model which corresponds to Figure 2C. In this case the curve shapes are the same and the two curves are nearly superimposed. Model data from the end of cure in Figure 6 were not available, however, the cure index indicated continuing residual reaction at 250 minutes, nearly 100 minutes into the final hold at 178 C.

It is interesting to note that the cure temperature profiles shown in Figure 2A and 2C are very similar except for the ramp rates. The resulting cure index data in Figures 4 and 6 are also very similar. However, the degree of cure predicted by the Springer and Landuyt models are significantly different. At the beginning of the second ramp one model indicates 35% cure while the other indicates 20% cure. The former sample was at the intermediate hold for about 20 minutes longer and may account for some of this difference, however, there appears to still be some discrepancy between the models. The dielectric cure index cannot be used at this time to verify either model since it is only expected to be monotonic with cure state.

SUMMARY

The dielectric properties of 3501-6 graphite epoxy were monitored during three different cure cycles. The temperature dependence of the ionic conductivity was determined for uncured and cured material. Using these data the dielectric cure index was calculated and compared to two different model predictions for degree of cure. The cure index curves exhibited similar shapes in all cases and showed good superposition onto the Landuyt modeled degree of cure data. In all cases the cure index was very sensitive to small changes in cure state near the end of cure. Future work will include further comparison of log resistivity data with modeled viscosity and comparison of viscosity and degree of cure predictions between various models.

ACKNOWLEDGEMENTS

The authors would like to thank Pete Cirisciole of Stanford for supplying the Springer model predictions and Pierre Landuyt of the University of Liege, Belgium, for supplying his model predictions.

REFERENCES

1. Micromet Instrument Product Literature, Cambridge, MA

2. W.E. Baumgartner and T. Ricker, SAMPE Journal, 19, 4, pg. 6 (1983)

3. D.R. Day, Poly. Eng. Sci., 26, 5, pg. 362 (1986)

4. D.R. Day, Qualitative Nondestructive Evaluation, Plenum Press, NY, 7B, pg. 1573 (1988)

5. D.E. Kranbuehl, M. Hoff, S. Weller, and P. Haverty, Qualitative Nondestructive Evaluation, Plenum Press, NY, 7A, pg. 525 (1988)

6. D.E. Kranbuehl, M. Hoff, D. Eichinger, and R. Clark, 34th Int. SAMPE Symp. Proc., 34, pg. 416 (1989)

7. Micromet Product Bulletin, "Critical Point Control Module," Cambridge, MA (1989)

8. S.D. Senturia and N.S. Sheppard, Jr., Advances in Polym. Sci., 80, pg. 1 (1986)

9. D.R. Day, A. Wall, and D. Shepard, Soc. Plast. Eng. Annual Tech. Conf. Proc., 34, pg. 964 (1988)

10. W.I. Lee, A. Loos, and G.S. Springer, J. of Composite Materials, 16, pg. 510 (1982)

11. J. Denoel, J.M. Liegeois, and P. Landuyt, "Composites," Liege's University, Belgium, May-June 1989

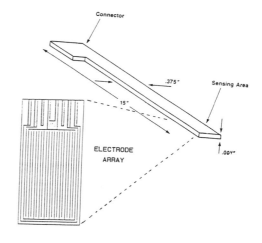

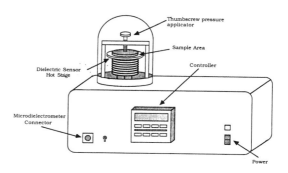

Fig. 1. A) Schematic diagram of Filtered IDEX sensor. B) Schematic diagram of programmable oven.

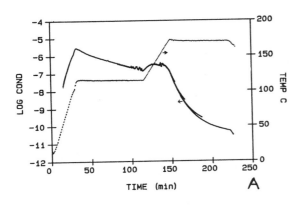

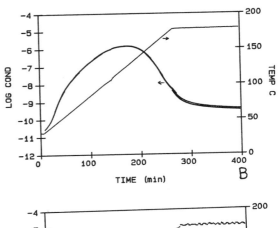

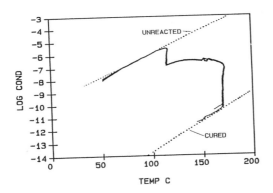

Fig. 2. Conductivity and temperature vs time for 3501-6 graphite-epoxy. (Three different cure schedules A, B, and C)

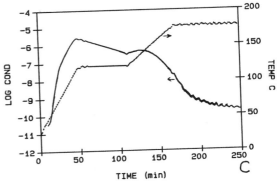

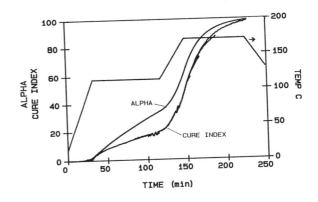

Fig. 3. Conductivity vs temperature (from Fig. 2A). Dotted lines represent temperature dependence estimates for 0% and 100% cured resin.

Fig. 4. Cure index (data from Fig. 2A) and degree of cure (Springer model) vs time. (Degree of cure is plotted along the same scale as cure index.)

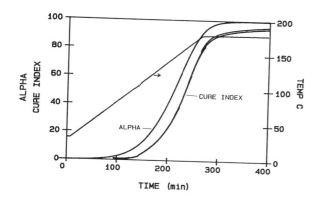

Fig. 5. Cure index (data from Fig. 2B) and degree of cure (Springer model) vs time. (Degree of cure is plotted along the same scale as cure index.)

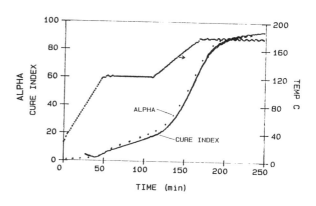

Fig. 6. Cure index (data from Fig. 2C) and degree of cure (Landuyt model) vs time. (Degree of cure is plotted along the same scale as cure index.)

MICROWAVE PROCESSING OF POLYMER COMPOSITE MATERIALS

Martin C. Hawley
Department of Chemical Engineering
Michigan State University
East Lansing, MI 48824 USA

John D. DeLong
Composite Materials and Structures Center
Michigan State Univeristy
East Lansing, MI 48824

ABSTRACT

Microwave processing of polymer composite materials is being studied as an alternative to conventional thermal processing. These studies are being carried out using a resonant microwave cavity technique. The inherent advantages to microwave processing have led to several significant developments in microwave processing technology. These include isothermal curing in both space and time of epoxy/amine resins, production of epoxy/amine samples with tensile properties equivalent to thermally processed materials, and the ability to process isothermally both unidirectional and crossply composites. The ability to cure carbon fiber/epoxy composites using microwave energy will allow production of uniform thick materials with equal or superior properties as compared to thermally processed materials.

MICROWAVE PROCESSING OF POLYMERS and polymer composites is an alternative to conventional thermal processing. Microwave processing offers several advantages over thermal processing, including 1) selective and controlled heating of the material due to absorption of microwave energy by polar functional groups of the polymer and conductive fibers, 2) decreased thermal degradation of the material due to rapid uniform bulk heating, and 3) increased control of the material time-temperature profile and cure cycle. The ability to heat rapidly and controllably offers the potential to virtually eliminate temperature gradients during cure of thick section materials and thus reduce thermal stress. By rapidly pulsing the input of electromagnetic energy, the temperature in a curing composite can be precisely controlled both in space and time. This paper discusses microwave processing of polymer composites using a resonant microwave cavity applicator system.

MECHANISM OF MICROWAVE HEATING

The level of electromagnetic heating of a material exposed to microwave radiation is determined by the frequency and intensity of the radiation, the orientation of the material in the electromagnetic field, the field pattern, and the dielectric loss factor of the material. The power density, P, dissipated as heat by a material at a given position in the field is given by

$$P = 1/2\ \epsilon''\ \omega\ |E|^2$$

where ϵ'' is the dielectric loss factor, ω is the frequency, and E is the electric field intensity at a point in the field. The dielectric loss factor is the imaginary part of the complex dielectric constant, ϵ^*, of a material, given by

$$\epsilon^* = \epsilon' - j\ \epsilon''$$

where ϵ' is the real part of dielectric constant, related to the capacitance of the material. The complex dielectric constant expresses the ability of a material to store and dissipate electromagnetic energy.

Two mechanisms contribute to the dissipation of electromagnetic energy as heat in a material, energy loss due to electrical conduction and loss due to dipole relaxation. This is shown by the following expression for the loss factor:

$$\epsilon'' = \sigma/\omega\epsilon_o + \epsilon''_d$$

where σ is the conductivity of the material, ϵ_o is the permittivity of vacuum, and ϵ''_d is the contribution due to dipole relaxation. At microwave frequencies, dielectric loss by conductive materials such as metals or carbon fiber is dominated by conduction, and loss by nonconductive materials such as polymers is dominated by dipole relaxation.

For a composite material consisting of a polymer matrix reinforced with fiber, the tendency for each component to heat in a microwave field depends on the magnitude of its loss factor. Table 1 shows representative values of the loss factor for several matrix and fiber materials used in composites. If the loss factors for a matrix and fiber are very dissimilar, the component with the higher loss factor will heat significantly faster than the other component. This selective heating results in locally high temperatures which can be exploited to tailor interfacial adhesion.

Table 1. Dielectric Loss Factor for Selected Materials at 2.45 GHz

Material	ϵ''
DGEBA/DDS uncured	.323
nylon	.038
Thermid IP-600 polyimide	.00326
teflon	.0003
AS4 carbon fiber	conductor; high loss
E-glass fiber	.002-.005
K-49 aramid fiber	.0002-.001

Conventional thermal heating of composite materials is controlled by the heat transfer characteristics of the material and autoclave. The rate at which heat is transferred into the material depends on the thermal conductivity of the material, the convective heat transfer coefficient, and the heating rate in the autoclave. Such a heating strategy is often slow, with little control available to alter the process. Microwave heating, on the other hand, is inherently controllable with the heating rate nearly independent of the heat transfer characteristics of the material. By raising the microwave power input the heating rate changes nearly instantaneously followed by a rapid temperature rise. This rapid heating feature is illustrated in Figure 1 for nylon [1].

MICROWAVE RESONANT CAVITY APPLICATOR

Development of processing systems for the coupling of microwave energy into polymers and polymer composites has been ongoing for more than a decade. Researchers have investigated the use of conventional multimode microwave ovens [2] and waveguides [3] on the curing of epoxies. These studies found that microwave processing offers rapid heating, faster cure, and properties at least equal to those of materials cured using conventional thermal curing. Microwave processing using multimode ovens or waveguides in not efficient, however, as much of the input energy is not utilized in the cured of the thermoset. In a waveguide, energy passes through the material continuously and any energy not absorbed exits the system. In a multimode oven such as a typical household microwave oven, much of the input energy is dissipated to the walls of the oven or is relected back into the circuit. In order to cure a thermoset, then, using either of these microwave techniques, a fairly high input power is required to compensate for the inefficiency of the energy coupling.

A more energy efficient design for microwave processing is the single mode resonant cavity applicator developed at Michigan State University [4]. A diagram of a single mode cylindrical cavity is shown in Figure 2. A circuit diagram for the experimental microwave system is shown in Figure 3. The single-mode applicator as part of an overall microwave circuit maintains a resonant microwave field and concentrates energy in regions where the material is located. The resonant field pattern is achieved by tuning the cavity to compensate for disturbance in the electromagnetic field by the loaded material due to its dielectric properties. This tuning is done by varying the axial dimension of the cavity and the depth of the microwave coupling probe until a resonant field is found. Resonant field patterns for empty cavities and cavities loaded with low loss materials are predictable. Knowledge of these field patterns is used to locate the sample to be processed in the region of maximum field intensity. Coupling efficiencies (absorbed power/input power) as high as 95% have been achieved with this technique [4].

MICROWAVE CURING OF EPOXY/AMINE

The resonant microwave cavity system operating at a frequency of 2.45 GHz has been successfully used for the curing of an epoxy/amine resin. The materials used were DER 332 (Dow Chemical Co.), a high purity DGEBA epoxy, and diaminodiphenyl sulfone, DDS, as shown in Figure 4. The materials were prepared as described previously [5]. An uncured epoxy/amine mixture was poured in a cylindrical teflon mold of approximately 2 cm^3 and suspended coaxially in the microwave cavity. The cavity was set to operate in a TM_{012} mode in which the electric field is directed along the major axis of the cavity as shown in Figure 5. Location of the sample in the center of the cavity operating in a TM_{012} mode ensures maximum absorption of electromagnetic energy. Other modes can be used, the choice of which depends upon the sample geometry.

Three stages are identifiable during heating of the epoxy/amine mixture as shown by the dielectric loss factor of the material. The dielectric loss factor was measured using a cavity perturbation method [5]. Figure 6 shows ϵ'' as a function of time during cure of DER332/DDS for a constant input power level of 5 W. Initially ϵ'' shows a rapid increase due to increased thermal motion by the heated monomer. This monomer heating is the first stage of the process. ϵ'' reaches a maximum then rapidly falls off to an asymptotic value. The rapid decrease in ϵ'' corresponds to formation of an increasingly cross-linked network during reaction, which is the second stage of the process. This indicated by the extent of cure vs time curve in Figure 6. Network formation restricts movement of the dipoles causing a decrease in ϵ''. Approach to an asymptotic value for ϵ'' corresponds to the third stage of the process which is formation of a fully cross-linked network.

A characteristic feature of the epoxy/amine reaction is the exothermic temperature excursion illustrated in the time-temperature curve of Figure 6. With an exothermic heat of reaction of approximately 400 J/g, isothermal curing in a constant power microwave process or a convention thermal process is not possible due to relatively slow convective heat transfer at the sample/air boundary. The cure temperature in such a process is limited such that the maximum temperature achieved during the reaction does not exceed the degradation temperature of the material. It is usual then to ramp to higher temperatures at successive points during cure to avoid thermal degradation. The exothermic reaction can also lead to uneven cure in a material contributing to internal stress.

To eliminate the need for temperature ramping, isothermal processing can be achieved using a pulsed microwave input. Figure 7 shows the temperature-time profile for epoxy/amine cured isothermally at 190 C with a pulsed microwave input. The material is initially heated to the set-point temperature. As the reaction proceeds, the exothermic heat is utilized to maintain the reaction temperature. In response to a temperature feedback control loop, microwave power is pulsed at such a rate as to compensate for convective heat transfer loss to the surroundings. Upon reaching the set-point temperature, the control system operates an on/off switch that regulates the input of microwave energy and maintains isothermal conditions.

The ability to cure isothermally allows the use of much higher cure temperatures. Cure temperatures above the 230 C ultimate glass transition temperature for DER 332/DDS have been used in a pulsed microwave process with no thermal degradation of the material [1]. Similar temperatures in a continuous power microwave process or thermal process have lead to degradation due to the exothermic temperature excursion.

The two temperature profiles shown in Figure 7 represent temperatures monitored at the center and near the boundary of the cylindrical epoxy/amine sample. The difference in temperature shown is due to convective heat loss at the boundary since the surrounding air in the cavity remains at nearly ambient temperature. The processed samples had a relatively high surface area/volume with consequent high heat loss. Larger samples can be expected to have a less significant temperature gradient. For practical composite applications, addition of a conductive reinforcement such as carbon fiber reduces the temperature gradient [6]. In Figure 8, pulsed-microwave process temperature-time profiles for epoxy and epoxy with ground AS-4 carbon fiber are shown. Upon each power pulse the boundary temperature matches the center temperature, but falls again when the power is turned off. A more rapid pulse would maintain a nearly constant temperature in space as well as in time.

Microwave processed materials show equivalent or superior material properties when compared to thermally processed materials. As will be discussed in the following section, the selective heating of carbon fibers in a carbon fiber/epoxy composite promotes improved adhesion at the fiber/matrix interface as reflected in some improved strength properties [7]. In the case of epoxy/amine resin, tensile strength and tensile modulus of thermally cured, microwave cured, and hybrid microwave/thermally cured specimens have been measured [8]. Figures 9 and 10 illustrate the tensile strength and tensile modulus, respectively, for thermally, microwave, and hybrid processed epoxy/amine resin. These curves show that equivalent tensile properties are achieved for microwave processed resin. Microwave curing of polyurethane using a similar cavity system [9] has shown an increased hardness as compared with thermally cured material.

MICROWAVE PROCESSING OF CARBON FIBER COMPOSITES

Microwave processing of composites is of particular interest due to the difficulty in processing thick section materials thermoset composites. As discussed earlier, the large mass of a thick section thermoset composite allows a heat buildup during the exothermic reaction, with consequent temperature gradients, thermal stress, and possible thermal degradation. The rapid, controllable, and selective heating nature of microwave processing allows uniform curing of thick section materials without degradation.

Successful microwave curing of carbon continuous-fiber composite is extremely dependent on the location of the sample in the resonant cavity and the cavity dimensions. Identification of a resonant field is done using a sweep frequency generator and an oscilloscope as described previously [10]. For an empty cavity a resonant mode is identifiable as a peak in the power absorption vs frequency trace as shown in Figure 11. Typically, the cavity length is adjusted to bring the resonant peak to 2.45 GHz, and the cavity penetration depth of the coaxial coupling probe is varied until the peak intensity is maximized. With a material of high dielectric loss such as a carbon fiber composite, however, the electromagnetic field is disturbed to such an extent that the usual resonant modes are not recognizable. Identification of cavity conditions necessary for successful curing becomes a matter of experience and, although the curve of absorbed power vs. frequency for the best cavity configuration is reproducible from sample to sample, the curve does not appear as a single resonance peak but as several peaks. This suggests that the composite is cured in an as yet undefined hybrid electromagnetic mode.

Carbon fiber/epoxy composites of two different fiber orientation types have been cured in the resonant cavity system. The first type was a 24 unidirectional-ply laminate, the second was a 24 cross-ply laminate. Various 3 inch x 3 inch samples were cured for 90 minutes in the resonant cavity. Similar samples were cured thermally for 90 minutes with an oven temperature of 180 C. Each sample was cured under vacuum without externally applied pressure as discussed elsewere. [11,12]

Figures 12 and 13 illustrate the sensitivity of the sample surface temperature to the cavity length, indicating the effect of cavity tuning on cure of cross-ply samples. In Figure 12, at a cavity length of 9.9463 cm the temperature is relatively uniform across the sample. Figure 13 illustrates the effect of slightly de-tuning the cavity away from the position of uniform cure. By changing the cavity length by only 1.17 mm the electromagnetic field is sufficiently disturbed to cause an extreme variation in temperature. Such variation leads to uncured areas as well as overcured, degraded areas in the sample. The most significant feature of both Figures 12 and 13 is that they demonstrates microwave curing of crossply carbon fiber composite material. Previous work [2] suggested that crossply curing was impossible.

Table 2 is a comparison of the physical and mechanical properties of microwave and thermally cured composites having unidirectional or cross-ply orientations [11,12]. For unidirectional samples, flexural modulus and flexural strength show a strong dependence on the orientation of the fiber direction with respect to the microwave coupling probe. An orientation of 45° gives significantly higher flexural properties than 90° or 0°. In addition, heating rates and temperature uniformity are also dependent on orientation, with the 45° position giving best results. This orientation dependence is due to the anisotropic nature of the material and the high dielectric loss of the carbon fibers. Variation in the position of the material leads to disturbance in the electromagnetic field [13] so that correct positioning, along with the correct cavity length, is required for successful curing.

Comparison of the flexural properties for microwave and thermally cured samples as shown in Table 2 indicates superior properties for the microwave processed material. This is misleading, however. The thermally cured material was processed for 90 minutes in a 180 C oven. Temperature vs time curves for these runs showed that the sample temperature slowly increased and reached the oven temperature only at the end of the run. This is reflected in the lower extents of cure for the thermally cured crossply and unidirectional samples as shown in Table 2. The lower cure extents resulted in lower flexural properties for thermally cured material processed with a time-temperature cycle similar to the microwave cycle. It is apparent from the extents of cure that the microwave processed material is cured faster than the thermally cured material due to the faster heating of the microwave process.

The flexural properties given in Table 2 for the microwave cured unidirectional samples with 45° orientation compare with manufacturer's specifications of 128 GPa modulus and 2105 MPa strength [13]. The recommended process pressure is 100 psig but no external pressure was applied for either the microwave or thermally cured samples. The means of pressure application in the microwave cavity is currently being developed. It is anticipated that with application of pressure, equivalent or superior mechanical properties will be achieved in a shorter processing time for microwave cured materials. Superior properties for microwave processed carbon fiber composites can be anticipated since the interfacial shear strength for microwave processed single carbon fiber /epoxy specimens has been shown to be 70% higher than for thermally processed materials [7].

SUMMARY

The inherent advantages to resonant cavity microwave processing of polymer composite materials has led to significant developments in several areas. These include
- isothermal curing of epoxy/amine resins in both space and time
- production of epoxy/amine samples with tensile properties equivalent to thermally processed samples
- the ability to cure isothermally both unidirectional and cross-ply carbon fiber composites.

Future work in the area of microwave will focus on scale-up of the cavity technology for use in both batch and continuous processes with means to apply pressure during cure.

Table 2. Comparisions of Microwave and Thermally Cured
24-Ply Cross and Unidirectional AS4/3501-6 Composites

	Cross Ply		Unidirectional			
			Microwave			Thermal
	Microwave	Thermal	90° *	45° *	0° *	
Cure Condition:						
Cavity Length (cm)	9.95	----	9.30	17.08	9.46	----
Input Power (W/g)	2	----	1	1	1	----
Physical properties:						
Thickness (mm)	4.35	3.53	4.26	3.61	4.58	3.74
Density (g/cm^3)	1.33	1.28	1.40	1.41	----	1.60
Fiber Content (v.%)	47.27	64.92	55.43	54.76	----	80.39
Void Content (v.%)	10.38	25.48	6.91	7.26	----	6.88
Extent of Cure						
FTIR	0.95	0.34	1.0	1.0	----	0.60
DSC	0.94	0.60	0.84	0.89	----	0.78
Mechanical Property						
Flexural Strength (MPa)	724.9	313.7	913.5	1136.1	615.5	789.6
Flexural Modulus (GPa)	40.46	40.51	61.18	99.97	30.37	63.30

* Angle between fiber orientation and coupling probe

Note: 1. Pressure was not applied during the cure
2. Same temperature/time cycles were applied for the comparing samples

REFERENCES

1. Jow, J., J. D. DeLong, and M. C. Hawley, Polymer Processing Society, Fourth Annual Meeting, Orlando, FL, May, 1988.

2. Lee, W., and G. S. Springer, J. Comp. Materials 18, 357-386 (1984).

3. Le Van, Q., and A. Gourdenne, Eur. Polym. J. 10, 777-780 (1987).

4. Asmussen, J., H. H. Lin, B. Manring, and R. Fritz, Rev. Sci. Instrum. 58, 1477-1486 (1987).

5. Jow, J., M. C. Hawley, and M. C. Finzel, Rev. Sci. Instrum. 60, 96-103 (1989).

6. Jow, J., M. C. Hawley, and J. D. DeLong, American Society for Composites, Proceedings of the Third Technical Conference, Seattle, 305-312 (1988).

7. Agrawal, R., and L. T. Drzal, J. Adhesion in press (1989).

8. Singer, S. M., J. Jow, J. D. DeLong, and M. C. Hawley, SAMPE Quarterly 20, 16-18 (1989).

9. Jullien, H., and H. Valot, Polymer 26, 506-510 (1985).

10. Jow, J., M. C. Hawley, M. Finzel, and T. Kern, Polym. Eng. and Sci. 22, 1450-1454 (1988).

11. Wei, J., J. Jow, J. D. DeLong, and M. C. Hawley, presented at the 7th International Conference on Composite Materials, Beijing (August, 1989).

12. Vogel, G., J. Jow, J. D. DeLong, and M. C. Hawley, to be presented at the Fourth Technical Conference of the American Society for Composites, Blacksburg, VA (October, 1989).

13. Chen, Y. F., and C.Y.C. Lee, Polym. Materials Sci. Eng. (ACS) 60, 680-684 (1989).

14. AS4/3501-6 Materials Specifications, Hercules Inc.

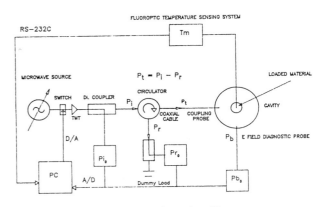

Figure 3. Microwave circuit diagram.

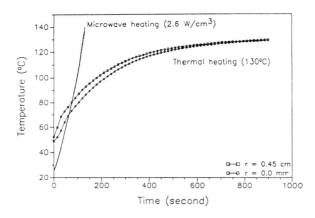

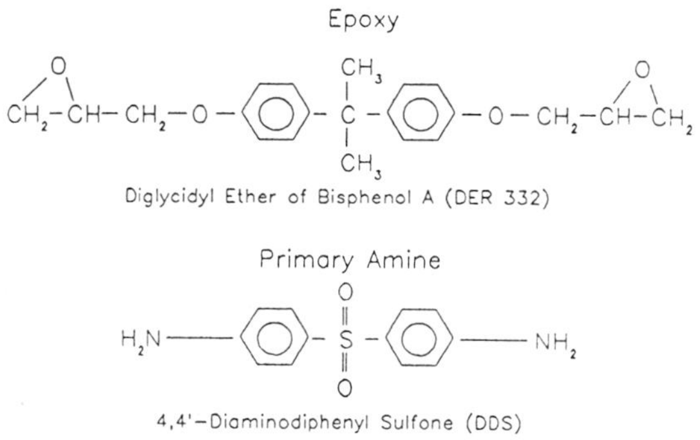

Figure 1. Microwave and thermal heating of Nylon 66

Figure 4. Chemical structures of reactants

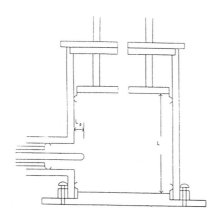

Figure 2. Resonant microwave cavity; L: cavity length, Lp: probe depth

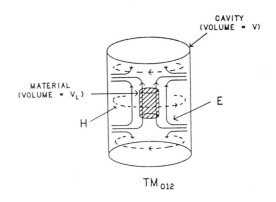

Figure 5. Representation of TM_{012} resonant mode in microwave cavity

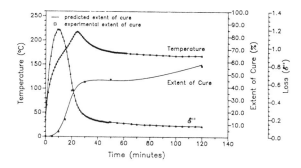

Figure 6. Dielectric loss, temperature, and extent of cure vs time for DER 332/DDS

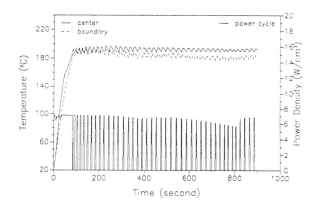

Figure 7. Pulsed microwave cure of DER332/DDS

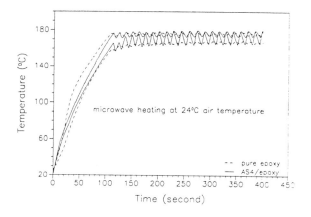

Figure 8. Pulsed microwave cure of epoxy/carbon powder

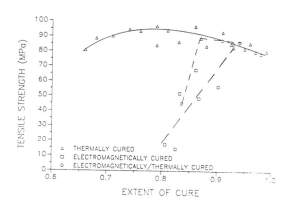

Figure 9. Tensile strength of microwave and thermally cured DER 332/ DDS

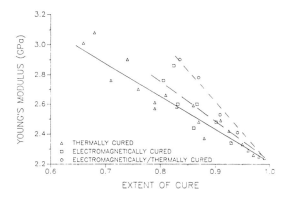

Figure 10. Young's modulus of microwave and thermally cured DER 332/ DDS

159

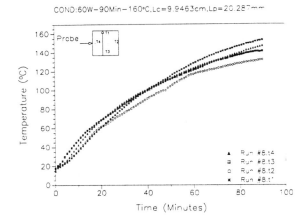

Figure 12. Temperature profiles for cross-ply cure in well tuned cavity.

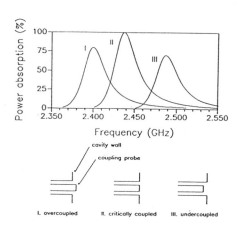

Figure 11. Power absorption curves indicating cavity resonance condition

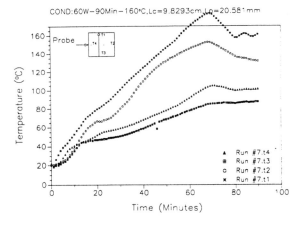

Figure 13. Temperature profiles for cross-ply cure in poorly tuned cavity.

THERMAL ANALYSIS OF COMPOSITE TOOLING MATERIALS

Nancy M. Ham, Michael S. Molitor
Ciba-Geigy Corporation
East Lansing, MI USA

Abstract

Technologies are now available which offer improved methods for making parts using intermediate and high temperature tooling. The technology employed in this improved scenario involves room temperature setting, intermediate high temperature use tooling prepregs and CAD/CAM technology. Thermal analysis of the cure characteristics of some of these materials will be investigated by DuPont thermal analysis using isothermal DSC kinetics. Theoretical kinetics will then be compared to actual applications.

ADVANCES IN TWO DISTINCTLY DIFFERENT TECHNOLOGIES have resulted in major improvements in the manufacture of composite tools. By incorporating computer aided design/computer aided manufacturing techniques it is possible to replace an inherently inaccurate, hand built model with a master machined directly from engineering data.

The use of room temperature curing tooling prepreg systems has allowed the construction of high quality composite tools to be built directly from the master model. For prototype or demonstration and validation programs this shortened tooling sequence provides substantial reductions in lead time as well as cost.

DISCUSSION

Historically, the traditional composite tooling sequence[1] consisted of six distinct steps (Figure 1, Method A).

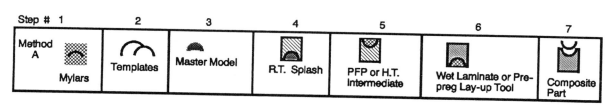

Figure 1

[1] Complete chart of sequences in composite tooling is shown in SME paper EM88-539, Unique Sequences in Composite Tooling by G. Sollner and M. Molitor.

The first three steps involve the transfer of engineering data to a physical working model. Full size mylars are used to lay out templates. These templates are then used to create lofting lines for the splining of either plaster or specially formulated epoxy resin systems. This method of master model fabrication is not only very labor intensive, but generally not repeatable due to the human element required in the transformation of data to model. The Coefficient of Thermal Expansion is transferred at each step causing considerable error by the end of the sequence.

In Method B, however, the transfer of engineering data to physical model is accomplished by the use of cutter path generating software downloaded to a numerically controlled machining center (Figure 2, Method B). This system of master fabrication not only reduces labor but allows exact duplications of models to be generated. One potential benefit of this feature would be multiple masters for offload purposes.

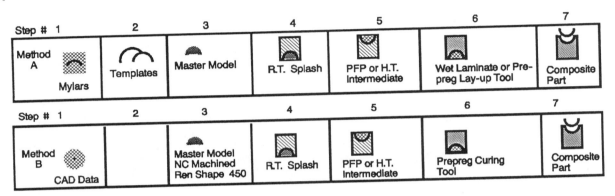

Figure 2

Three inherently inaccurate operations have now been replaced by two precise manufacturing steps, computer aided design and computer aided manufacture of master models. The true master is now the mathematical data and the physical representation of that data is now a tooling aid that can be considered disposable. The elimination of physical masters would offer a number of benefits including a reduction of capital assets and removing the need for large storage facilities (Figure 3).

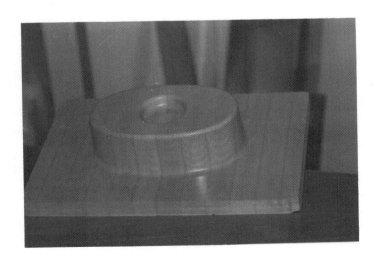

Figure 3

Another technology having a major impact in the processing of composite tools is the emergence of room temperature curing prepreg systems. The unique feature of these materials is their ability to cure to a "self-supporting" condition at room temperature, allowing an unsupported postcure in an oven. This advancement in materials offers numerous advantages over traditional, room temperature curing, wet lay-up systems.

The traditional hand lay-up systems have the potential problems of inaccuracy in weighing and mixing the two components of a system, uneven resin content, and a high void content. These problems have been essentially eliminated with vacuum bag compaction and curing of the room temperature prepreg systems.

Resin distribution and mixing ratios are precisely controlled with preimpregnated fabric, thereby eliminating much of the human error in hand lay-up. Lower void content is achieved by the uniform, predictable bleed out of resin under vacuum compaction. Safety is improved by the lack of worker exposure to amines in the mixing process and the elimination of known carcinogenic materials in the room temperature curing prepreg system (Figure 4).

Figure 4

Room temperature curing prepreg was developed to be self-supporting after a four-day room temperature cure. A prepreg needed to be developed that would have a minimal rate of cure when frozen and yet cure in four days at room temperature. Systems have been developed to meet this criteria through the use of isothermal DSC kinetics.

The rate of cure of room temperature curing prepreg is temperature and time dependent. Kinetics studies have been done to determine the degree of cure at various temperatures and times. Kinetics were compared with actual experience and it was determined that a 55% cure was sufficient for self-support. A tool could take as long as 6 to 8 days at 20°C to cure to a free standing state or only 3 days at 27°C. A faster cure can be achieved by adding heat when curing under vacuum. Eight hours at 49°C is sufficient for self-support if time is critical.

The "self-supporting" nature of room temperature prepregs allows the manufacturer to eliminate both the negative "splash" and the high temperature intermediate tool required when a heat cured tooling prepreg is utilized.

This further reduces the number of steps to a total of three. (Figure 5, Method D). This reduction of steps helps improve accuracy by reducing accumulated error and reducing cost by eliminating raw materials and labor.

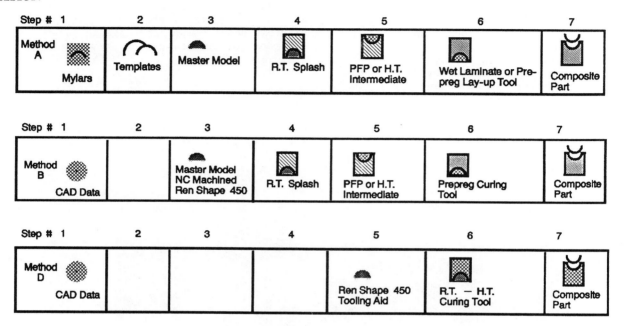

Figure 5

Room temperature prepreg has room temperature properties equivalent to a typical wet lay-up system or a 100°C curing prepreg. The high temperature properties are better than the wet lay-up system. Typical physical properties of a wet lay-up system, room temperature curing prepreg, and 100°C curing prepreg are as follows:

Physical Property Comparisons

Flexural Strength	Wet System[1]	Pre-Cat™[1]	Aerolam® Prepreg[2]
@ RT	77,000	79,800	76,500
@ 177°C	20,000	33,000	43,000
@ 204°C		22,000	27,000
Flexural Modulus			
@ RT	4.4×10^6	5.7×10^6	5.8×10^6
@ 177°C	1.7×10^6	4.4×10^6	4.7×10^6
@ 204°C		3.4×10^6	3.8×10^6

---

[1] 8 ply, 6K graphite, 0° rotation laminate, cured 4 days @ RT, postcured to 177°C.

[2] 18 ply, 6K graphite, 0°, 45° rotation laminate, cured 8 hrs. @ 100°C, postcured to 177°C.

Physical Property Comparisons (continued)

Aged 1,000 hrs. @ 177°C

	Wet System[1]	Pre-Cat™[1]	Aerolam® Prepreg[2]
Flexural Strength			
@ RT	79,000	81,000	77,900
@ 177°C	18,000	40,700	60,600
@ 204°C		31,000	35,200
Flexural Modulus			
@ RT	4.3×10^6	6.0×10^6	5.8×10^6
@ 177°C	1.5×10^6	5.0×10^6	5.2×10^6
@ 204°C		3.6×10^6	3.7×10^6

---

[1] 8 ply, 6K graphite, 0° rotation laminate, cured 4 days @ RT, postcured to 177°C.

[2] 18 ply, 6K graphite, 0°, 45° rotation laminate, cured 8 hrs. @ 100°C, postcured to 177°C.

CONCLUSION

By utilizing two new technologies, CAD/CAM and room temperature curing prepregs, multiple steps can be removed from the typical 6-stage composite tooling sequence. This trimming of steps results in labor and material savings as well as improved accuracy. Additional benefits include the use of temporary N.C. machined tooling aids to reduce storage and the elimination of wet lay-up systems.

MOEN HEATING SYSTEMS FOR HIGH VOLUME PRODUCTION OF THERMOPLASTICS

Robert W. Aukerman
Heat Transfer Technologies, Inc.
Sun Valley, CA USA

ABSTRACT

In the development of cost effective tooling to produce advanced thermoplastic structures, a key procedural step is controlled heating of the matrix material to its 'Tg' melt point. Processing time and product economy is dependent on the ability of a tooling heat transfer system to rapidly deliver enough energy into a predesigned area (large or small) to achieve the melting of the thermoplastic resin.

One such heating method utilizes high velocity gas jets. This technique demonstrates the ability to rapidly transfer and control the required energy to reach the melt phase in these high temperature thermoplastic resins. This hot gas jet heat transfer system is a forced convection concept relying on a small volume of heated gas delivered to the surface to be heated at high velocities.

This system impinges hot gas onto the tooling and optimizes the ability of a heat transfer medium (i.e., air, nitrogen) to provide efficient, rapid, controlled heat transfer onto the desired tool area. (viz. filament winding, autoclave augmentation, heated press platens, other integrally heated tooling applications).

INTRODUCTION

The MOEN high velocity hot air impingement heating system using a multi-orifice, jet stream delivery network is a heating concept which greatly improves heat cycle controllability and shortens time duration of production schedules. This concept provides faster and more accurate heating and cooling cycles at temperatures from -300oF to +2000oF.

These high velocity hot air impingement heating techniques were developed by Mr. Walter Moen, a Research Engineer, at North American Aircraft Division, Rockwell International for use as a production and testing procedure on the B-1 Bomber, and the Space Shuttle. These same techniques are now being marketed and produced for general industrial use by HEAT TRANSFER TECHNOLOGIES, INC., of Sun Valley, CA under U.S. Patent NO. 4,386,650; foreign and other U.S. patents are pending.

THERMOPLASTIC COMPOSITE STRUCTURES

The MOEN System and its unique high velocity convection delivery concepts provides an integral heating and cooling system which is operated by shop manufacturing personnel to produce <u>High Quality</u> dense advanced composite parts. The system maintains an accurate heat rate profile with uniform part temperatures held within close tolerances and a controlled cool down cycle, at low cost, with superior repeatability.

The system is easily controlled either manually or by computer. The part assumes the hot air high velocity jet stream temperature of the heating gas which is computer controlled very accurately, assuring no temperature over-shoot.

TYPICAL APPLICATIONS USING MOEN HEATING CONCEPT

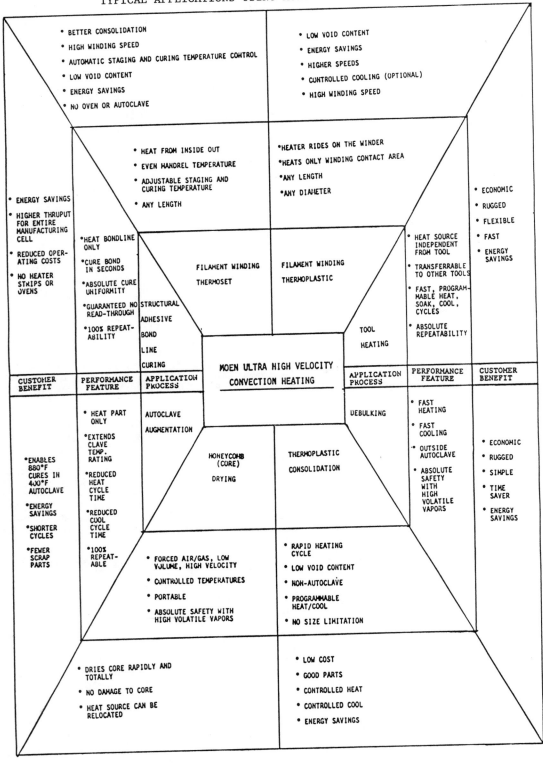

CUSTOMIZED HEATING DISTRIBUTION NETWORK

The photograph below is a typical illustration of a customized heating/cooling delivery system network. In this example a full scale mock up of the B-1 Bomber Cockpit Windshield area uses plant air and forced convection heating/cooling principles to create the required environmental effects for hot-cold and moisture stress and fatigue conditions. The applications provide the necessary thermal shock & thermal fatigue testing data for flight certification requirements.

A needed insulation blanket which covers the entire distribution and heating area is missing from this photo.

COCKPIT OF B-1 BOMBER

- Inexpensive tubular network with orifices drilled on the inner perimeter of the tube wall.

- Network provides frame for inexpensive insulated enclosure.

- The distribution network is formed to the configuration of the part usually 1 to 2 inches from the part providing <u>excellent heat transfer and temperature uniformity.</u>

- The distribution network can be made portable or installed in existing ovens, furnaces or process heating enclosures. It can be installed around tanks, petrochemical lines, valves, bulk containers used in chemical and food processing.

- Heating and cooling applied without moving the part.

- Temperature ranges beyond the limits of other technologies (-200°F to +1600°F).

- High reliability and repeatability of heating/cooling cycles.

- Adability to a wide range of applications.

BASIC OPERATING PRINCIPLES

The physical principles of the "MOEN" Heating System are uniquely different from most conventional industrial heating/cooling concepts. These principles are:

* Heating is accomplished by passing compressed air through a tublular resistance heating element.

* Improved heating efficiencies are achieved via a low pressure tubular delivery system using heated gases at high velocity through small orifices closely positioned adjacent to the part or process to be heated or cooled.

* The high velocity impact concept utilizes the advantages of improved heat flux densities (rate of heat transfer) obtained by the increased high velocity scrubbing action of the heated gas blowing against the part or process: (attainable from a carefully designed tubular heating delivery network).

* The delivery networks (with appropriate insulation techniques) utilizes the best advantages of Conductive, Radiant and Convection heat transfer techniques.

* Electronically controlled temperatures are accomplished via thermocouple feedback from the heated areas: (response time to needed changes is almost instantaneous).

* Additional temperature control flexibility is achieved via throttling the rate of flow of the gas through the tubular heater: (response time to needed changes is minimal).

ADVANTAGES OF HIGH VELOCITY JET STREAM

High thermal transfer efficiency produces savings in energy consumption and provides improved time savings. The MOEN heating concepts are from five to thirty times faster than convection ovens. It is three times faster than infrared heating. This faster heating is accomplished by exploiting the high velocity, high impact advantages of convective heating. The heat transfer coefficient between the heating medium (air or other gas) and the product are geometrically enhanced in using the high velocity close proximity distribution networks.

The "MOEN" System is unique in combining the unequaled advantage of <u>high heat flux rates</u> of conductive heating, the rapid heat transfer of <u>radiant heat</u> and the exceptional economy and <u>utility of</u> near super sonic high velocity <u>convection heating</u> all from the same singular <u>system</u>.

The system is most useful in meeting unusual requirements for controlled elevated heating and cooling within the same process. In a "less than ambient temperature mode" a cryogenic gas medium is used.

Open or closed loop delivery system applications can be devised. The system operates on a conventional 480 Volt single or three phase power source feeding into a controller which regulates a step down transformer. The low voltage secondary is connected to a tubular resistance heater element. Gases or fluids pass through the tubular heater providing a heated medium. This is delivered to the part(s) or process via a close proximity tubing network.

These high velocity, high impact concepts can be designed for use in existing ovens, presses, furnaces, process heating enclosures, or work cell environments. These concepts are particularly useful for pre-heat or postcure processes which call for a finite control and also where there is a requirement for rapid temperature change influxes.

CONTROL FLEXIBILITY

The ability to easily control the input power and/or the volume or velocity of heated gases provides processing flexibility and energy conservation advantages. In addition, the "MOEN" System encompasses the use of a combination of Conductive, Radiant and Convection Heating Concepts. The maximization of all three modes of heat transfer are utilized in the design of properly devised heat delivery networks and improved insulation techniques.

Systems control for the heating and cooling unit can be manual, semiautomatic, or fully computerized depending on the application requirements.

DELIVERY SYSTEMS

The materials used to build a tubular close proximity delivery network are low in cost, allowing for flexibility in the design and application.

A typical distribution network provides the framework on which to install a covering of insulation material suitable to enclose the part or product to be heated. This assists in maximizing the efficiency of the heat transfer process.

SUMMARY OF NOTEWORTHY FEATURES

This self contained heater is a major improvement within industrial heating/cooling temperature control systems. Among its useful characteristics are:

- heating uniformity
- faster turnaround response on heating/cooling cycle requirements (uses maximum heat flux density principles of super high velocity convection heating)
- sharp or flat temperature ramps (speed of heat-up cycle) easily achieved
- ---uniquely controlled

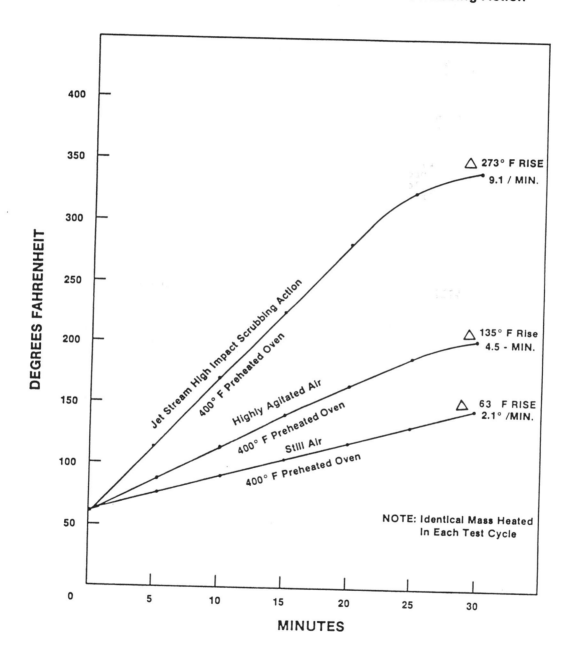

- ---easily programmed
- cost savings in energy consumption (50 to 75% energy savings)
- high reliability and repeatability of heating/cooling cycles
- positive temperature control to +/- 5oF
- overshoot of desired temperatures eliminated
- inherent safety features
- low maintenance costs
- adapatability to a wide range of applications
- can be used with pressurized systems
- adaptable to existing ovens, presses, furnaces, or other process equipment

PERFORMANCE SPECIFICATIONS

The "MOEM" System furnishes excellent operational efficiencies by providing superior heat transfer characteristics. These improved techniques can be achieved inside and existing oven, press, furnace, conveyor, pipe or other apparatus used in any heating process. This is accomplished by using a close proximity distribution network with orifices conveying heated gases at high velocities against the entire heating surface. This close high velocity impingement of the heating medium upon the product produces higher operational efficiencies in both energy consumption and time required to achieve the desired level of heating. In a like manner, the same proximity delivery system network provides savings in the cooling cycle efficiency and time requirements.

PREHEATING OF THERMOPLASTIC PREFORMS

The MOEN System is used to shorten the time for preheating and cooling Reinforced Plastic Composites (RP/C) materials within a RP/C Manufacturing Cell.

In the heating of polypropylene thermoplastic materials prior to compression molding, the MOEN Heating concepts have distinctive advantages

- The heating cycle is three times faster than conventional heating techniques(i.e. infra red lamps).
- Precision in-line production heatup timing is coordinated to the exact press molding cycle time.
- Multiple layer heating chamber lay-ups are available to allow for correct part thickness variations.
- Energy savings of 50 percent or better are achieved using the MOEN Heating concepts.
- Heated RP/C materials can be adapted to automated loader transfer devices.
- MOEN Heating/Cooling logic can be integrated into the newest and most sophisticated automated composites manufacturing cells.
- Minimizes or eliminates manual operational costs and offers total payback in less than a one year production period.
- If cooling of parts or tools is needed, the same MOEN energy transfer system(s) can be utilized to impinge ambient or cryogenic air to parts or tool faces requiring rapid cooling.

BONDING OF SMC PARTS

MOEN bonding fixtures are now being supplied for numerous high speed bonding applications. The results have been excellent. Final production data confirms that two thermoset parts, .100" and .120" thick, can be bonded in 45 seconds. Sections up to .225" requires 60 second bonding cycle time for bonding two sections of SMC together.

There are six major advantages associated with bonding using the MOEN impingement air system:

1. There is no objectionable bond line read-through on finished parts.
2. The MOEN impingement heating keeps pace with a 50 to 60 second molding cycle.
3. There are no heat losses when the fixture is opened, therefore, reducing kilowatt consumption.
4. A single MOEN heating unit can be used with many different bonding fixtures. Subsequently, the stand alone investment of each individual bonding machine is considerably reduced.
5. The control of heat transfer to the part is infinitely adjustable through the computerized controller permitting a wide range of temperatures within seconds. This is a contributing factor for controlling bonding line read-through and provides maximum flexibility for any type of bond fixture or part thickness.
6. The computerized controller permits entering several programs suitable for various bond applications when a single MOEN unit is used for different bonding fixtures.

CONCLUSION

The "MOEN" Heat Transfer System provides a superior method of heat transfer and control. The system provides the following;

-Greater efficiency as a heat transfer medium. More of the heat generated is effectively transferred into the target part. The result is low energy consumption for a given process or part, less waste heat generated to be dissipated in to the surrounding work environment, and heat ramps potentially three times faster than by using traditional means.

-Even heating throughout parts within varyuing cross-sections and materials.

-Flexibility in forming and curing large, oddshaped parts, special projects, and research.

CONCLUSIONS (continued)

-Cooling capability, varying from shop air to liquid nitrogen, can employ the same distribution network, further speeding the cycle time. The cooling cycle can be a pre-programmed function of the control system and need not require operator input.

-Close and precise real time control over the complete cycle. The system can maintain the target temperature within plus or minus 5 degrees, with the elimination of temperature overshoot.

-High level of consistency and repeatability between production cycles. Once a distribution network has been developed, the heat characteristics it provides will not change. Quality is improved.

-Adaptable to high production situations. The system eliminates the waiting and scheduling inherent with an oven or autoclave. The system can be used on the shop floor and be designed as an integral part of the mold or tool.

The development of the "MOEN" High Velocity Hot Air Impingement Heating System represents a significant advancement in the state of available heating technology. The system not only assists in moving the pre-heating, curing and forming of composites from a piece meal approach to a linear production mode, but also allows the forming or curing process to adapt to the special requirements of the part rather than requiring the part to adapt to the available heating process. Further, and perhaps more importantly, the system allows real time control over the forming and curing cycle. The energy savings and lower initial captial investment are an additional step in lowering the cost of composite construction and thus widening the application base.

SURFACE WAVINESS AND OBSERVED REFLECTIONS —MATHEMATICAL MODELING FOR VARIOUS INSPECTION METHODS

Ching-Chih Lee
GenCorp Research
2990 Gilchrist Road
Akron, OH 44305 USA

ABSTRACT

The relation between the surface topography of a part and the observed reflections from the part has been studied for three surface waviness inspection methods: (1) reflection of a back-lighted grid, (2) reflection of a scanning laser beam, and (3) reflection of a point light onto and then returning from a retro-reflective screen. Computer programs for predicting the observed reflections were developed based on the physics of optical reflection and image formation. The surface profile of an SMC panel was measured and the reflection from the panel for each method was predicted. The predicted and photographed images were in close agreement. The reflections for typical long-term waviness were studied and their characteristics discussed. The inspection sensitivity of the methods was shown to be different and to depend on the part orientation.

SURFACE WAVINESS OF EXTERIOR BODY PANELS is an important consideration in the automotive industry. To date, waviness ratings are based on examination of surface profiles[1-4] or images reflected off the panels[2-11]. Common image rating methods use the (1) reflection of a back-lighted grid[1-5], (2) reflection of a scanning laser beam[6-8], and (3) reflection of point light sources onto and then returning from a retro-reflective screen[9-11]. Establishing the relation between the surface topography of a panel and the reflected image is necessary for proper interpretation of the image.

It has been found that different inspection methods have different sensitivity to surface waviness. Even for the same inspection method, sensitivity may change with part surface orientation and viewing angle. By developing a mathematical model to relate the surface topography with the reflection, one can obtain quantitative measures of inspection sensitivity.

The surface waviness of an FRP (fiber reinforced plastics) molding is postulated to result from waviness in the tooling before and/or during molding and relief of the residual stresses after the molding. Residual stresses are caused by shrinkage during cure, if thermosets are used, and differential thermal shrinkage during post-molding cool-down, and are relieved or balanced by deformations. The deformations depend on both the distribution of residual stresses and the properties within the part, which are, in turn, related to material formulation, flow and fiber orientation patterns, and cure and heat transfer history. Surface deformations as affected by these factors can be estimated from solutions of stress-strain equations. A relation between surface topography and observed reflections may enable us to predict the effect of each material or processing parameter on the surface appearance.

This paper describes the mathematical modeling of light reflection that occurs in the three image rating methods. Although the emphasis is on FRP panels and long-term waviness, i.e., of wavelengths greater than 1cm[2,4,5], the modeling is equally valid for panels of other materials and for short-term waviness.

INSPECTION METHODS

A. BACK-LIGHTED GRID - A sketch of the back-lighted grid inspection most widely used and to be considered in this paper is shown in Fig. 1. It is composed of a bank of fluorescent lights, a simple grid pattern on a diffuse light transmitting panel, a reflecting surface, i.e., the part under inspection, and a camera or the inspector's eye.

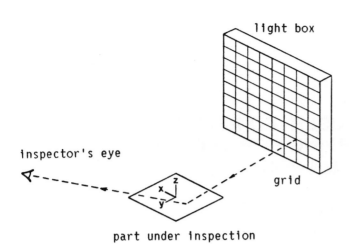

Fig. 1 Back-lighted grid inspection

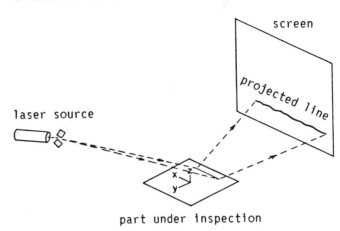

Fig. 2 Laser beam scanning method

The fluorescent light bank is used to project the grid pattern onto the part surface. Some light, after being reflected off the surface, enters the camera lens or the inspector's eye, and an image of the grid lines is formed. The grid lines on the light box are straight, of uniform width and are equally spaced. Due to the waviness on the part surface, the grid image is distorted. The surface waviness is rated based on the waviness and width variation of the lines in the image and is quantified by comparing to the standards previously judged by a group of experts. Because of the finite grid line spacing, an inspector usually moves his eyes around (up and down, and left and right) in order to "scan" over the entire part surface.

Some arrangements have the light box with the grid pattern hung overhead, roughly parallel to the part surface[5], but the principle of inspection is the same. An alternative set-up is the so-called "green room" inspection. A bank of green fluorescent light tubes are vertically placed inside the light box and no grid lines are employed[10]. The inspector looks for perturbations in the straight, light to dark transitions formed between the "edges" of the light tubes and their background. For analysis purpose, this set-up can be considered as a special case of that shown in Fig. 1, since the vertically placed fluorescent tubes are equivalent to the vertical grid lines.

B. LASER BEAM SCANNING - This method, developed by Ashland Chemical Co.[6-8] and named "Loria™", is schematically shown in Fig. 2. A laser is used as the light source and with moving mirrors is directed to scan linearly at a number of levels over the part under inspection. The laser beam reflects off the part surface, then strikes the screen and yields a projected line, which is wavy and/or fuzzy if the part surface is wavy. A typical analysis is limited to a 25.4 cm x 25.4 cm (10 inches x 10 inches) area, and has a total of 21 lines across the part.

The projected line on the screen is viewed with a video camera whose output is directed to a digitizing board residing in a microcomputer. To evaluate long-term waviness, each line is fitted with a low-order polynomial (second-order for a flat panel and third-order for a curved panel). The root-mean-square deviations are summed and averaged over all the lines giving a number as a measure of the long-term waviness across the part[6,7]. Short-term waviness, such as "orange peel", is characterized by the variation in the line width[8].

C. RETRO-REFLECTIVE SCREEN - A retro-reflective screen reflects an incoming light beam back along the incident light path. The inspection method involving such a screen was developed by Diffracto Limited[9-11] and is named "D-Sight™". Fig. 3 is a sketch of the set-up. The essential components are a light source, a part surface of interest, a retro-reflective screen, and an imaging lens or an inspector's eye. The light source radiates light in all directions, some of which is reflected by the part surface and strikes the screen. The screen consists of millions of small (25-76 microns or 0.001-0.003 inch diameter), tightly-packed, half-silvered glass beads. Due to the combination of reflection and refraction, the beads do not function as perfect retro-reflectors. Instead of sending a given light ray directly back along its incident path, a single bead tends to return a diverging cone of light. If the screen were perfectly retro-reflective, then all the rays emitted from the light source would be returned to it and an eye offset from the light source would receive no light and see nothing at all. Because of the slight divergence, an eye or a lens near the light source will receive enough light to produce an image in which the light intensity varies with the part surface waviness. In a D-Sight™ Audit Station, the image is captured by a video camera with a digitizing board, which is connected to a microcomputer for image analysis. The surface waviness is evaluated by the light intensity or gray level variation in the image.

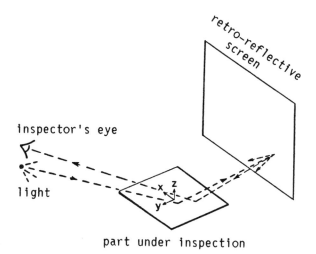

Fig. 3 Retro-reflective screen method

The retro-reflective screen may be flat or curved. We will consider a flat screen in this article, although the modeling procedure should be equally applicable for a curved screen.

SMC TEST PANEL

SURFACE TOPOGRAPHY - To study the relation between the surface profile of a part and the observed reflection in each inspection method, a flat SMC (sheet molding compound) panel with a high level of waviness was chosen. The panel, measuring 40.64 cm x 40.64 cm x 0.292 cm (16" x 16" x 0.115"), was glued onto a piece of glass plate, 40.64 cm x 40.64 cm x 0.635 cm (16" x 16" x 0.25"), for rigidity using Devcon® 5-Minute® epoxy. The surface profile was measured by a Sheffield CORDAX® RS-30 DCC Coordinate Measuring Machine with a touch probe which had a ruby tip of 0.196 cm (0.077") diameter. The sampling points were spaced 0.254 cm (0.1") in x and y directions (see Fig. 1 for coordinate definition). Linear regression was performed on the measured data to compensate for any tilting of the part during measurement. The result for the central 25.4 cm x 25.4 cm (10" x 10") area is plotted in Fig. 4. The scale in the z direction is magni- fied relative to the scale in the x and y directions to show the surface waves. For clarity, only one out of every two measured points in both the x and y directions is plotted.

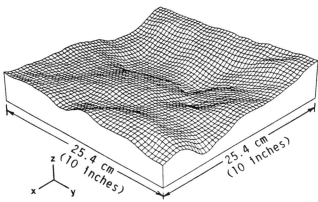

$z_{max} = 1.46 \times 10^{-2}$ cm (5.73×10^{-3} inch)

$z_{min} = -1.39 \times 10^{-2}$ cm (-5.48×10^{-3} inch)

Fig. 4 Surface topography of the profiled SMC panel

REFLECTION IMAGES - The surface waviness of the profiled SMC panel was inspected using all three methods. In each inspection, only the central area of the panel, as plotted in Fig. 4, was exposed for reflection. Referring to the coordinates defined in Figs. 1, 2 and 3, the exposed part surface lay on the area -12.7 cm $\leq x \leq 12.7$ cm, -12.7 cm $\leq y \leq 12.7$ cm ($-5" \leq x \leq 5"$, $-5" \leq y \leq 5"$) with z = 0 being the nominal surface plane. The positions of other components are given below.

A. <u>Back-Lighted Grid</u> - The grid pattern was in the plane y = -33.3375 cm (-13.125"). The vertical grid lines were at $x = \pm 0.635n$ cm ($\pm 0.25n$ inch), and the horizontal grid lines at $z = 3.048 + 0.635n$ cm ($1.2 + 0.25n$ inches), where n = 0, 1, 2, 3,... Note that the spacing between vertical lines and between horizontal lines was 0.635 cm (0.25 inch). The line width was 0.05 cm (2×10^{-2} inch). The camera for photographing the grid image had its lens center at (x, y, z) = (0, 176.812 cm, 45.784 cm) or (0, 69.611", 18.025") and was aimed at the center of the part. A photograph of the grid image is shown in Fig. 5(a).

B. <u>Laser Beam Scanning</u> - A Loria™ Surface Analyzer was used. The scanning laser beam was projected from the point (0, 167.0 cm, 106.68 cm) or (0, 65.75", 42"). The first scan was along y = 12.065 cm (4.75") on the part surface. For a nominally flat panel, the location of the i-th scan on the part can be approximated by the equation

$$y_i = 12.065 \text{ cm} - 1.130 \frac{r^{i-1} - 1}{r - 1} \text{ cm},$$

where r = 1.012096 and i = 1, 2,..., 21. Note that the spacing between consecutive scans on the part increased slightly with i, but all spacings were close to 1.27 cm

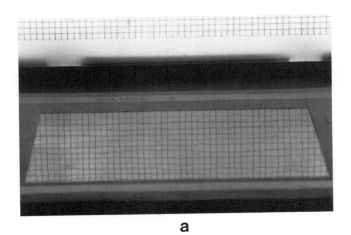

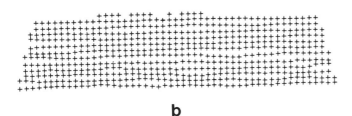

Fig. 5(a) The reflected image of grid from the profiled SMC panel
(b) Prediction.

(0.5"). The screen was at y = - 137.795 cm (-54.25"), and the camera lens at (- 16.51 cm, 156.337 cm, 112.522 cm) or (- 6.5", 61.55", 44.3"). Fig. 6(a) shows the projected lines on the screen, viewed by the camera and reconstructed by the computer. For clarity, only odd-numbered lines from line 1 to line 19 are shown. Also note that in the computer reconstruction, the lines were displaced in the vertical direction, and therefore the spacing between lines in this figure do not correspond to that in the real projections on the screen.

C. <u>Retro-Reflective Screen</u> - A Diffracto hand-held device with a flat retro-reflective screen was used. The hand-held device had multiple light bulbs; all except one on the top were blocked so an approximate point light source was obtained. The light source was at (0, 171.206 cm, 35.966 cm) or (0, 67.404", 14.16"). The camera lens was at (0, 172.463 cm, 40.020 cm) or (0, 67.899", 15.756"), above the light source. The screen was at y = -53.34 cm (-21"). Fig. 7(a) is the photographed D-Sight image.

MATHEMATICAL MODELING -
PREDICTION OF OBSERVED REFLECTIONS

Computer programs were developed to predict the reflections for the different inspection methods for a part with a known surface profile. A numerical grid of 101 x

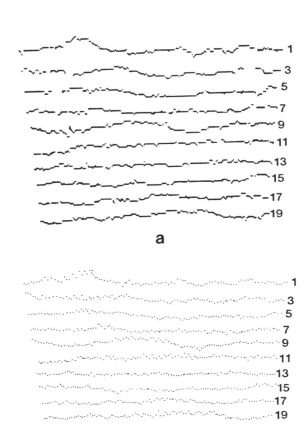

Fig. 6(a) The projected lines on the screen for the profiled SMC panel in the laser scan inspection.
(b) Prediction.

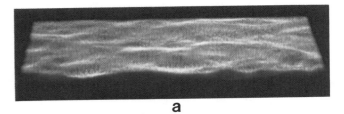

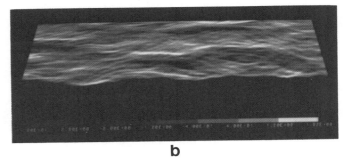

Fig. 7(a) The D-Sight image of the profiled SMC panel.
(b) Prediction.

101 points on the surface with Δx = Δy = 0.254 cm (0.1") was used in the mathematical modeling. Whenever a point other than a numerical grid point was considered during the calculation, its height (z) value and the associated slopes were found by two-dimensional linear interpolations from the neighboring grid points. Basically, each program traced the light paths in the reflection process and predicted the image or projection according to the physics of optical reflection. Specular reflection from a part surface was assumed. The predicted image or projection was then displayed using SUPERTAB software from SDRC[12]. Some unique features pertinent to the modeling of each inspection method are described as follows.

A. BACK-LIGHTED GRID - Each grid line intersection was treated as a point light source from which light rays were emitted in all directions. However, only the light rays reflected off a small area on the part surface would enter the inspector's eye and form an image. To determine what area on the surface would reflect light into the eye (or camera lens), a numerical iteration was carried out. The iteration started with an initial guess, which would have been the reflecting point on the part surface if the surface had been perfectly flat. Since the surface was not flat, the reflected light ray was unlikely to enter the eye. If the reflected ray went to, say the left of the eye, the next guess of reflecting point would be adjusted to the right. If one guess led to a reflected ray to the left of the eye, and the next guess led to a reflected ray to the right of the eye, then the following guess would be in the middle of the two preceding guesses.

To determine whether a particular reflected light ray would enter the eye, a finite eye opening was used. The diameter of the aperture of a human eye may be varied by unconscious, muscular action, according to the total flux of light entering the eye, between approximately 0.15 cm and 0.6 cm[13]. In the present work, an aperture diameter of 0.3 cm is used. According to the physics of optical reflection and image formation, the grid intersection and its image are geometrically symmetric with respect to the tangential plane at the reflecting point on the part surface. Therefore, once the reflecting point is found, the image location of a grid intersection can be calculated.

It is very likely that there are many reflecting points for a point light source, but ordinarily they are clustered in a very small area on the part surface. If the part is perfectly flat or the surface wavelength is sufficiently long, the multiple reflecting points would be on the same plane, and mathematically we would find a unique image point regardless of the reflecting point chosen, which corresponds physically to a distinct image. However, for a part surface with short-term waviness, the multiple reflecting points would lie on different planes, each plane producing an image. Physically a blurred image of the light source is then observed. In the present work, the surface is considered free of such short wavelength irregularities so that only one reflecting point is found for each light source. The program can be easily modified for multiple reflecting points in order to study the effect of short-term waviness.

Fig. 5(b) shows the predicted images of the grid intersections by reflection off the SMC panel. Very good agreement with the photographed image is apparent.

B. LASER BEAM SCANNING - Knowing the source of the laser beam and the points on the part scanned by the beam, one can find, for each point, the local tangential plane and the associated incident light path. Then the direction of the reflected laser beam and its intersection with the screen are calculated. In modeling we sampled 101 discrete points along each line of scan; the points were uniformly spaced at intervals of 0.254cm (0.1"). The program gave the projection on the screen of the laser beam reflected off each sampled surface point. Fig. 6(b) is the predicted projections on the screen of the reflected laser scans for the SMC panel. Again, only the odd-numbered lines from line 1 to line 19 are displayed for clarity. It can be seen that the prediction compares very well with Fig. 6(a); some slight discrepancy may be due to the difference between the actual and the modeled reflecting points and the fact that the part surface contained some short-term waviness.

C. RETRO-REFLECTIVE SCREEN - For this inspection method, we predict the light intensity distribution in the image formed by double reflection off a part surface with a known profile. One feature essential to this method is the divergence of light returning from the retro-reflective screen. For mathematical modeling, one needs to know the cone angle as well as the light intensity distribution inside the cone. To this end, a simple experiment was done. A laser beam was directed to the screen, and a piece of white paper was placed in front of the laser source to intercept the light returning from the screen. There was a small hole at the center of the paper, allowing the laser beam emitted from the source to pass through. The projection on the paper had a circular bright region with a dark concentric ring inside, as shown in Fig. 8. The light intensity varied only in the radial direction, highest at the center, decreasing into and then increasing out of the ring, and finally decreasing again towards the surroundings. There were no distinct boundaries for the ring and the circle. It was speculated that the ring was some diffraction pattern due to the nature of the retro-reflective screen, but the exact cause and its effect on the D-Sight image remain to be investigated. In our modeling, we neglected the ring and assumed a radial Gaussian distribution of light intensity inside the cone. The planar measurement of the cone angle was approximately 2.9°.

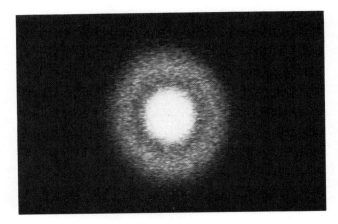

Fig. 8 The projection of a retro-reflected laser beam on a piece of paper.

The computer program traced the light path from the point light source to each sampled point on the part surface, and the first reflection onto the screen. The intersection with the screen then became a secondary point light source, from which a diverging cone of light was emitted. Using an iteration scheme similar to that for the back-lighted grid method, the program searched for a point on the part surface from where the retro-reflected light would be reflected into the eye or camera lens, and calculated the contribution of light intensity in the overall image based on the incremental solid angles associated with the sampled light rays and the corresponding incremental projection area in the image. Since the light intensity at a particular point in the image might have contributions from reflections with different paths, a numerical summation was used for the intensity accumulation. The resulting light intensity distribution was then normalized with respect to the average intensity value in the whole image area. Since for a human eye the perceived brightness of an object is related to the light intensity approximately to a logarithmic scale[14], the natural logarithm value of the normalized intensity was taken and then displayed using seven gray levels. Fig. 7(b) shows the predicted D-Sight image for the SMC panel. Good agreement with the actual image (Fig. 7(a)) was observed. Some discrepancy may be due to the nonuniform reflectivity over the part surface and some degree of diffuse reflection.

TYPICAL SURFACE WAVINESS OF FRP PARTS AND THE REFLECTIONS

Having developed and verified the computer programs to predict reflections for each inspection method, we can now study the typical surface waviness on FRP parts and the associated reflection image. Table 1 lists eight surface profiles examined along with their mathematical expressions. The locations of the components in each method are assumed to be the same as in the inspection of the profiled SMC panel. The reflecting surface is assumed to occupy the area $-12.7 \text{ cm} \leq x \leq 12.7 \text{ cm}, -12.7 \text{ cm} \leq y \leq 12.7 \text{ cm}$ ($-5" \leq x \leq 5", -5" \leq y \leq 5"$) in each simulation.

Table 1
Surface Profiles Used to Simulate
Typical Surface Waviness of FRP Parts

Surface Profile	Reflection Images
(i) Flat panel $z = 0$	Fig. 9
(ii) Square dome $z = .05" \cos(\frac{2\pi x}{20"}) \cos(\frac{2\pi y}{20"})$	Fig. 10
(iii) Bi-directional sinusoidal wave, smaller amplitude $z = .00025" \sin(\frac{2\pi x}{3"}) \sin(\frac{2\pi y}{3"})$	Fig. 11
(iv) Bi-directional sinusoidal wave, larger amplitude $z = .001" \sin(\frac{2\pi x}{3"}) \sin(\frac{2\pi y}{3"})$	Fig. 12
(v) Circular blister $z = .001" [\cos(\frac{2\pi r}{2"}) + 1]$ if $r \leq 1"$ $= 0$, otherwise where $r = (x^2 + y^2)^{1/2}$	Fig. 13
(vi) Circular indent (sink mark) $z = -.001" [\cos(\frac{2\pi r}{2"}) + 1]$ if $r \leq 1"$ $= 0$, otherwise where $r = (x^2 + y^2)^{1/2}$	Fig. 14
(vii) Bond line along y-axis $z = -.001" [\cos(\frac{2\pi x}{2"}) + 1]$ if $|x| \leq 1"$ $= 0$, otherwise	Fig. 15
(viii) Bond line along x-axis $z = -.001" [\cos(\frac{2\pi y}{2"}) + 1]$ if $|y| \leq 1"$ $= 0$, otherwise	Fig. 16

FLAT PANEL - Fig. 9 shows the reflections from a flat panel as a reference. The grid line images (Fig. 9(a)) are straight and regularly spaced. The square part surface area becomes a trapezoid as the inspector sees because of a perspective effect. The projected line on the screen of each laser bean scan (Fig. 9(b)) is straight and the points are regularly spaced. The spacings between points and between lines, and the line lengths vary slightly according to the geometrical relations. The D-Sight image for a flat panel (Fig. 9(c)) has a nearly uniform light intensity except in the front area. Since the eye is above the light source, the light reflected off the front portion of the panel does not enter the eye, and therefore that portion becomes completely dark in the D-Sight image. The size of the dark area increases with the distance from the light source to the eye. For a flat panel, the dark band is straight; for a wavy panel, this band would have a wavy boundary, as will be shown by the following examples.

SQUARE DOME - Automotive body panels usually have some smooth design contours. Here we look at a dome-shaped panel with 0.127 cm (0.05") height at the center. Fig. 10 gives the predicted reflections. The grid line images and the projected lines of reflected laser scans are curved; the line spacings and curvatures vary according to the slope on the part surface. In the D-Sight image, the front and the rear portions of the panel become completely dark, while the rest of the panel has light intensity decreasing from the central area towards the sides.

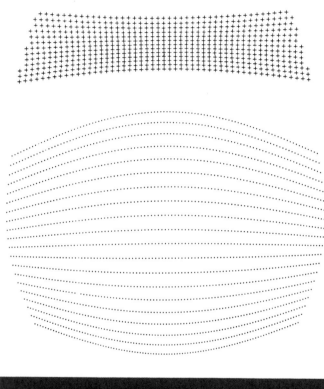

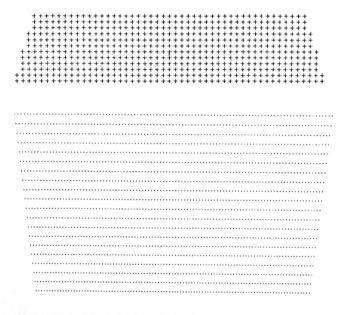

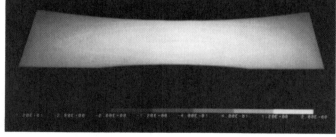

Fig. 10 Predicted reflection images for a dome-shaped panel, profile (ii) in Table 1. From top to bottom: (a) Grid, (b) Laser scan, (c) D-Sight.

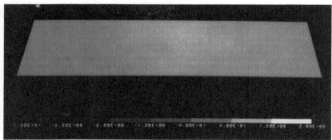

Fig. 9 Predicted reflection images for a flat panel, profile (i) in Table 1. From top to bottom: (a) Grid, (b) Laser scan, (c) D-Sight.

BI-DIRECTIONAL SINUSOIDAL WAVE, SMALLER AMPLITUDE - The surface waviness of an FRP panel can, in principle, be represented by a suitable series of sine and cosine waves, i.e., a Fourier series. Here we consider a representation of the surface waviness in terms of a bi-directional sine wave with a wavelength of 7.62 cm (3") in both x and y directions. The assumed amplitude coefficient is 6.35×10^{-4} cm (.00025"). Fig. 11 shows the predicted reflections. With grid inspection, the waviness is barely visible. If Fig. 11(a) is viewed sideways, the deviation of a horizontal grid line image from a straight line becomes clear. However, an inspector can never view the image this way since the reflected image changes with the viewing direction. With the laser beam scanning, the wavy projected lines reveal the waviness, but the most vivid impression of waviness is obtained from the D-Sight image, where the light intensity variation gives a hill-and-valley feel. Note that a small amplitude of surface waviness already gives quite an objectionable impression.

BI-DIRECTIONAL SINUSOIDAL WAVE, LARGER AMPLITUDE - As shown in Fig. 12, when the amplitude of the bi-directional sinusoidal wave is quadrupled, the waviness becomes very obvious under the grid inspection. The projected lines of laser beam scans also become wavier as the amplitude is increased. The light intensity pattern in the D-Sight image remains the same, but the gradient becomes larger. The most interesting is the boundary of the front dark band - the boundary is like the surface waviness profile of a cross section of the part, magnified in the part thickness direction. The same thing was observed in the image for the SMC panel (see Figs. 4 and 7), and the image for the square dome (Fig. 10).

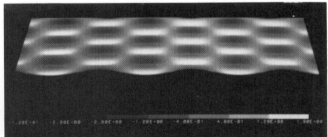

Fig. 12 Predicted reflection images for a bi-directional sinusoidal wave with larger amplitude, profile (iv) in Table 1. From top to bottom: (a) Grid, (b) Laser scan, (c) D-Sight.

CIRCULAR BLISTER - The reflection from a flat panel with a circular blister or "outdent" of 5.08×10^{-3} cm (2 mils) height in the center is predicted in Fig. 13. In grid inspection, the blister causes a disturbance in the otherwise regular grid image pattern.

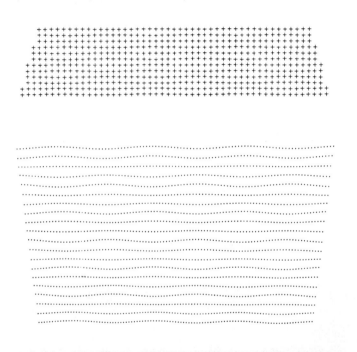

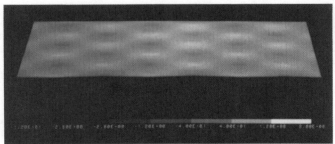

Fig. 11 Predicted reflection images for a bi-directional sinusoidal wave with smaller amplitude, profile (iii) in Table 1. From top to bottom: (a) Grid, (b) Laser scan, (c) D-Sight.

Along the viewing direction, the grid spacing becomes larger, then smaller, and then larger again before returning to the regular spacing. The projected laser scans have a bulb shape in the middle.

When a panel surface has a local defect such as a pimple, the D-Sight image would contain two images of the defect if a single point light is used. One image, called the primary image, is caused by the light coming from a uniformly illuminated area on the screen and reflected by the defect area. The other, called the secondary image, is caused by the light coming from the area on the screen with light intensity disturbed (due to the first reflection off the defect region) and reflected by a smooth surface area[10]. For a small defect, the two images can be easily separated by moving the eye relative to the point light. However, the blister under consideration is 5.08 cm (2") in diameter and with the point light and the eye locations assumed, the two images are not separated; in fact, there are some interactions, as shown in Fig. 13(c). Using multiple lights would usually smear the primary image[10].

CIRCULAR INDENT (SINK MARK) - Fig. 14 is the predicted image for a circular indent or sink mark in the center of a flat panel. The sink causes the grid image spacing to become smaller, then larger, and then smaller again before returning to the otherwise regular spacing in the viewing direction. The projected laser scans have a hourglass shape in the middle. The D-Sight image has, again, two "defect" images with interactions.

Fig. 14 Predicted reflection images for a circular indent (sink mark) in the center of a flat panel, profile (vi) in Table 1. From top to bottom: (a) Grid, (b) Laser scan, (c) D-Sight.

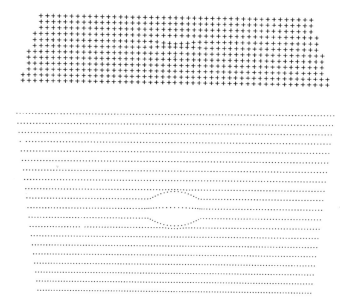

Fig. 13 Predicted reflection images for a blister in the center of a flat panel, profile (v) in Table 1. From top to bottom: (a) Grid, (b) Laser scan, (c) D-Sight.

BOND LINE READ-THROUGH - A common surface defect when two FRP parts are adhesively bonded is the so-called bond line read-through, which is a line of depression due to the shrinkage of adhesive. Here we consider a bond line with a depth of 5.08×10^{-3} cm (0.002") and a width of 5.08 cm (2"). In Fig. 15, the bond line is oriented along the viewing direction. The defect is invisible under the grid inspection. In the laser beam scanning inspection, the projected lines are nearly straight, but the spacing between projected points shows a disturbance due to the bond line. The D-Sight image has a very slight intensity variation in the bond line area, but the variation is too small to be visible with the scale in Fig. 15(c). It should be noted that as the depth of the depression increases, the defect would gradually become visible, i.e., the grid line image would become wavy, so would the projected lines of the laser scans. However, the inspection sensitivity is lowest when the defect is so oriented.

has two bond line images with interactions. This orientation gives the highest inspection sensitivity.

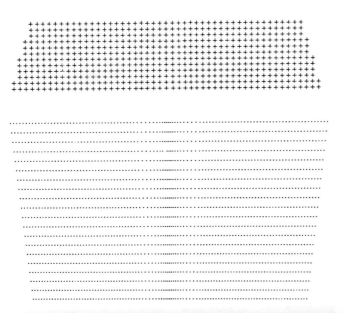

Fig. 15 Predicted reflection images for a bond line oriented along the viewing direction, profile (vii) in Table 1. From top to bottom: (a) Grid, (b) Laser scan, (c) D-Sight.

In Fig. 16, the bond line is oriented perpendicular to the viewing direction. The images of grid lines have their spacings disturbed due to the bond line. Along the viewing direction, the spacing becomes smaller, then larger, and then smaller again, just like that for a isolated indent, but with a bond line, the spacing change is across the part width. The projected lines of laser scans are all straight, but their spacings are disturbed. The D-Sight image

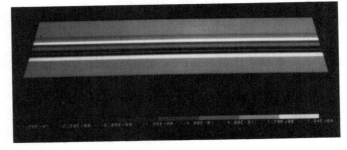

Fig 16. Predicted reflection images for a bond line oriented perpendicular to the viewing direction, profile (viii) in Table 1. From top to bottom: (a) Grid, (b) Laser scan, (c) D-Sight.

CONCLUSIONS AND DISCUSSION

The relation between the surface topography of a part and the observed reflections has been established. Computer programs have been developed to predict the images for three inspection methods, namely, back-lighted grid inspection, laser beam scanning, and double reflection using a retro-reflective screen. The surface profile of an SMC panel was measured. Based on the measured profile, the reflection images were predicted and compared very well with actual ones. The reflections for some representative types of FRP long-term surface waviness are predicted and their characteristics identified. As expected, the distortion in the reflection image in each method is related to the slope of the surface profile with larger slopes giving more severe distortion. The sensitivity of different inspections and with respect to wave orientation and viewing direction are compared.

Although long-term waviness is the emphasis in this study, short-term waviness such as orange peel can be detected by the fuzziness of the image lines in the grid method and the laser beam scanning method, but becomes invisible under D-Sight inspection. Within the range of applicability, waviness is least visible by the grid method, and highest by D-Sight inspection. However, the grid method is the closest to what a car buyer would judge for smoothness of a body panel. Because of the image sensitivity to waviness orientation, one should inspect a part with at least two quite different orientations unless the waviness direction is known a priori and the part is oriented accordingly.

Various proposals have been made to quantify the surface waviness of a part from the projected lines in the laser scanning[6,7] and the D-Sight image[11]. However, caution is needed to avoid misleading results. For example, the laser scans for a bond line oriented perpendicular to the viewing direction yield straight projected lines; the straightness is the same as for a perfectly flat panel. The D-Sight images for the two bi-directional sinusoidal waves in Figs. 11 and 12 have similar patterns although light intensity gradients do differ. What algorithm is appropriate to give a number or numbers to indicate the difference? The computer programs developed in this work may provide a guide for better quantification of surface waviness from the reflection image as well as assist in relating material and processing parameters to part surface appearance.

ACKNOWLEDGMENT

The author would like to thank R. M. Griffith, J. F. Roach, and L. F. Marker for helpful discussions. J. F. Roach also carried out the experiment to measure the cone angle in the D-Sight reflection. D. Millward and B. Henry helped in the graphic display. The Engineering Research Center at Ohio State University made the Sheffield coordinate measuring machine available and C.-L. J. Wong helped carry out the surface profile measurement. Their assistance is greatly appreciated.

REFERENCES

1. A. T. Hurst, "Measurement Aspects and Improvement on Surface Profile in Thin Gauge Molded SMC," 37th Annual Conference, Reinforced Plastics/Composites Institute, SPI, Session 7-E, New York, N.Y., 1982.
2. Jurgen Gunter, "Method for Determining the 'Long Term Waviness' of Large SMC Panels," paper no. 830142, SAE International Congress and Exposition, Detroit, Michigan, 1983.
3. K. A. Iseler and R. E. Wilkinson, "A Surface Evaluation System for Class A Applications," 39th Annual Conference, RP/C, SPI, Session 2-D, 1984.
4. G. Georg, "PALAPREG SMC, A Case of New Resin Systems Being Used for Automotive Parts," pp. 129-132, Proceeding 14th Reinforced Plastics Conference, British Plastics, February, 1984.
5. W. M. Kralovec, "Optical Evaluation of Long Term Surface Waviness," 23rd Annual Technical Conference, RP/C Division, SPI, Session 1-C, 1968.
6. S. S. Hupp, "A Quantitative Method for Analysis of Surfaces of Molded SMC Parts," 43rd Annual Conference, Composites Institute, SPI, Session 10-A, 1988.
7. S. S. Hupp and T. B. Hackett, "Quantitative Analysis of Surface Quality for Exterior Body Panels," paper no. 880358, SAE International Congress and Exposition, 1988.
8. S. S. Hupp and T. B. Hackett, "The Characterization of Painted Surfaces for Exterior Body Panels," 44th Annual Conference, Composites Institute, SPI, Session 18-B, 1989.
9. D. A. Clarke, R. L. Reynolds, and T. R. Pryor, "Panel Surface Flaw Inspection," U.S. Patent 4,629,319, December 16, 1986.
10. R. L. Reynolds and O. L. Hageniers, "Optical Enhancement of Surface Contour Variations for Sheet Metal and Plastic Panel Inspection," International Symposium on Optical Engineering and Industrial Sensing for Advanced Manufacturing Technologies, Dearborn, Michigan, June 26-30, 1988.
11. W. Pastorius, "A New Machine Vision Technique for Surface Inspection," 44th Annual Conference, Composites Institute, SPI, Session 18-D, 1989.
12. I-DEAS Supertab Pre/Post Processing Engineering Analysis Version 4.0 User's Guide, Structural Dynamics Research Corporation, Milford, Ohio, 1988.
13. Max Born and Emil Wolf, "Principle of Optics," Pergamon Press, New York (1959), p. 233.
14. G. A. Baxes, "Digital Image Processing," Prentice-Hall, New Jersey (1984), p. 10.

ALUMINUM ALLOYS MATRIX COMPOSITES USING PARTICLE DISPERSION

Hideo Ohtsu
Nippondenso Co., Ltd.
1-1, Showa-Cho, Kariya-City
Aichi-Pref., Japan

ABSTRACT

Particle dispersion composite Aluminium alloys in which particles of SiC are dispersed in Aluminium can be obtained by firstly mixing and dispersing SiC particles in molten Aluminium using the vortex method and then forming this melt using a squeeze casting method.

The characteristic of this technique is that near-net-shape MMC can be obtained without the necessity of making a preform, which means that the same level of productivity as with conventional Aluminium alloys is possible. MMC using A390 as the matrix obtained by this method is extremely strong and resistant to wear and also has a coefficient of thermal expansion which is close to that of steel (12.5 x 10^{-6}/K).

This makes this type of MMC ideal for pump parts etc. which require lightness, good wear resistance and a low coefficient of thermal expansion for high speed sliding.

RECENTLY, in line with the trend for automobile parts to become more compact and lighter, the demand for MMC (which is very strong, has excellent resistance to wear and a low coefficient of thermal expansion) with Aluminium matrices is increasing. However, such MMC have not yet reached the stage of practical application. There are three major reasons for this.

The first is that the main reinforcements which have been researched to date (continuous and discontinuous fibers and whiskers) are expensive. The second reason is that the main MMC production techniques — the powder metallurgy method, the liquid infiltration method and the hot pressing method — all require complex production processes and are characterized by low productivity. The third reason is that it is difficult to obtain near-net-shape MMC using these methods and that many post-production stages are required.

In this paper we will describe a production method we have developed which solves the problems described above by efficiently providing near-net-shape MMC which has excellent strength and wear resistance properties and a low coefficient of thermal expansion. The first half of the paper concerns the production method and the second half concerns the properties of this MMC.

PRODUCTION METHOD FOR MMC

PREPARATION OF MMC — The particles used for MMC in this research were three types of α-SiC particles obtained from Showa Denko Co,LTD. Their average diameters and compositions are indicated in Table 1. Two

Table 1 Average diameters and chemical compositions of SiC particles for reinforcements

Average diameters of particles (μm)	Chemical Compositions (wt%)			
	SiC	C	SiO$_2$	Fe
10				
28	98.5	0.1	1.0	0.2
57				

Table 2 Chemical compositions of Aluminium alloys for matrix

Aluminium Alloys	Chemical Compositions (wt%)				
	Si	Mg	Fe	Cu	Al
pure Al	0.15	0.02	0.15	0.03	99.65
A390	17.0	0.6	0.2	4.6	77.6

compositions of Aluminium alloys were used for the matrix metal. These were pure Aluminium and a hyper-eutectic Aluminium alloy (A390). Their compositions are indicated in Table 2.

The reason for using α-SiC in this research is that it has a high hardness and a low coefficient of thermal expansion[1], and it is reported that it has good wetting characteristics with Aluminium[2]. These SiC particles are very inexpensive — about $10/kg, which is less than one tenth the price of SiC whiskers.

The reason for using pure Aluminium as a matrix metal is that it obviates the need to consider constituents other than aluminium in the basic consideration and assessment, which is convenient. The reason for using A390 is that is a conventional alloy with a low coefficient of thermal expansion and good resistance to wear.

PRODUCTION PROCESSES FOR MMC — Two production processes for MMC were used in this research; a process in which SiC particles were dispersed in molten Aluminium using the vortex method and a process in which the resulting molten material (hereafter called "MMC melt") was formed by squeeze casting. These processes are described below:

1) Dispersion Of SiC Particles Using The Vortex Method — SiC particles were mixed and dispersed in the molten Aluminium using the vortex method. A schematic view of the apparatus involved in this process is presented in Fig.1.

The procedure is as follows. First, the Aluminium is melted in a crucible and then stirred at high speed to create a vortex by stainless steel agitators which have been coated with Molybdenum using the plasma spray method. The SiC particles are then gradually added and stirred in. At this time the surrounding atmosphere is replaced by Argon gas in order to prevent oxidation of the Aluminium. The conditions for the vortex method are presented in Table 3.

The important points in these conditions are the temperature of the molten Aluminium and the speed of the agitators. If the temperature of the molten Aluminium is too low, it will not be possible to create a vortex and if it is too high, there is the possibility that the SiC and Aluminium will react with each other; the temperature of the molten Aluminium was therefore set at 150K above the melting point. If the stirring speed is too low, it will not be possible to create a vortex and if it is too fast, the added SiC particles will be liable to be scattered; the optimum speed is 800 - 1,000 rpm.

2) Forming The MMC Melt Using The Squeeze Casting Method — After dispersing the SiC particles in the molten Aluminium using the vortex method, the resulting MMC melt is poured into the required mold and quickly formed under high pressure and solidified to give near-net-shape MMC. A schematic view of the apparatus used in this process is presented in Fig.2 and the conditions for squeeze casting are presented in Table 4.

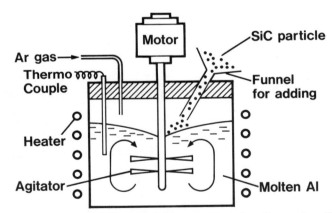

Fig. 1 Schematic view of the apparatus for dispersing SiC particles in molten aluminium

Table 3 Experimental conditions of dispersing SiC particles in Aluminium alloys by vortex method

Aluminium Alloys	Temperature of Melt (K)	Rotation Speed (r.p.m)	Adding Rate (g/min)	Pre-heat Temperature of Particles (K)
pure Al	1073	900	50	773
A390	1023	800		

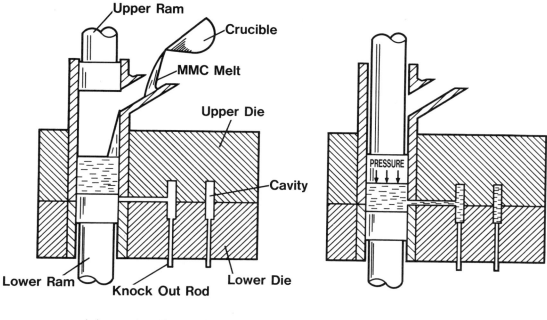

Fig. 2 Schematic view of the apparatus for squeeze casting

Table 4 Experimental conditions of forming MMC melt by squeeze casting method

MMC Melt	Mold Temperature(K)	Pouring Temperature(K)	Squeeze Pressure(MPa)	Pressure Holding Time (sec)
SiCp/pure Al	553	1073	98	60
SiCp/A390	523	1023		

RESULTS AND CONSIDERATIONS

The results and detailed considerations for this production method are described below.

MIXING AND DISPERSION TIME FOR SIC PARTICLES IN THE VORTEX METHOD — The stirring time required for mixing and dispersing the SiC particles in the molten Aluminium in the vortex method is an important factor in MMC production.

The effects of SiC particle diameter on the mixing and dispersion time were therfore clarified by using pure Aluminium as the matrix. Additionally, to clarify the effects of adding Calcium, which is reported to improve the wetting properties of SiC and Aluminium[3][4], Calcium was added within the range 0.5 - 5 wt% and effects on mixing and dispersion time were clarified. Finally, these effects were compared with the results obtained using an A390 matrix.

To assess the state of mixing in and dispersion, the melt was scooped out with a ladle at various times and after prompt solidification, was polished and the polished section examined using an optical microscope.

The results obtained on mixing and dispersing SiC particles 28 μm in diameter into pure Aluminium are shown in Photo 1. As can be seen from these results, the stirring time required for mixing in the particles is about 30 minutes after addition. The stirring time required to disperse the particles evenly was about 90 minutes.

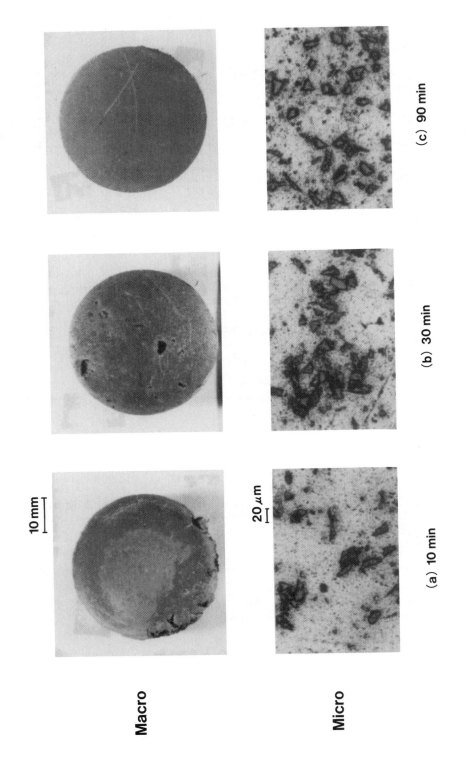

Photo 1 Typical photographs from the polished sections of MMC of each stirring time ($28\,\mu m$-SiC$_P$/pure Al)

Using particles with a diameter of 57 μm, all the particles were mixed in after about 20 minutes and it was possible to disperse the particles evenly in about 60 minutes of stirring, as shown in Photo 2. In the case of particles 10 μm in diameter, it was possible to mix in all the particles but, as shown in Photo 3, even after 120 minutes of stirring the particles had agglomerated and even dispersion could not be confirmed. This is because the larger the diameter of the SiC particles, the smaller is their surface area in comparison with their volume, the smaller is their combined surface energy and the more easily they can be dispersed in the molten Aluminium.

The effects of the addition of Calcium are shown in Fig.3 and the dispersion of SiC particles after stirring for 30 minutes with 1% Calcium added are shown in Photo 4. As shown in Fig.3, it has been clarified that the addition of Calcium considerably reduces the mixing in and dispersion time. The results of adding Calccium at about 1% by weight are considerable and adding 3% by weight is sufficient. Photo 5 shows the results of electron probe micro analyses of the MMC after adding 3% by weight of Calcium. On the basis of these results it is thought that the effect of Calcium is that it accumulates in high concentrations in the vicinity of the surface of the SiC particles, reducing the surface tension of the Aluminium and increasing the wetting properties of the Alominium and SiC, thus reducing the mixing in and dispersion time.

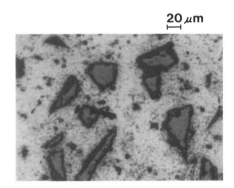

Photo 2 Typical microphotographs from the polished section of 57 μm-SiCp/pure Al after 60 minute stirring

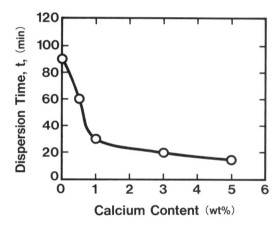

Fig. 3 Effect of calcium content on the dispersion time of SiC particles in the molten aluminium (28 μm-SiCp/pure Al)

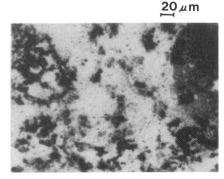

Photo 3 Typical microphotographs from the polished section of 10 μm-SiCp/pure Al after 120 minute stirring

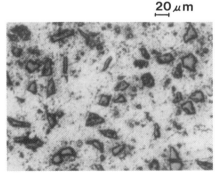

Photo 4 Typical microphotographs from the polished section of 28 μm-SiCp/pure Al after 30 minute stirring when 1wt% Ca is added

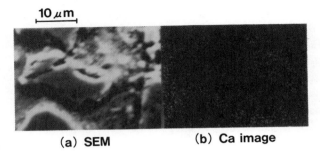

(a) SEM   (b) Ca image

Photo 5 Ca distribution in Matrix by Electron Probe Micro Analyses (28 μm-SiCp/pure Al-3wt%Ca)

Finally, we will consider the results when A390 is used as the matrix. As shown in Fig.4, the dispersion time is shorter than with pure Aluminium. This is because its constituents Silicon and Magnesium have an effect in reducing the surface tension of the molten Aluminium, though their combined effect is not as great as that of Calcium[5][6]. It is possible to reduce the time required for mixing in and dispersion still further by adding Calcium to A390. However, if too much Calcium is added, there is a tendency for the size of the precipitated primary silicon grains to increase and this has an adverse effect on the mechanical properties; the quantity of Calcium added should be 1 wt% or less.

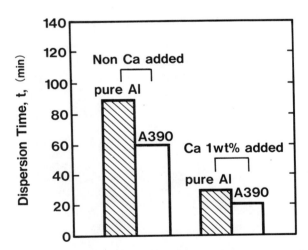

Fig. 4 Effect of difference of aluminium matrix on the dispersion time of SiC particles (28 μm)

REACTION BETWEEN SIC AND ALUMINIUM — it has been reported that at high temperatures, SiC and Aluminium undergo the reaction described by equation (1)[7][8] and SiC particles are lost.[9]

$$3\ SiC + 4\ Al \rightarrow Al_4C_3 + 3\ Si \qquad eq.(1)$$

This was confirmed by electron probe micro analyses and X-ray diffraction of MMC made using a matrix of pure Aluminium with 1% by weight of Calcium. The results are shown in Photo 6 and Fig.5.

The results of the electron probe micro analyses show that although there was practically no diffusion of Si into the MMC after 30 minutes of stirring, a small amount of Si was seen to have diffused into the matrix after 240 minutes of stirring. However, the size of the SiC particles had hardly changed at all, and it is therefore thought that the extent of the reaction is extremely slight. This is backed up by the X-ray diffraction results in Fig.5 which show that the presence of the reaction product Al_4C_3 could not be confirmed.

In the case of MMC using A390 as the matrix, since there is already a large amount of Si in the matrix from the start, it was thought that equation (1) would proceed with more difficulty than in the case with a pure Aluminium matrix. The results obtained show that no loss of SiC particles was observed after stirring for 480 minutes.

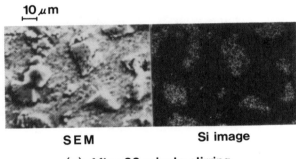

SEM   Si image
(a) After 30 minute stirring

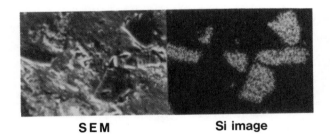

SEM   Si image
(b) After 240 minute stirring

Photo 6 Si diffusion in Matrix by Electron Probe Micro Analyses (28 μm-SiCp/pure Al-1wt%Ca)

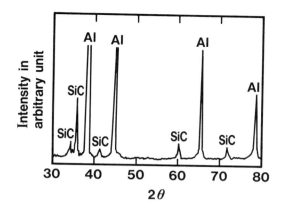

Fig. 5 Xray diffraction pattern of MMC after 240 minute stirring at 1073K (28 μm-SiCp/pure Al-1wt%Ca)

SETTLING AND SEPARATION OF SIC PARTICLES IN THE MOLTEN MMC — it was assumed that, due to the difference in the specific gravity of the SiC particles and Aluminium, settling or separation would occur in molten MMC in which the SiC particles had been dispersed using the vortex method. This would place a restriction on the time from the MMC production process to pouring, forming and solidification.

The settling of SiC particles after stirring molten MMC in which SiC particles 28 μm and 57 μm in diameter had been dispersed in A390 matrix was therefore observed. The method used was to maintain the molten MMC at the same temperature for a fixed time after stirring, then quickly cool it while still in the crucible and cut samples from the top, middle and bottom of the crucible. The assessment was made by polishing these cut pieces, examining the polished sections under an optical microscope and calculating the number of particles and the volume fraction. The results SETTLING AND SEPARATION OF SIC PARTICLES IN THE MOLTEN MMC are shown in Photo 7 and Fig.6.

The results show that in the case of 28 μm particles, about 3 minutes after cessation of stirring hardly any settling of the particles is observed, after 10 minutes there is some tendency toward setting and after 20 minutes there is considerable settling. With the 57 μm particles however (see Fig.6) a tendency for faster settling is seen and a considerable amount of particles had settled after 10 minutes.

In the light of these results, it is necessary to pour, form and solidify within several minutes after the cessation of stirring in order to obtain MMC in which SiC particles are evenly dispersed.

FLOWABILITY OF MMC MELT — The flowability of the molten MMC is an important factor when it is formed by the squeeze casting method to make MMC.

A comparative assessment was therefore performed by measuring the flow length using the simple apparatus shown in Fig.7. The flowability of pure Aluminium was taken to be 100 and the results are shown in Fig.8.

The flowability of MMC melt with a matrix of pure Aluminium was reduced to approximately half at a particle volume fraction of 20%. At a volume fraction of 35%, there was almost no flow. On the other hand, with MMC melt with a matrix of A390, the flowability was approximately 50% of that of pure Aluminium at a particle volume fraction of 25%, which means that the flowability is higher than in the case when pure Aluminium is used as the matrix. However, as might be expected, with a volume fraction of 35%, there is almost no flow.

As can be seen from these results, if the actual casting process is considered, the limit for the particle volume fraction is 30% and it should preferably not exceed 25%.

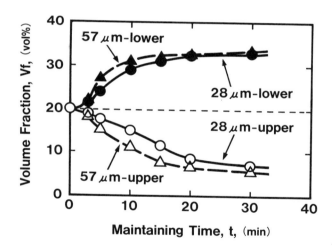

Fig. 6 SiCp volume fraction changes in top, middle, bottom fields dependence on maintaining time after cessation of stirring (28, 57 μm-SiCp/A390)

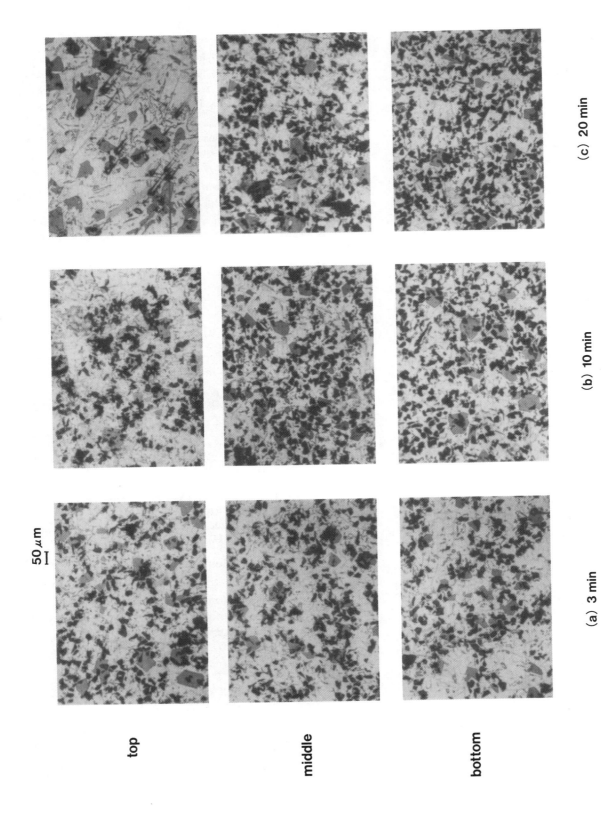

Photo 7 SiC particle distribution in top, middle and bottom fields of MMC at each maintaining time after cessation of stirring (28 μm-SiCp/A390)

(a) 3 min (b) 10 min (c) 20 min

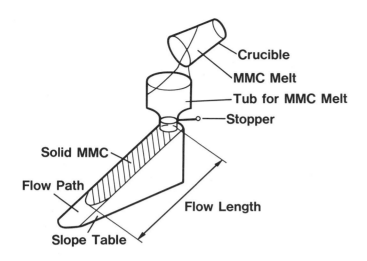

Fig. 7 Schematic view of the apparatus for measurement of flowability

The MMC tensile strength starts to decrease somewhat when the particle volume fraction is 20%, and there is a large amount of dispersion in tensile strength values. This is because when the volume fraction exceeds 20%, the particles agglomerate easily and the material itself becomes brittle, stress concentration becomes more liable to occur and strength decreases. The reason that the bigger the particles are, the lower the strength is that, in the case of particles 57 μm in diameter, breakdown occurs not just at the particle/matrix interface but inside the particles as well.

VICKERS HARDNESS — The relationship between the particle volume fraction and MMC hardness is shown in Fig.10.

The hardness increases almost linearly with increasing particle volume fraction, regardless of the type of matrix. When A390 is used as the matrix and the SiC volume fraction is 25%, the hardness is approximately 1.5 times that with A390 alone. No difference in MMC hardness due to differences in particle diameter was observed.

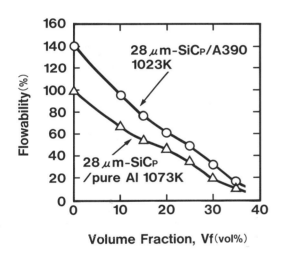

Fig. 8 Effect of volume fraction on flowability for pure Al and A390 matrix composites (28μm-SiC$_P$/pure Al, A390)

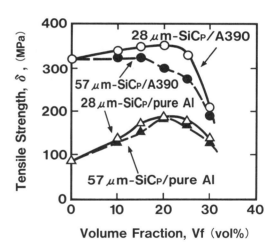

Fig. 9 Effect of volume fraction and SiC particle size on tensile strength for pure Al and A390 matrix

MMC PROPERTIES

The properties of the MMC obtained using this method are described below.

TENSILE STRENGTH — The relationship between the SiC particle volume fraction and the tensile strength of MMC is shown in Fig.9.

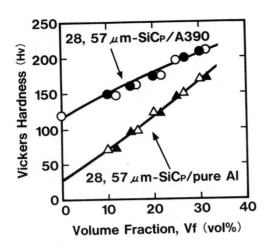

Fig. 10 Effect of volume fraction on vickers hardness for two different particles in pure Al and A390 matrix composites (28, 57 μm-SiCp/pure Al, A390)

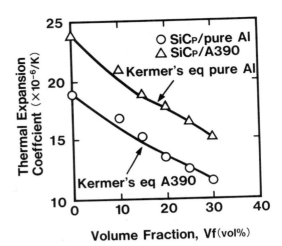

Fig. 11 Effect of volume fraction on thermal expansion coefficient for pure Al and A390 matrix composites. (28 μm-SiCp/pure Al, A390)

COEFFICIENT OF THERMAL EXPANSION — The relationship between particle volume fraction and the coefficient of thermal expansion is shown in Fig.11.

The coefficient of thermal expansion falls almost linearly with increasing particle volume fraction, regardless of the type of matrix, in approximate correspondence with the Kermer's equation (equation (2)).

$$\alpha_{MMC} = \alpha_p + (1 - V_p)(\alpha_m - \alpha_p)$$
$$\times \frac{K_p(3K_m + 4G_p)^2 + (K_m - K_p)(16G_p^2 + 12G_pK_m)}{(4G_p + 3K_m)[4V_mG_p(K_m - K_p) + 3K_pK_m + 4G_pK_p]}$$

eq.(2)

α: thermal expansion coefficient   m: matrix
V: volume fraction (0 to 1)   p: particle
K: Bulk modulus
G: rigidity

The coefficient of thermal expansion for MMC with A390 as the matrix and with 25 vol% SiC particles dispersed in it is $12.5 \times 10^{-6}/K$, which is close to the coefficient of thermal expansion for steel.

WEAR RESISTANCE — The wear resistance of MMC was assessed using a rotating disk on disk type abrasion tester as shown in Fig.12 under the test conditions indicated in Table 5.

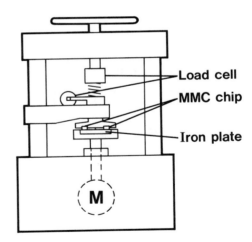

Fig. 12 Schematic view of the apparatus for measurement of wear

Table 5 Experimental conditions of measuring wear resistance

Materials of Chips	Materials of Sliding plates	Sliding Speed (m/sec)	Pressure (kPa)	Condition of Lubrication
MMC A390	Mo plasma sprayed SUJ2	1.5	49	in Air at 293K dry

As a control, a material which is both an alloy in practical use and the matrix used for these MMC – A390 — was also assessed. The results are shown in Fig.13.

As shown in Fig.13, the wear resistance of the MMC is considerably better than that of the A390. Moreover, if we consider the difference in wear resistance attributable to differences in particle diameter; the MMC containing SiC particles 57 μm in diameter had better wear resistance than the MMC containing SiC particles 28 μm in diameter. This is because in MMC in which smaller diameter SiC particles are dispersed, soft Aluminium can easily move over other Aluminium at sliding surfaces by plastic flow, increasing the amount of wear of the Aluminium which constitutes the matrix. Particles with smaller diameters also fall out of the matrix more easily. It is thought to be the result of these factors that MMC in which larger diameter composites are dispersed have better wear resistance.

were dispersed, using the vortex method, in molten A390 to which 1 wt% Calcium had been added. The MMC melt was then poured into the mold for a cylindrical product and true near-net-shape MMC was obtained by squeeze casting. The external appearance of this MMC is shown in Photo 8 and its polished section is shown in Photo 9 and properties in Table 6.

Photo 8 External appearance of Near-Net-Shape MMC by squeeze casting method (25vol%-28μm-SiCp/A390-1wt%Ca)

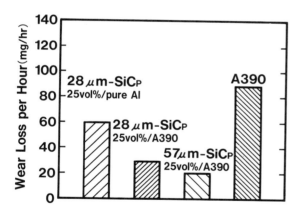

Fig. 13 Comparison of wear resistance between MMC and A390

NEAR-NET-SHAPE MMC

Next we will consider near-net-shape MMC.
PRODUCTION OF NEAR-NET-SHAPE MMC — Enough SiC (28 μm) particles to give a 25% volume fraction

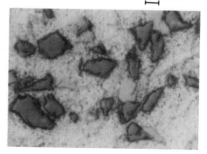

Photo 9 Optical micrograph from the polished section of Near-Net-Shape MMC (25vol%-28μm-SiCp/A390-1wt%Ca)

Table 6 Properties of Near-Net-Shape MMC(28μm-SiCp/A390-1wt% Ca)

MMC Type	Tensile Strength(MPa)	Vickers Hardness(HV)	Wear Resistance	Thermal Expansion Ceofficient(×10⁻⁶/K)	Density (g/cm³)
28μm-SiCp / A390 25vol% / −1wt%Ca	310	175	>A390 ×3	12.5	2.84

APPLICATION OF NEAR-NET-SHAPE MMC — Near-net-shape MMC with properties like those indicated in Table 6 can be obtained with good productivity. They are ideal for products such as those given as examples in Fig.14 — compressor vanes and vacuum pump rotors — which, in addition to lightness and wear resistance suited to high speed sliding, require a coefficient of thermal expansion close to that of steel to improve performance by ensuring a minimal change in clearance due to temperature changes.

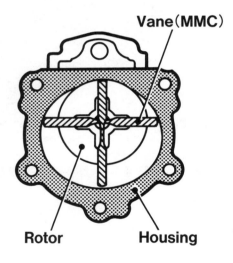
(a) Compressor Vane

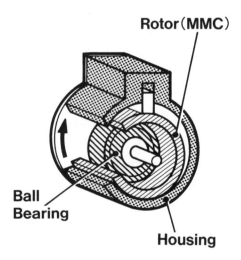
(b) Vacuum Pump Rotor

Fig. 14 Examples of Near-Net-Shape MMC Application

CONCLUSION

The conclusions can be summarized as follows:
1. It is possible to achieve good productivity in the manufacture of near-net-shape MMC by dispersing SiC particles into molten Aluminium using the vortex method and then forming by squeeze casting.
2. By adding Calcium to the matrix in the vortex method, the stirring time required to disperse the SiC particles in the molten Aluminium is considerably reduced.
   This is because the Calcium is highly concentrated in the vicinity of the interface between the SiC particle and Aluminium, reducing the surface tension of Aluminium and improving the wetting characteristics of the Aluminium and SiC. It is advisable to add about 1 wt% of Calcium.
3. With regard to the reaction between SiC and Aluminium which is a possibility in the vortex method and would lead to the loss of SiC particles; when particles about 28 μm in diameter were used, only a slight quantity of Si was diffused into the matrix and even after 240 minutes of stirring, there was no loss of SiC particles.
4. When squeeze casting is performed, provided that the molten MMC is poured, formed and solidified within a few minutes after cessation of stirring, there is no settling or separation of the SiC particles due to the difference in the density of SiC and Aluminium and MMC in which the SiC particles are more or less evenly dispersed can be obtained.
5. Molten MMC with particle volume fractions up to 25% can be assured to have sufficient flowability for squeeze casting.
6. MMC made by this production method with 25vol% of SiC particles dispersed in A390 not only have excellent strength, wear and lightness properties but also have a coefficient of thermal expansion close to that of steel (12.5×10^{-6}/K).
7. MMC with these properties are ideal for application to pump parts etc.

REFERENCE

1) G.V.SAMSONOV, "High Temperature Materials" pages 128 and 184, Plenum Press, NEW YORK (1964)
2) Warren, R., Anderson, C.H, Composites, 15(1984), page 101
3) Takao Cho, Giro Ebihara, Takeo Oki, The Japan Institute of Metals summary of Spring Lectures (1987), page 321
4) Sato, IMONO (Casting), 53, (1981), page 42
5) Koichiro Kubo, Masumi Koisi, Mitsuo Kakuta, "Composites and Interfaces" page 123, Society for Materials Technology Research, Tokyo (1983)
6) U.T.S. Pillai, R.K. Pandey, K.D.P. Nigam, Fifth International Conference on Composite Materials Proceedings, (1985)
7) Takayoshi Izeki, Second survey of Trends for Forthcoming Technology relating to Composites, (1983) page 147
8) Sinichi Towata, Senichi Yamada, Journal of The Japan Institute of Metals No.47, (1983) page 159
9) Takeo Hikosaka, Kaoru Miki, Naoki Kawamoto, ALUTOPIA, 2, (1988) Page 56

EFFECTS OF REACTION PRODUCTS ON MECHANICAL PROPERTIES OF ALUMINA SHORT FIBER REINFORCED MAGNESIUM ALLOY

Harumichi Hino, Mikiya Komatsu
Materials Research Laboratory,
Central Engineering Laboratories
Nissan Motor Co., Ltd, Yokosuka, Japan

Yoshikazu Hirasawa
Technical Department,
Light Metal Division
Ube Industries, Ltd., Ube, Japan

Masato Sasaki
Material Research Section,
Research and Development Dept.
Atsugi Motor Parts Co., Ltd.,
Atsugi, Japan

Abstract
An alumina short fiber reinforced Mg-Nd alloy, referred to here as SAFRMg-Nd, has been fabricated by squeeze casting and the relation between its properties and metallurgical structure has been examined with the following experimental results. (1) The high temperature strength of SAFRMg-Nd with a sound metallurgical structure is superior to that of alumina short fiber reinforced AZ91. (2) Mg_2Si produced by the reaction between the molten magnesium metal and silica binder in the fiber preform is preferentially present in the area where the molten metal finally infiltrates. (3) The deformation behavior of SAFRMg-Nd containing Mg_2Si shows fracture of Mg_2Si in contact with the fiber in the heavily deformed matrix. (4) The properties of SAFRMg-Nd containing Mg_2Si scatter considerably, with some of them being inferior to those of the matrix. Examination of the metallurgical structure of SAFRMg-Nd containing Mg_2Si reveals that the Mg_2Si fractures at the beginning stage of plastic deformation because of stress concentration in the matrix, resulting in degradation of the properties of SAFRMg-Nd.

AN ALUMINA SHORT FIBER reinforced magnesium alloy is thought to have the potential to provide high strength and great durability when applied to automotive components. Reducing the weight of pistons through the application of an alumina short fiber reinforced magnesium alloy is one important way of achieving higher engine performance and lower energy consumption. We reported at ACCE '88 that an AZ91 alloy could be reinforced using alumina short fiber[1]. In order to improve the high temperature properties, especially creep, we have developed an alumina short fiber reinforced Mg-Nd alloy referred to as SAFRMg-Nd in this report. The mechanical properties of SAFRMg-Nd have displayed higher mean values than those of SAFRAZ91, i.e., an alumina short fiber reinforced AZ91 alloy, however, individual values tended to scatter greatly. One of the causes of this scattering of properties in the alumina short fiber reinforced magnesium alloy was thought to be the reaction product, for example Mg_2Si, between the magnesium melt and silica binder in fiber preform[2]. The reaction process and the effects of reaction products on the mechanical properties have not been well understood, however. This report presents the results of a detailed examination of the reaction process and the relationship between the metallurgical structure and the behavior of reaction products during deformation. It also discusses the mechanism causing deformation and fracture of SAFRMg-Nd.

EXPERIMENTAL PROCEDURE

A Mg-5mass%Nd alloy was selected as the matrix alloy. Since the high temperature properties of magnesium alloys are improved by the addition of rare earth elements[3], we selected neodymium as the alloying element. This element allows good solubility into magnesium, and was therefore expected to promote solid-solution hardening at high temperature. Alumina short fiber was chosen as the reinforcement because of its superior tensile strength, high Young's modulus and excellent thermal and chemical stability. The properties

Table 1 - Properties of alumina fiber

Chemical composition	mass %
Al_2O_3	96 - 97
SiO_2	3 - 4
Mean fiber diameter	3 μm
Mean fiber length	150 μm
Tensile strength	2 GPa
Young's modulus	300 GPa
Crystal phase	mainly delta alumina

of the alumina fiber reinforcement are summarized in Table 1. The squeeze casting method used to fabricate the composite is illustrated schematically in Fig. 1. The method of fabricating a prototype piston and a columnar casting for use as specimens is shown in (a) and (b) in the figure, respectively. The molten matrix alloy infiltrated under pressure into a preheated fiber preform set in the die. The volume fraction of fibers was about 9% in the preform. Creep strength, tensile strength at room and elevated temperatures, and hardness of SAFRMg-Nd were examined. The metallurgical structure was also analyzed using an optical microscope, a transmission electron microscope and a scanning electron microscope.

RESULTS AND DISCUSSION

METALLURGICAL STRUCTURE- one half of a vertical section of the prototype piston and typical metallurgical structures observed under an optical microscope are shown in Fig. 2. Magnifications of the microstructure in the regions denoted by the letters b and c in figure (a) are shown in (b) and (c), respectively. In Fig. 2 (b), alumina short fibers, seen as black dots or lines, are uniformly dispersed in the Mg-Nd alloy matrix. This is a typical sound structure. On the other hand, many polygonal precipitates in contact with the fiber are observed in Fig. 2 (c), for example, in the area indicated by P with an arrow in the micrograph.

The microstructures of the vertical section of the columnar casting (Fig. 1 (b)) were also observed under an optical

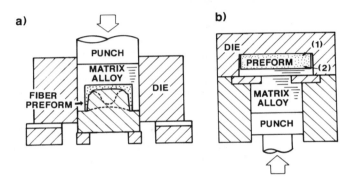

Fig. 1 - Schematic diagrams of fabrication process for SAFRMg-Nd
(a) Prototype piston (b) Columnar casting

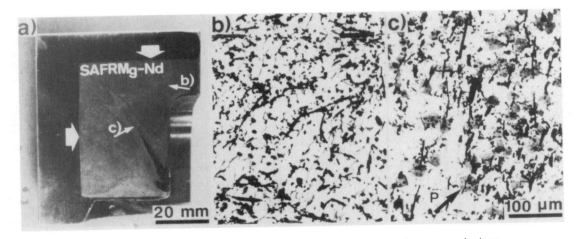

Fig. 2 - Metallurgical structure of prototype piston
(a) One-half of vertical section of piston
(b) Microstructure of region denoted by (b) in (a)
(c) Microstructure of region denoted by (c) in (a)

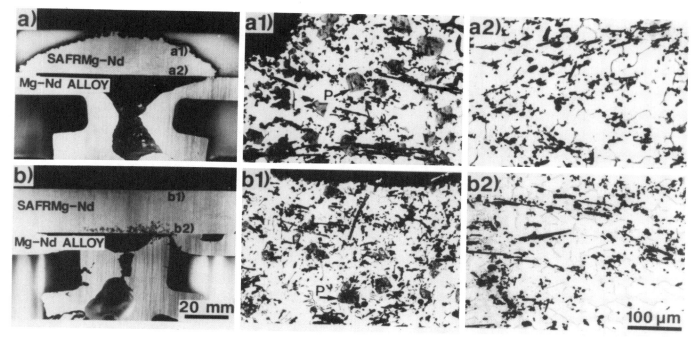

Fig. 3 - Metallurgical structures of columnar castings
(a),(b) Vertical section of casting
(a1),(b1) Microstructure of solidification front
(a2),(b2) Microstructure near bottom surface of fiber preform

microscope. In order to examine the microstructures as a function of the infiltrated distance of the molten metal, two kinds of castings were prepared. The results obtained are shown in Fig. 3. Photographs of the vertical section of the columnar castings are shown in (a) and (b) for infiltration distances of about 12 mm and 20 mm, respectively. In this figure, the area where the molten metal infiltrated and solidified in the fiber preform and the area where it did not are denoted by the notations SAFRMg-Nd and Mg-Nd alloy, respectively. The white areas seen in the upper corners of Fig. 3 (a) are the fiber preform. The large black areas seen at the center of Fig. 3 (a) and (b) are shrinkages resulting from solidification without the application of high pressure. Fig. 3 (a1) and (b1) show the microstructures of the solidification front in the fiber preform, denoted by (a1) and (b1) in (a) and (b); (a2) and (b2) show the microstructures near the bottom surface of the fiber preform, denoted by (a2) and (b2) in (a) and (b). It is seen in (a1) and (b1) that many polygonal precipitates, for example, notation P with an arrow, are present at the solidification front in the fiber preform. However, such precipitates are not observed in (a2) and (b2).

These precipitates were then observed under an analytical electron microscope. The results are presented in Fig. 4, where (a) shows a typical bright-field image of the specimen. The precipitate, the matrix alloy and alumina short fiber are indicated by the notations P, M and ALUMINA, respectively in the micrograph. The results of energy dispersive X-ray spectro scopy (Fig. 4-b)) suggested that the precipitate was Mg_2Si.

The matrix was then deeply etched and the precipitates were observed under a scanning electron microscope. The results presented in Fig. 5 indicate that some Mg_2Si precipitates were in contact with the fiber, for example, the notation P with an arrow in the micrograph.

REACTION PROCESS- Based on the results shown in Fig. 3, the reaction process can be understood as follows. The silica binder is reduced to silicon by the molten magnesium alloy and the silicon dissolves in the magnesium alloy melt. This elementary reaction occurs successively at the front edge of the molten magnesium alloy as it infiltrates into the fiber preform, thereby increasing the silicon concentration at the melt front. As a result, many Mg_2Si precipitates segregate in the area where the molten metal finally infiltrates.

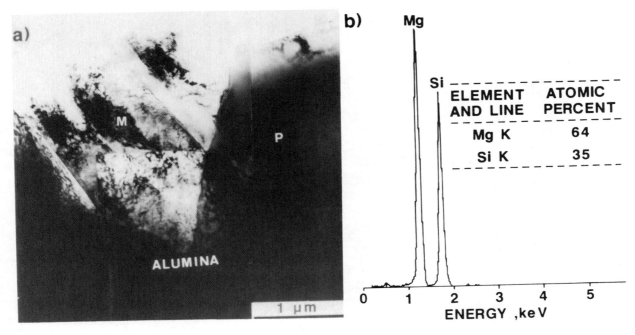

Fig. 4 - Precipitates observed under an analytical electron microscope
(a) Typical bright-field image (b) Analysis of region P in (a)

Fig. 5 - Mg_2Si precipitated in contact with fiber

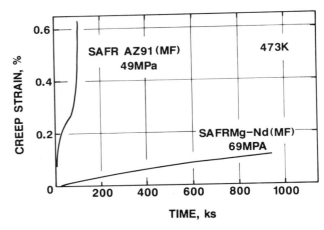

Fig. 6 - Creep strength of SAFRMg-Nd and SAFRAZ91

In Fig. 2 a), it is thought that the molten metal infiltrates in two directions, vertically and horizontally, as indicated by the two large white arrows in the micrograph, and that a lot of Mg_2Si is preferentially present at the center region denoted by (c).

MECHANICAL PROPERTIES-The mechanical properties of SAFRMg-Nd, which was not heat treated, are shown in Figs. 6 through 9. Test pieces were cut from columnar castings fabricated under optimized conditions. Two kinds of specimens were prepared. One was SAFRMg-Nd containing Mg_2Si, referred to as specimen IM in the following, which was cut from the region denoted by (1) in Fig. 1 (b). The other was cut from a region of SAFRMg-Nd that was free of Mg_2Si, denoted by (2) in Fig. 1 (b), and is referred to as specimen MF in the following.

Creep strength of specimen MF is superior to that of the specimen cut from a region of SAFRAZ91 that was free of Mg$_2$Si, as seen in Fig. 6.

The tensile strength of SAFRMg-Nd as a function of temperature is shown in Fig. 7. The strain rate during testing was 8.3×10^{-4} s^{-1}. In this figure, the tensile strength of specimen MF is superior to that of SAFRAZ91. However, individual values for specimen IM scatter largely and some of them are inferior to those of the matrix alloy.

In order to compare the fracture elongation of specimen MF with that of specimen IM, typical stress-displacement curves are shown in Fig. 8. It is seen that the fracture elongation of specimen IM is smaller than that of specimen MF, and that specimen IM fractures in a brittle manner.

A difference in properties between specimens MF and IM can be seen in the results of a hardness test presented in

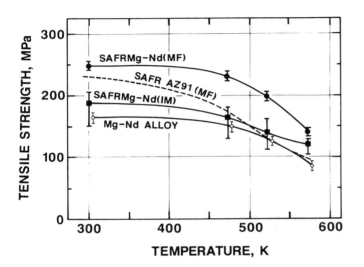

Fig. 7 - Tensile strength of SAFRMg-Nd(FM and IM), Mg-Nd alloy and SAFRAZ91 as a function of temperature

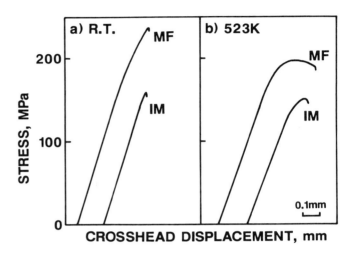

Fig. 8 - Typical stress displacement curves of specimens MF and IM
(a) At room temperature (b) At 523 K

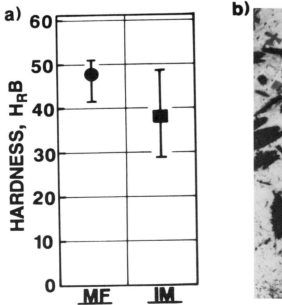
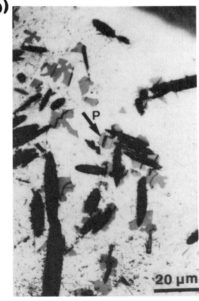

Fig. 9 - Hardness test and microstructure observed following test
(a) Hardness of specimens MF and IM
(b) Fracture of Mg$_2$Si in specimen IM

Fig. 9. Individual hardness values of specimen IM scatter largely as seen in (a), exhibiting the same tendency as mentioned earlier for tensile strength. The deformation structure near the indentation in specimen IM following the hardness test is shown in Fig. 9 (b). The large black area seen in the upper right corner in (b) is the indentation. It is seen that many Mg_2Si precipitates fractured in a cleavage manner in the neighborhood of the indentation, for example, notation P with an arrow in the micrograph.

The fracture pattern of Mg_2Si was believed to be closely related to the tensile strength of SAFRMg-Nd. In order to confirm this point, the tensile deformation behavior was continuously observed under an optical microscope. The results obtained are shown in Fig. 10. The specimen was notched in advance and deformed in the direction denoted by T.D. with an arrow in the micrograph. The black area of each micrograph in (a) is the notch. This series of micrographs in (a) shows the typical deformation behavior of specimen MF. The process from the beginning stage of deformation (a-1) through fracture (a-4) is followed in the same region. The same position is indicated by the white arrow in each micrograph. When tensile stress is initially applied, slip deformation, for example, notation A with an arrow in (a-2), occurs in the matrix the yield stress of which is lower than that of the fiber. As the amount of deformation increases, this deformed region expands and a deformation band is formed, for example, region B between the two dotted lines in (a-3). Regions where deformation is concentrated locally occur simultaneously at the interface between the fiber and matrix, for example, notations C and D in (a-3). It is thought that these regions are induced by stress concentrations due to the discontinuous strain. As the amount of deformation increases further, fracture occurs when these deformed regions are successively combined, as seen in (a-4). It is thought that the fiber fractures due to cracking induced by such stress concentrations and that the crack propagates in the matrix producing new stress concentrations at the crack tip. It has been clarified by this experiment that the tensile deformation behavior of specimen MF is determined essentially by stress concentrations arising from the discontinuous strain occurring at the interface between the fiber and matrix.

The effects of such stress concentrations on the deformation behavior of specimen IM is discussed next based on the results shown in Fig. 10 (b). The deformation behavior observed under a microscope is followed successively in (b-1) through (b-6) in the same way as in (a). The black lump structure, for example, notation E indicated in (b-1), is alumina fiber, and the polygonal structure is Mg_2Si. When tensile stress is initially applied in the direction denoted by T.D. in the micrograph, slip deformation, for example, notation F with an arrow in (b-3), occurs in the matrix. This occurs in the same manner as shown in (a). In this stage, the change in contrast of some Mg_2Si precipitates, which is thought to be due to bending, is already recognizable, for example, notation P1 with an arrow in (b-3). As the amount of deformation increases, the signs of deformation appear, as seen in the region denoted by G in (b-4), and some Mg_2Si precipitates fracture simultaneously, for example, notations P1, P2 and P3 with an arrow. It is thought that Mg_2Si fractures because of the stress concentrations in the matrix. As the amount of deformation increases further, the crack in Mg_2Si is propagated into the matrix, as shown by notation H with an arrow in (b-5), and specimen IM fractures when some heavily deformed regions combine, for example, notations I and J in (b-5) and (b-6).

Based on the observation results shown in Fig. 10, the brittle fracture behavior of specimen IM seen in Fig. 8 is discussed in the following. In Fig. 8, plastic deformation occurs in each specimen. This corresponds to the plastic deformation observed in the matrix, as seen in Fig. 10. As the amount of deformation increases, stress concentrates at the interface between the fiber and matrix. In specimen IM, Mg_2Si fractures at the beginning stage of plastic deformation because of stress concentrations in the matrix. As seen in Fig. 5, since the Mg_2Si precipitates are frequently in contact with the fiber, the crack in Mg_2Si propagates rapidly accompanied by fracture of neighboring fiber and/or Mg_2Si. As a result, specimen IM fractures in a brittle manner, as seen in Fig. 8. The deformation and fracture behavior in specimen IM is closely related to fracture of Mg_2Si, i.e., it depends on the distribution of Mg_2Si precipitates. Therefore, it is thought that the scattering of values for the mechanical properties of specimen IM depends on the discontinuous distribution of Mg_2Si precipitates.

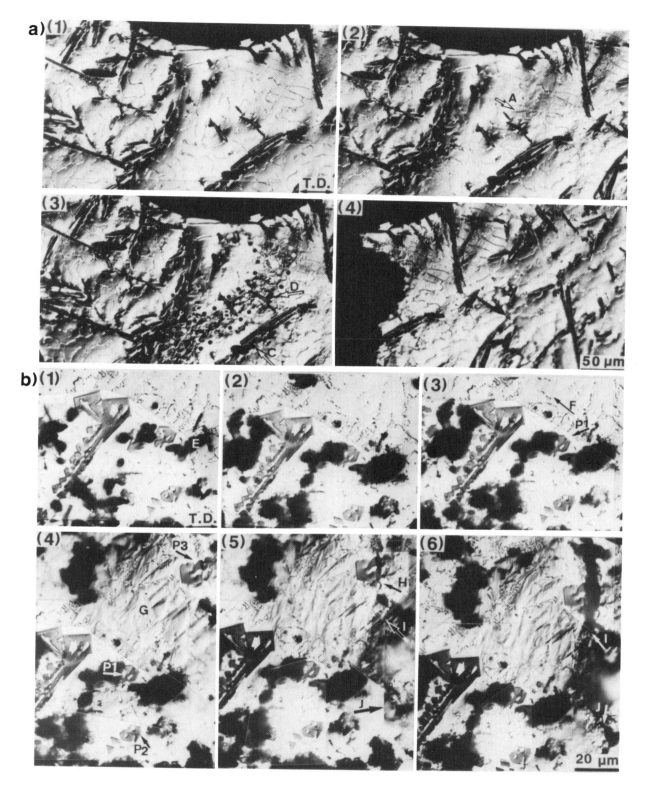

Fig. 10 - Successive stages of tensile deformation behavior in SAFRMg-Nd
  (a) Deformation and fracture behavior of specimen MF
  (b) Fracture of Mg_2Si caused by stress concentration in matrix and crack propagation in specimen IM

SUMMARY AND CONCLUSIONS

An alumina short fiber reinforced Mg-Nd alloy(SAFRMg-Nd) was fabricated by squeeze casting and the relation between the properties and the metallurgical structure was examined with the following results.

(1) The mechanical properties of SAFRMg-Nd having a sound metallurgical structure are superior to those of alumina short fiber reinforced AZ91 alloy.

(2) In the fabrication process, the silica binder in the fiber preform is reduced to silicon by molten magnesium melt. As a result, many Mg_2Si precipitates segregate in the area where the molten metal finally infiltrates.

(3) When tensile stress is applied to SAFRMg-Nd, slip deformation is observed in the matrix. As the amount of deformation increases, stress concentration induced by discontinuous strain occurs at the interface between the fiber and matrix. Ultimately the fiber fractures because of the stress concentration.

(4) In the deformation process of SAFRMg-Nd containing Mg_2Si, Mg_2Si fractures at the beginning stage of plastic deformation because of the stress concentration in the matrix.

The results obtained in this work have clarified that the deformation and fracture mechanism of SAFRMg-Nd is closely related to stress concentration in the matrix and fracture of Mg_2Si, i.e., the crack in Mg2Si at the beginning stage of deformation degrades the properties of SAFRMg-Nd.

REFERENCES

1. Sayashi, M., Hino, H., Komatsu, M. and Sasaki, M., Proc. Annual Conf. Advan. Compos. ACCE-4th, 479-485(1988)
2. Dinwoodie, J. and Horsfall, I., Proc. Inter. Conf. Compos. Mater. ICCM-6th, 2390-2401(1987)
3. Unsworth, W., Metals. Mater. 2, 83-86(1988)

CONTINUOUS SiC FIBER REINFORCED METALS

Melvin A. Mittnick
Textron Speciality Materials
Lowell, MA USA

ABSTRACT

Continuous silicon carbide (SiC) fiber reinforced metals (FRM) have been sucessfully applied on numerous and varying DoD and aerospace development programs fulfilling primary design objectives of high specific strength over baseline monolithic materials.

Programs have included:

   Rocket motor cases
   Army Assault Bridging
   Projectile fins and sabots
   Small and large caliber gun barrels
   Turbine engine disks, shafts and blades
   Tank Track Pins for the M-1
   Wing stiffners
   MK50 Torpedo shells

THE ABILITY TO PRODUCE low cost SiC FRM is fostered by the capacity of the fiber to be easily wet by the respective metal and to resist degradation of strength and modulus during consolidation and high temperature secondary processing. Strong metallurgical bonding at the fiber-matrix interface is readily achieved for several matrix materials through tailoring of the continuous fiber surface chemistry during the chemical vapor deposition (CVD) process. The results of this tailoring is the ability to consolidate aluminum composites using high temperature, less costly processes such as investment casting and low pressure (hot) molding. For titanium composites the fiber can withstand long time exposure (7 hours or more) at diffusion bonding temperatures. This capability makes possible the fabrication of complex shapes by the innovative superplastic diffusion bonding and hot isostatic pressing (HIP'ing) processes.

Secondary metal-working operations of machining, joining and attaching methods have remained compatible with structural alloys used in baseline designs. For aluminum, the composite is weldable and requires no outgassing procedure, thus offering a good potential for fix forward capability. The inherent environmental resistance over broad temperature ranges and field ruggedness of these lighter weight, stronger and stiffer fiber reinforced metals, continues to present enhanced performance, long-term survivability and improved logistics for defense material.

This presentation will review the current state-of-the-art in silicon carbide fiber reinforced metals through discussion of their application on DoD-sponsored and IRAD programs at Textron. Discussion will include a review of mission requirements, program objectives and accomplishments to date employing these FRM's.

Briefly highlighted will be fabrication, producibility, and quality assurance issues associated with the manufacture of continuous silicon carbide fiber and the consolidation processes for producing end item metal matrix components.

COMMERCIALIZATION OF *DURALCAN* ALUMINUM COMPOSITES

William R. Hoover
Dural Aluminum Composites Corporation
San Diego, CA USA

ABSTRACT

Dural Aluminum Composites Corporation, a subsidiary of Alcan Aluminum Corporation, produces ceramic particle-reinforced aluminum composites by mixing the reinforcing ceramic powder into the aluminum melt. Dural is now strongly focused on the commercialization of *DURALCAN* aluminum composites by ensuring a reliable product material and by preparing for large-scale, low-cost production. An extensive mechanical property database has been developed and will be discussed, with an emphasis on relating the observed behavior to the composite structure. In addition, the status of the technology necessary to support widespread MMC usage, such as shape casting, extruding, forging, machining, and welding, will be reviewed.

CERAMIC-REINFORCED METALS, or metal matrix composites (MMCs), have been studied and evaluated for more than 20 years. Although these engineered materials typically have outstanding mechanical properties and, often, desirable physical properties as well, they have never become widely used. The reason for this is their excessive cost; only in rare instances have they been found to be cost-effective.

Until recently, it was widely believed that the reason for the high price of metal matrix composites was the small market, and the reason for the small market was the high price. It was assumed that if a sufficiently large market could be developed, the price would fall sufficiently to justify the demand. This was an illusion, however, because the high <u>absolute</u> <u>cost</u> of manufacturing almost all of these materials precludes a price anywhere near that of unreinforced aluminum. Even in large-scale production, they would be much too expensive to be widely competitive in the commercial and automotive markets.

Most ceramic particle-reinforced aluminum composites are manufactured by the blending of ceramic and aluminum powders, followed by hot pressing to achieve consolidation. By contrast, *DURALCAN* particulate-reinforced aluminum composites are produced by incorporating the ceramic powder directly into the molten aluminum. This ingot metallurgical approach was developed in 1986 by Dural Aluminum Composites Corporation (San Diego), a subsidiary of Alcan Aluminum Corporation. It is inherently less expensive—by an order of magnitude—than the powder metallurgical approach.

Dural currently manufactures two basic types of composites. *DURALCAN* foundry composites are reinforced with SiC and are produced as foundry ingot (or pig) for subsequent remelting and shape casting. *DURALCAN* wrought composites are reinforced with alumina and are produced as extrusion billet of up to 12 inches in diameter and 50 inches in length; by the end of 1989, we could be producing rolling slab as well. Our production capacity is 4 million lb/yr, and our estimated total production for 1989 is about 2 million lb.

In mid-1988, Alcan Smelters and Chemicals Ltd. began construction of a commercial manufacturing plant in Quebec for *DURALCAN* aluminum composites. This $30 million (U.S.) plant, with an annual production capacity of 25 million lb, is expected to produce at a cost close enough to that of ordinary, unreinforced aluminum to stimulate a very wide demand for the material.

Alcan's action has had a profound effect on Dural's process and materials development activities. These are now focused on three vital ingredients required for successful commercialization: a low-cost, large-capacity production process, a desirable, consistent product, and a large, broad-based market. This paper will discuss these aspects of Dural's commercialization efforts, the level of which is unprecedented within the metal matrix composite community.

A LOW-COST, LARGE-CAPACITY PRODUCTION PROCESS

The central technical issue in any commercialization effort is the manufacturing process itself. It must be inexpensive, reliable, and amenable to scale-up. Alcan's commitment to the large-scale commercialization of *DURALCAN* aluminum composites led Dural to scale up its production process, change one of the raw materials used in some of its products, and adopt a better casting process for manufacturing extrusion billet. Let us examine each of these in turn.

THE *DURALCAN* PROCESS - In the *DURALCAN* process, ceramic particles are incorporated into molten aluminum; the composite melt is then cast into extrusion billet or foundry ingot. This process was being used in mid-1988 to manufacture 100-lb composite batches at Dural's pilot plant in San Diego. Since then, the process has been scaled up through 250-lb batches to its current level of 1500-lb batches.

This effort was the prelude to commercial-scale production (in Quebec) of 15,000-lb batches of *DURALCAN* composites, which will begin late in 1989. During the scale-up, equipment and casting practices were developed and quality control practices were modified to make them adaptable to the commercial-scale operation. Fortunately, the *DURALCAN* process is both inexpensive and fully compatible with commercial aluminum production practices.

RAW MATERIALS - Originally, only silicon carbide (SiC) was used as the ceramic reinforcing agent for *DURALCAN* composites. It is not readily available, however, in the particle size, quantity, and price range desirable for a 25,000,000-lb/yr composite plant.

In searching for a suitable substitute, Dural found that alumina (Al_2O_3) is also an effective reinforcing agent for many alloys. Although its density is higher than that of SiC, alumina has the advantages of being inexpensive and available in large quantities. Furthermore, it is more chemically stable than SiC in many aluminum-alloy melts. The initial disadvantage of alumina was the lack of a broad database on alumina-reinforced aluminum composites.

Under pressure to develop a low-cost composite, Dural began to make wrought *DURALCAN* composites using the alumina reinforcement. SiC was retained as the reinforcement for the foundry *DURALCAN* composites primarily because of its chemical stability in high-Si alloys and its lower density.

CASTING PROCESS - Prior to Alcan's commercialization decision, Dural was manufacturing extrusion billet by low-pressure casting into steel book molds. This process was considered unacceptable for commercial production, however. Consequently, Dural modified the direct-chill (DC) casting technique used for the production of wrought aluminum alloys and applied it to the casting of *DURALCAN* composites. This technique has the advantages of low cost and of producing a superior microstructure, since the composite solidifies rapidly.

In the DC process, shown schematically in Figure 1, the molten composite flows across a casting table and into a water-cooled mold. As the composite solidifies within the mold, the bottom block is lowered at a rate that maintains a stable liquid-solid interface. Once the supply of molten metal is exhausted, the billets are removed from the casting table, inspected, and cropped. A sample from each heat goes to quality control to ensure consistent chemistry and particle distribution and volume fraction. After passing QC, the billets are shipped to the extruder; no surface preparation is required.

A DESIRABLE, CONSISTENT PRODUCT

After scaling up the *DURALCAN* process and selecting a new raw material and a more efficient casting process, it became necessary to characterize the *DURALCAN* composites thoroughly. Here the focus has been on obtaining a statistically significant database that will allow Dural to quote 99%-confidence-interval minimum values for the primary mechanical properties of the various *DURALCAN* composites. This effort has revealed a very consistent set of properties that are attractive for many applications of cast and wrought products.

DURALCAN FOUNDRY COMPOSITES - Dural's primary foundry composite is A356 aluminum reinforced with 10, 15, or

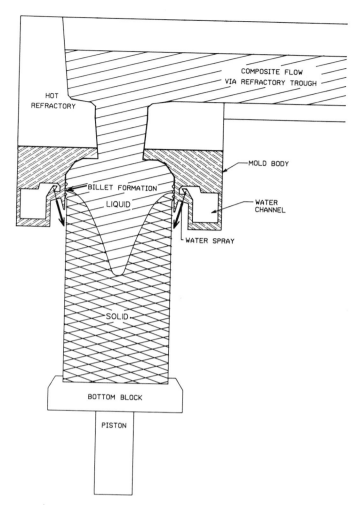

Figure 1. Schematic of the direct chill (DC) casting process.

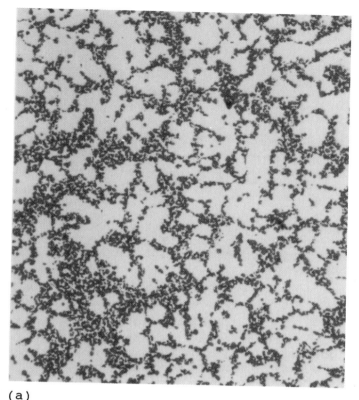

Figure 2. Microstructure of <u>DURALCAN</u> A356-20% SiC castings. (a) Investment casting (43X). (b) Permanent mold casting (85X).

20 vol% SiC (A357 is also used, in smaller quantities). This material is produced in standard 35-lb notched ingots intended for remelting and shape casting in our customers' foundries. It has been successfully cast using most of the major casting techniques: investment (shell and plaster), sand, permanent mold, and low-pressure die casting.

The microstructure of the cast composite is dependent on the casting technique used and the solidification rate achieved. When very slow solidification is allowed, as in investment casting, the microstructure shows a less than optimal distribution of the SiC particles (Figure 2a). This is caused by the formation of primary aluminum dendrites that "push" the SiC into the interdendritic spaces. This microstructure is significantly different from that produced when the composite is more rapidly solidified, as in a permanent mold casting (Figure 2b).

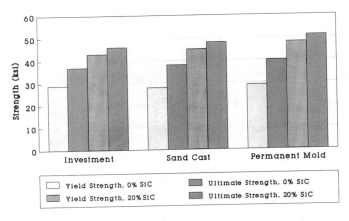

Figure 3. Effects of casting method on the tensile strengths of DURALCAN A356-SiC composite castings (cast-to-size tensile bars).

Table 1. Room-temperature tensile properties of DURALCAN A356-SiC composites (T6 condition) made by the permanent mold casting process. Typical and (minimum) values.

Volume Percent SiC	Yield Strength (ksi)	Ultimate Strength (ksi)	Elongation (%)	Modulus (Msi)
0	29	40 (37)	6.0	10.9
10	41 (40)	44 (41)	0.6	11.7
15	47 (44)	48 (45)	0.3	13.0
20	48 (46)	51 (48)	0.4	14.0

Minimum values represent 99% confidence interval.
Cast-to-size tensile bars.

More uniform microstructures would be expected to produce composites with higher strength. Figure 3 shows a comparison of the yield and ultimate tensile strengths of unreinforced and reinforced A356 as a function of casting technique. As expected, the yield strength of the composite increases as a more uniform microstructure is produced by more rapid solidification. Note, however, that even the microstructure of the investment cast composite produces a composite yield strength that is almost 50% greater than that of the unreinforced investment casting.

The mechanical property data for the cast DURALCAN composites show that both the yield and ultimate tensile strengths increase as the SiC volume fraction increases (Table 1). Surprisingly, most of the strength increase is achieved with only 10% SiC. The elastic modulus, however, increases linearly with increasing SiC content. The wear resistance also increases with SiC content, and the machinability decreases.

The tensile ductility of these composites is lower than desired; this is the price to be paid for increased stiffness and strength. Research is being conducted to improve the tensile ductility through the use of modifiers. It is important to note, however, that this very low ductility is _not_ accompanied by very low fracture toughness. The plane-strain fracture toughness of the composites with 15% and 20% SiC content has been determined to be 16.6 and 14.6 ksi·in.$^{1/2}$, respectively. These data indicate that the material's ability to resist crack growth is quite tolerable. The relatively high fracture toughness despite low tensile ductility indicates that the latter is not a reflection of poor casting quality, but is an inherent property of the material.

Figure 4 shows that the increases in room-temperature yield strength achieved by the addition of SiC are not only retained at elevated temperatures, they are _expanded_. In the 400-600°F range, the composite yield strengths are more than twice that of the unreinforced alloy. Table 2 reveals that the advantages in ultimate strength and elastic modulus are also retained at elevated temperatures.

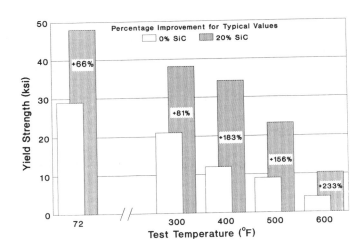

Figure 4. Typical high-temperature yield strength of DURALCAN A356-SiC composite castings (permanent mold casting process).

214

Table 2. High-temperature strength of DURALCAN A356-SiC composites (T6 condition) made by the permanent mold casting process. Typical values.

Temperature (°F)	Yield Strength with x% SiC (ksi)		Ultimate Strength with x% SiC (ksi)		Elongation with x% SiC (%)		Modulus with x% SiC (Msi)
	0%	20%	0%	20%	0%	20%	20%
72	29	48	40	51	6.0	0.4	14.0
300	21	38	24	39	15	1.0	11.8
400	12	34	15	35	30	1.0	10.7
500	9	23	11	24	50	1.5	7.8
600	3	10	4	11	60	6.3	4.5

DURALCAN A356-SiC: Average of two cast-to-size tensile bars, held at temperature for 30 minutes.

A356-0%: Handbook values.

In summary, DURALCAN A356 composites show increased yield strength, ultimate strength, elastic modulus, wear resistance, and thermal stability over the unreinforced alloy. Since not all of these properties increase linearly with SiC content, trade-offs can be made to provide the best material for a specific application. For example, a low SiC content may be desirable for applications requiring increased strength but not maximum elastic modulus, while maintaining acceptable machinability.

DURALCAN WROUGHT COMPOSITES - Dural produces wrought composites consisting of 6061, 2014, and 2219 aluminum alloys reinforced with 10, 15, and 20 vol% alumina. These composites are produced as DC extrusion billet in sizes up to 12 inches in diameter and 50 inches in length. As with the DURALCAN foundry composites, the DURALCAN wrought composites demonstrate increased strength, stiffness, wear resistance, and thermal stability when compared with their unreinforced counterparts.

Extrusion of the wrought DURALCAN composites produces microstructures such as that shown in Figure 5. Although some microstructural alignment of the particles is caused by the extrusion process, their distribution remains basically isotropic, and a strong dependence of mechanical properties on extrusion direction is not observed.

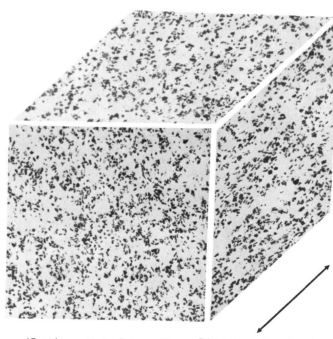

15 volume % Al$_2$O$_3$ in 6061   Direction of Extrusion

├──┤
100 μm

Figure 5. The microstructure of a DURALCAN 6061-15% Al$_2$O$_3$ composite extrusion.

Table 3. Room-temperature tensile properties of extruded DURALCAN 6061-Al_2O_3 and 2014-Al_2O_3 composites (T6 condition). Typical and (minimum) values.

Volume Percent Al_2O_3	Yield Strength (ksi)	Ultimate Strength (ksi)	Elongation (%)	Modulus (Msi)
6061				
0	40 (35)	45 (38)	20	10.0
10	43 (37)	49 (44)	7.6	11.8
15	46 (42)	52 (49)	5.4	12.7
20	52	55	2.1	14.3
2014				
0	60 (55)	70 (64)	13	10.6
15	69 (63)	73 (68)	2.3	13.3

Minimum values represent 99% confidence interval.

Table 4. Plane-strain fracture toughness of extruded DURALCAN composites (T6 condition), demonstrating significant toughness retention even at the 20% reinforcement level.

Volume Percent Al_2O_3	K_{IC} (ksi-in.$^{1/2}$)	
	6061	2014
0*	27.0	23.0
10	21.9	16.4
15	20.0	17.1
20	19.6	—

CTS per ASTM E399 and B645. Typical values with 3 - 11 valid tests each.

* Handbook values.

Table 5. Wear behavior (block-on-ring test) of extruded DURALCAN composites, showing excellent wear-resistance improvement over the unreinforced alloys.

Volume Percent Al_2O_3	Volume Loss	
	(10^{-6} in.^{3})	(mm^3)
6061		
0	534	8.75
10	1.64	0.0269
15	1.06	0.0173
20	0.91	0.0149
2014		
0	683	11.2
10	1.16	0.019
15	0.60	0.0099
20	0.58	0.0095

ASTM G77; ring material 4140 steel; Rc = 50-60. Tests performed in 10W40 motor oil for 2 hrs with a 150-lb load.

The mechanical properties obtained by extruding the DURALCAN wrought composites are summarized in Table 3. These data show that the composites have approximately 15% higher yield strength and 30% higher elastic modulus than their unreinforced matrix alloys. Table 4 demonstrates that the plane-strain fracture toughness of these composites is above the 15-ksi·in.$^{1/2}$ level, even for 20% alumina reinforcement. This is a respectable fracture toughness level for a metallic composite and is quite useful for design purposes.

The addition of alumina to the wrought alloys causes a dramatic increase in wear resistance, as shown by the data in Table 5. Preliminary fatigue data indicate that the fatigue strength of these composites is quite similar to that of the unreinforced matrix alloys, as is the corrosion resistance.

Analysis of the data for both the wrought and foundry composites indicates a very small difference between the typical (or average) strength values and the 99%-confidence-interval minimum values. The small spread indicates that these materials have quite consistent mechanical behavior.

A LARGE, BROAD-BASED MARKET

Even with a low-cost, large-capacity production process and a desirable, consistent product, the development of a broad-based market is dependent on the ability to cast or work the material into useful shapes. Consequently, Dural has spent considerable effort in working with semifabricators to develop appropriate techniques and approaches.

CASTING - The DURALCAN foundry composites can be shape-cast by most ordinary aluminum foundries, using standard casting techniques without the need for expensive new equipment and procedures. Except in most induction furnaces, where convection currents prevent the SiC particles from settling, a mixing impeller is needed to keep the SiC suspended in the molten aluminum.

This stirring, coupled with extreme care to avoid hydrogen pickup (since degassing is generally not recommended), is the only significant modification to normal procedures that most foundries must undertake before successfully casting *DURALCAN* composites. In addition, conservative casting design utilizing well-gated and well-risered molds is recommended because the composite does not feed quite as readily as the unreinforced alloy.

EXTRUSION - A wide variety of shapes has been extruded using the *DURALCAN* wrought composites. They extrude similarly to the hard aluminum alloys, with two exceptions. One is that the increased strength of the composites requires a roughly 20% higher extrusion pressure than for unreinforced aluminum. The other is that the abrasive nature of these materials causes substantial wear on conventional tool-steel dies. This problem can be avoided by using ceramic or carbide dies. Certain die coatings and some new die materials have provided encouraging wear results.

FORGING - Forging studies on the *DURALCAN* wrought composites have just begun. Preliminary results indicate that the material demonstrates a good die-filling capability if both it and the die are sufficiently hot prior to forging. Test results on open-die forgings indicate that no significant change in mechanical properties occurs during open-die forging. Closed-die forging studies are currently in progress.

WELDING - Difficulty is often encountered in welding metallic composite materials because of the gas dissolved in the composite during manufacture. This gas is released during fusion welding, resulting in weld porosity. Because *DURALCAN* composites are made in the molten state, there is no significant dissolved gas in the metal. Consequently, weld porosity is not a problem in welding these materials. Both wrought and cast *DURALCAN* composites have been successfully welded using standard aluminum welding procedures and filler alloys.

MACHINING - The ceramic content of metal matrix composites makes these materials more difficult to machine than their underlying alloys. It is apparent, however, that they can be readily machined if the correct tooling and procedures are utilized. Carbide tooling is required for most operations, although polycrystalline diamond is preferred. To maximize cutting rate and minimize tool wear, slow speeds and large feeds are generally recommended. A thorough machining study is being conducted by Cleveland Twist Drill under contract to Alcan International Ltd. The results of this study will provide a comprehensive guide to machining these materials.

SCRAP - Foundry scrap (gates and risers) from *DURALCAN* composites can be recycled if sound foundry practices have been followed to avoid undesirable hydrogen pickup during the initial melting and casting. The oxides and inclusions that would be picked up by the gates and risers can be filtered out using special feeding and gating techniques. Furthermore, preliminary research indicates that some degassing and gas fluxing can be accomplished if very specific procedures are followed.

Scrap from the wrought *DURALCAN* composites is more problematic. The aluminum can be reclaimed if special procedures are followed, but the recycling of wrought scrap as a composite is still a technical challenge. In view of the current difficulties with wrought scrap recycling, Alcan has announced that it will purchase wrought *DURALCAN* scrap at a price that is tied to the current price of aluminum.

SUMMARY

Progress is being made toward the commercialization of *DURALCAN* aluminum composites. A low-cost, large-capacity manufacturing process has been developed and is being readied for full-scale production. Desirable, consistent products have been developed and extensively characterized. Finally, significant effort has been devoted to developing the fabrication technology necessary to make useful products from these metal matrix composites. It is anticipated that the results of these efforts will be the first metallic composite to be produced and sold on a large commercial scale.

NUMERICAL STUDY OF THERMAL CONDUCTIVITY OF FIBER-MATRIX COMPOSITE MATERIALS

Wei Cha
Department of Mechanical Engineering
Michigan State University
East Lansing, MI 48824 USA

James V. Beck
Heat Transfer Group and Composite
Materials and Structure Center
Michigan State University
East Lansing, MI 48824 USA

ABSTRACT

The conductive heat transfer characteristics of a fiber-matrix composite material are studied in this paper. The effects on the effective thermal conductivity of a composite material due to 1) fiber thermal conductivity, 2) matrix thermal conductivity, 3) contact conductance at the fiber-matrix interface, and 4) fiber density are investigated. A method is suggested for using this information to estimate the thermal conductivities of fiber and matrix from the measured effective thermal conductivities of a group of composite materials made of the same fiber and matrix materials and measured fiber volume fractions. Some preliminary data are used to estimate the conductivities of the components.

A two-dimensional finite element heat conduction program, TOPAZ2D, is used in this study to solve the heat conduction model of a composite material. The comparison between the numerical results from this study and Hasselman and Johnson's analytical results for a dilute composite shows very good agreement for volume fractions up to 70%.

COMPOSITE MATERIALS have become important structural elements for a variety of advanced devices and vehicles. One reason for the importance of these new materials is that some of their properties can be tailored by choosing appropriate matrix and fiber materials as well as fiber volume fraction and fiber orientation. Some of the properties of interest are specific gravity, Young's modulus, electrical conductivity and thermal conductivity. The underlying principles for describing and prescribing the thermal conductivity of composite materials are not adequately understood and are investigated in this paper.

A purpose of this paper is to investigate the effects of the thermal conductivities of the fiber, k_f, and the matrix, k_m, on the effective thermal conductivity of a composite material, k_{eff}. The effect of thermal resistance at the fiber-matrix interface is included in the analysis. The results of the analysis can be used to a) calculate the effective thermal conductivity from basic components and b) aid in the design composite materials with desired effective thermal conductivities. It also can be employed to estimate the thermal conductivity of fiber and matrix from measured effective thermal conductivities of composite materials information.

Most papers containing studies of the effective thermal conductivity incorporate the assumption of perfect thermal contact between the constituents. Only a few papers take the interface contact resistance into consideration. Hasselman and Johnson developed an equation calculating the effective thermal conductivity of dilute concentrations with spherical, cylindrical, and flat plate geometries. They also considered the thermal contact conductance at the interface between the fiber and matrix. The results were considered to be valid when the fiber volume fraction ranges from 0 to 30%. Benveniste studied the effective thermal conductivities of non-dilute concentrations with the thermal contact resistance between the fiber and matrix. He presented two methods of modeling the thermal contact resistance at the interface and proved that both approaches have the same results. The provided methods do not cover a wide range of fiber volume fraction. Chiew and Glandt also conducted a numerical study of the effective thermal conductivity of composite materials in which the fillers are identical spherical particles. In their paper, the variation of effective thermal conductivity due to the interface resistance has been studied. In none of these papers is the subject discussed of estimating the individual components of thermal conductivity, k_f and k_m, from measurements k_{eff} and volume fraction, V. Another purpose of this study is to investigate the determination of k_f and k_m from measurement of k_{eff} and V.

The present study assumes fibers are distributed uniformly in epoxy with a triangular arrangement as shown in Figure 1. The advantages of studying this geometry include 1) experiments at

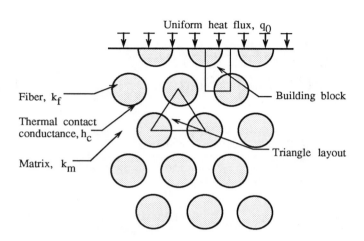

Figure 1 Problem geometry

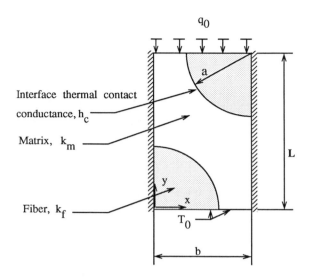

Figure 2 Basic building block

Michigan State University have been performed for this geometry and 2) the numerical solution is much easier and less uncertain than for other fiber arrangements, such as chopped fibers or different fiber orientations for each layer. With the uniform distribution assumption, the effective thermal conductivity of the composite material can be calculated by studying a basic building block as indicated in Figure 1.

It is very difficult, if not impossible, to measure directly the fiber thermal conductivity normal to its axis because of their small sizes; the fiber used in present study had a diameter of 7.2 μm. However, the derived relationship of k_f and k_m to k_{eff} has the potential to estimate k_f after k_{eff} is measured.

The following outlines each section of this paper. The mathematical formulation is presented in Section 2. Section 3 discusses numerical results. Comparisons between Hasselman and Johnson's results and the numerical results are presented in Section 4. A estimation procedure of k_f using numerical results and experimental results is presented in Section 5 and Section 6 contains the conclusions.

MATHEMATICAL FORMULATION

As mentioned in connection with Figure 1, a repeating geometrical pattern exists. Hence the problem can be studied by considering the basic building block shown in Figure 2. The effective thermal conductivity, k_{eff}, is calculated using this basic building block. The k_{eff} value for transient and steady state conditions can be approximated by considering only steady state.

This is a consequence of the fibers being very small in diameter, causing the temperature distribution to be locally in a near steady state condition. In the very small region near individual fibers, the heat flux leaving a small control volume such as in Figure 2 is nearly that entering but the temperature gradients are changing rapidly in position, caused by the large changes in thermal properties in the control volume. Hence a near steady state is approached locally. Also the particular boundary conditions, such as prescribed temperature or prescribed heat flux, are not critical because the same temperature pattern is repeated one or two fiber diameters away from the boundary.

Since the actual types of boundary conditions are not critical, a uniform heat flux, q_0, is applied to the top boundary of the building block, and the bottom boundary has a uniform temperature, T_0. These boundary conditions give very nearly the same effective conductivity as for many more elements than shown in Figure 2. Both sides of the body are insulated because there is symmetry at these boundaries. The mathematical formulation of the problem is

$$\frac{\partial^2 T_f}{\partial x^2} + \frac{\partial^2 T_f}{\partial y^2} = 0, \quad \text{in fiber} \qquad (1)$$

$$\frac{\partial^2 T_m}{\partial x^2} + \frac{\partial^2 T_m}{\partial y^2} = 0, \quad \text{in matrix} \qquad (2)$$

$$-k_f \left.\frac{\partial T_f}{\partial n_f}\right|_f = h_c (T_f - T_m), \quad \text{at interface} \qquad (3)$$

$$k_m \frac{\partial T_m}{\partial n_m}\bigg|_m = h_c(T_f - T_m), \quad \text{at interface} \quad (4)$$

$$-k_f \frac{\partial T_f}{\partial y}\bigg|_{y=L} = q_0, \quad b-a<x<b \quad (5)$$

$$-k_m \frac{\partial T_m}{\partial y}\bigg|_{y=L} = q_0, \quad 0<x<b-a \quad (6)$$

$$k_f \frac{\partial T_f}{\partial x}\bigg|_{x=b} = k_m \frac{\partial T_m}{\partial x}\bigg|_{x=b} = 0, \quad 0<y<L \quad (7)$$

$$k_f \frac{\partial T_f}{\partial x}\bigg|_{x=0} = k_m \frac{\partial T_m}{\partial x}\bigg|_{x=0} = 0, \quad 0<y<L \quad (8)$$

$$T(x,0) = T_0, \quad 0<x<b \quad (9)$$

The domain of interest is shown in Figure 2, with the dimension of b in the x direction and L in the y direction. The radius of a fiber is a. The symbols T_f and T_m are the temperatures in the fiber and matrix, respectively. Equations (3) and (4) contain two interface conditions, one for continuity of the heat flux and the other modeling the temperature drop at the interface. The thermal resistance at the interface is treated by using the thermal contact conductance, h_c, multiplied by the temperature difference across the interface. The symbols n_f and n_m in Equations (3) and (4) are the outward pointing normals, that is, $n_f=r$ and $n_m=-r$ for r being measured radially from the center of a fiber. The effective thermal conductivity, k_{eff} is defined as

$$k_{eff} = \frac{q_0 L}{\overline{T_L} - T_0} \quad (10)$$

where $\overline{T_L}$ is defined by:

$$\overline{T_L} = \frac{1}{b}\int_0^b T(x,L)dx \quad (11)$$

The effective thermal conductivity, k_{eff}, the fiber thermal conductivity, k_f, the thermal conductance, h_c, and the fiber dimension, a, are normalized as

$$K_{eff} = \frac{k_{eff}}{k_m} \quad K_f = \frac{k_f}{k_m}, \quad B = \frac{h_c a}{k_f} \quad (12), (13), (14)$$

$$V = \frac{A_f}{A_c} = 0.907 \left(\frac{a}{b}\right)^2 \quad (15)$$

A solution of the model given by equations (1)-(9) can be used to predict the effective thermal conductivity of composite materials having the triangular fiber arrangement. The number 0.907 is a geometrical constant obtained for the triangular fiber arrangement. Moreover, this mathematical model can also be used to estimate k_f and k_m, given the appropriate measurements of K_{eff} and V. As indicated in equations (1)-(9), K_{eff} depends on K_f, B, and V.

NUMERICAL RESULTS AND DISCUSSION

Program TOPAZ2D, a finite element heat conduction code, was employed to solve the problem described in Section 2. Numerical results are presented in this section. The three dimensionless parameters affecting K_{eff}, the dimensionless effective thermal conductivity, are K_f, B and V. By varying one parameter at a time, the effects of these parameters on k_{eff} are displayed. Figure 3 shows the dependence of K_{eff} on K_f and B when the volume fraction, V, is the constant value of 64%. Recall that K_{eff} and K_f are normalized k_{eff} and k_f. In Figure 3 the range of K_f is between 1 and 10, because k_f is higher than k_m in our experiments. The range of B varies from 0.1 to 10. Figure 3 indicates that for each volume fraction, V, the normalized effective conductivity, K_{eff}, increases monotonically with the normalized fiber conductivity, K_f. As B is increased, the effective conductivity also increases, although negligible changes occur for B values larger than 10.

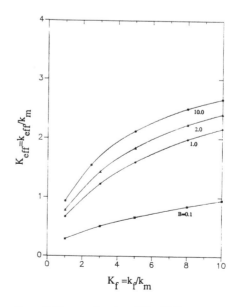

Figure 3 K_{eff} as a function of K_f with B as a parameter for the volume fraction, V, equal to 64.0%

From the curves in Figure 3, the effective thermal conductivity, K_{eff}, can be graphically found, given normalized fiber conductivity, K_f, and contact conductance, B. On the other hand, if the effective thermal conductivity, K_{eff}, is known, the normalized fiber conductivity, K_f, can also be estimated. This aspect will be addressed in detail in Section 5. Note that K_{eff} increases more rapidly at small K_f values than at large K_f values. The k_{eff} values asymptotically approach $(1+V)/(1-V)=1.56$ for $K_f=k_f/k_m$ going to infinity and for V=0.64.

The effect of volume fraction, V, on the normalized effective thermal conductivity, K_{eff}, is discussed next. Figure 4 shows K_{eff} as a function of K_f for different V values and for the fixed dimensionless contact conductance value of B=20. (This value of B gives results that are essentially the same for B values from 10 to ∞). The normalized effective thermal conductivity, K_{eff}, increases smoothly and monotonically with normalized fiber thermal conductivity, K_f, for each value of the volume fraction, V. Using the curves in Figure 4, the normalized thermal conductivity, K_{eff}, can be found graphically given normalized fiber thermal conductivity, K_f, and volume fraction, V. Note that as the volume fraction increases, the normalized effective conductivity becomes more sensitive to normalized fiber thermal conductivity for the range of $k_f=1$ to 10. In other words, as the density of fiber increases, the fiber thermal conductivity has more contribution to the effective thermal conductivity.

COMPARISON WITH HASSELMAN AND JOHNSON'S RESULTS

Hasselman and Johnson conducted an analytical study for the dependence of composite material thermal conductivity on fiber thermal conductivity, matrix thermal conductivity, interface thermal contact conductance, and fiber density. Their study presented the analytical expression

$$K_{eff}=[(K_f-B^{-1}-1)V+(K_f+B^{-1}+1)]/[-(K_f-B^{-1}-1)V+(K_f+B^{-1}+1)] \quad (16)$$

This equation was developed for the cases with volume fractions less than 30%, i.e., dilute composites. Equation (16) was not derived based on a particular geometrical arrangement and thus does not have any geometrical limitations for 0<V<30%. However, very good agreement between the results of equation (16) and the numerical results are found for even higher volume fractions for the geometry considered here. Notice that the current numerical study has a triangular arrangement of fiber as shown in Figure 1. Figures 5 and 6 give two examples of comparisons, one with V=64.0%, the other with V=89.0%. In Figures 5 and 6, the interface contact conductance, B, is treated as a parameter, and the effective thermal conductivity, K_{eff}, is given as a function of normalized fiber conductivity, K_f. Note that for V=64.0%, the numerical and analytical (equation(16)) results have less than 2% difference in K_{eff} values. The solid lines are the numerical values and the dashed lines are for equation (16).

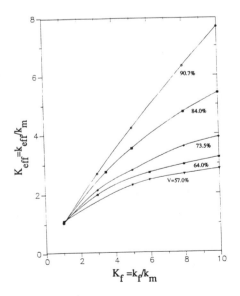

Figure 4 K_{eff} as a function of K_f with V as a parameter (B=20.0)

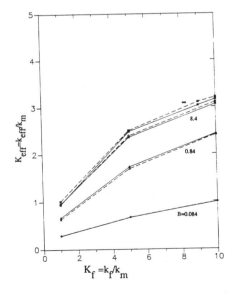

Figure 5 Comparison between numerical and analytical results (V=64.0%)
The solid lines are for the numerical solution values and
the dashed lines are for the Hasselman and Johnson analytical values

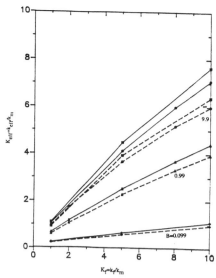

Figure 6 Comparison between numerical and analytical results (V=89.0%)

The maximum difference occurs at the curve of infinite contact conductance. For V=89.0%, Figure 6 shows that the analytical K_{eff} values are too low for all values of K_f with the differences being less for low contact conductances. This means that equation (16) can be applied to composites with volume fractions as high as 64% and the extrapolation of equation (16) should not be done for volume fractions higher than 70%, particularly for K_f considerably larger than one.

The differences between the analytical results of Hasselman and Johnson and the numerical results increase slightly as the contact conductance increases. In Figure 5, the maximum difference between the results from two methods varies from 0.01% for B=0.1 to 2% for B=infinity. The same trend is also observed in Figure 6. Since the agreement for low B values in Figure 6 is not as good as that in Figure 5, the agreement for high B values was not expected. This is due to the volume fraction effect discussed previously.

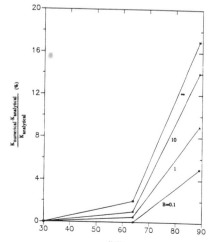

Figure 7 The effect of volume fraction on the difference between numerical and analytical results

In order to extrapolate Hasselman and Johnson's expression to higher volume fraction correctly, Figure 7 plots the difference between numerical and analytical results verses volume fraction. According to this plot, analytical expression can be used instead of the numerical approach, for volume fraction as high as 70%.

ESTIMATION OF PARAMETERS OF COMPONENTS

The above relations can be used with experimental values to estimate intrinsic parameters, such as fiber thermal conductivity, k_f, and thermal contact conductance, h_c. Measured values include the effective thermal conductivity, k_{eff}, fiber volume fraction, fiber radius and thermal conductivity of the neat matrix material, k_m. In the case of measuring k_m, one assumption is that its value is the same as for the composite material in its neat (that is, bulk) state; this assumption may not be completely justified, however. Two procedures are suggested, one assuming that k_m is known from measurements in its neat state and the other from measurements in the composite state.

The purpose of this section is to explore some possible experiments that can be used to estimate intrinsic parameter components (such as k_f and h_c) of thermal behavior of composite materials. It is shown that a certain function of the intrinsic parameters can be estimated if k_m, a, and V are known. Having measurements for various V values does not aid in determining that group. On the other hand, if measurements of k_{eff} are available for more than two V's, it is theoretically possible to estimate k_f and k_m provided B^{-1} is known or is small compared with unity.

In experiments it is possible to measure k_{eff} as a function of temperature, and more importantly for the present purpose, as a function of volume fraction, V, for given fiber radii. For values of V less than 0.70, equation (16) is a reasonable model to use for aligned fibers. For k_{eff}, k_m and V being measured, the dimensionless groups of K_{eff} ($=k_{eff}/k_m$) and V are known in equation (16). The group $B=h_c a/k_f$ involves the unknown k_f, which is the conductivity normal to the axis of the fiber. The B group can be written as

$$B=(h_c a/k_m)(k_m/k_f)=B_m/K_f \qquad (17a)$$
$$B_m=h_c a/k_m \qquad (17b)$$

Equation (16) can be written as

$$K_{eff}=(V+\beta)/(-V+\beta) \qquad (18a)$$

where

$$\beta=[K_f(1+B_m^{-1})+1]/[[K_f(1-B_m^{-1})-1] \qquad (18b)$$

223

The quantity β is a combination of K_f and B_m and can be estimated from one or more measurements of K_{eff} versus V, with the assumptions that K_f and B_m do not change with V. It also assumes that the matrix material and fiber radius remain constant. Suppose that several K_{eff} values have been made for different V's, then the least squares procedure can be used to estimate β. The method is to minimize

$$S = \sum_i^n (Y_i - K_{eff,i})^2 \quad (19a)$$

$$K_{eff,i} = (V_i + \beta)/(-V_i + \beta) \quad (19b)$$

with respect to β and where Y_i is a measured value of $K_{eff,i} = k_{eff,i}/k_m$. This is a non-linear problem and can be solved using a number of computer programs including NLINA (Beck and Arnold 1977).

For a single measurement the solution of equation (18a) for β is

$$\beta = V(1 + K_{eff})/(K_{eff} - 1) \quad (20)$$

One set of measurements at MSU at room temperature for a carbon based composite with 8 μm fibers and Epon 828 matrix material gave the values k_{eff}=0.8 W/m-C, k_m=0.24W/m-C and V=66.1%. The resulting values are $K_{eff} = k_{eff}/k_m = 0.8/0.24 = 3.33$ and β=1.23. If B_m is assumed to be large and thus disappears in equation (18b), K_f is given by

$$K_f = (1+\beta)/(\beta-1) = 9.8 \quad (21)$$

This is the minimum value of K_f given by equation (18b) for this case. The extreme case is for K_f going to infinity and then solving for B_m gives the similar result of

$$B_m = (1+\beta)/(\beta-1) = 9.8 \quad (22)$$

Finally, if K_f is assumed to be equal to B_m, one obtains

$$K_f = 2(1+\beta)/(\beta-1) = 19.6 \quad (23)$$

which is twice the above values. These values indicate that $k_f = K_f k_m = 0.24 K_f$ and is equal to or greater than 2.4W/m-C.

A method of estimating both K_f and B_m is to use a set of different diameters of fibers, with the same or different volume fractions V.

Some concepts for simultaneously estimating k_f and k_m are now suggested, assuming that B_m is very large. (If the thermal conductivity of the matrix, k_m, is unchanged in the composite material, it should be simply measured in its neat state). Then equation (16) can be written as

$$k_m = k_{eff,1}[(1-K_f)V_1 + (1+K_f)]/(K_f - 1)V_1 + (1+K_f) \quad (24)$$

for one experiment with a volume fraction of V_1. A similar equation can be written for a second volume fraction of V_2 and then equated to equation (24). The resulting equation can be equated to obtain the quadratic equation for K_f,

$$\lambda_1 K_f^2 + \lambda_2 K_f + \lambda_3 = 0 \quad (25)$$

where

$$\lambda_1 = (K_1 V_1 V_2 - K_2 V_1 V_2) - (K_2 V_2 + K_1 V_2 - K_1 V_1 - K_2 V_1) - (K_1 - K_2) \quad (26a)$$

$$\lambda_2 = 2[(K_2 - K_1) - (K_1 V_1 V_2 - K_2 V_1 V_2)] \quad (26b)$$

$$\lambda_3 = (K_1 V_1 V_2 - K_2 V_1 V_2) + (K_2 V_2 + K_1 V_2 - K_1 V_1 - K_2 V_1) - (K_1 - K_2) \quad (26c)$$

where

$K_1 = k_{eff,1}/k_m$ and $K_2 = k_{eff,2}/k_m$

The procedure is to solve equation (25) for K_f and then use this K_f value in equation (24) to obtain k_m. Finally k_f is found using $k_f = K_f k_m$. This procedure has been used but the present measurements are not sufficiently accurate to obtain satisfactory results. The procedure could also be improved by using more than two values of the volume fraction, this can be analyzed using the method of non-linear least squares (Beck and Arnold 1977).

Although the above estimation procedure pertains for a specific type of composite structure, the estimation concepts can be applied to other composite materials as well.

CONCLUSIONS

The effects of the thermal conductivity due to 1) fiber thermal conductivity, 2) matrix thermal conductivity, 3) contact conductance at the fiber-matrix interface, and 4) fiber density have been studied.

The numerical results have been compared with the Hasselman and Johnson's analytical results, and very good agreements for volume fraction less than 70% are observed.

The results then are employed to suggest methods for measuring fiber thermal conductivity and matrix thermal conductivity from effective thermal conductivity measurements for values of different volume fractions.

ACKNOWLEDGMENTS

We would like to thank Composite Materials and Structures Center, Michigan State University, who has sponsored our study. We also like to thank Dr. Elaine P. Scott who provided experimental data for this study.

NOMENCLATURE

a	fiber radius
A_f	cross sectional area of fiber in a basic building block of composite $[\pi a^2/2]$
A_c	cross sectional area of a basic building block of composite [bL]
b	width of a basic building block
B	normalized thermal contact conductance $[h_c a/k_f]$
h_c	thermal contact conductance
k_{eff}	effective thermal conductivity
K_{eff}	normalized effective thermal conductivity $[k_{eff}/k_m]$
k_f	fiber thermal conductivity
K_f	normalized fiber thermal conductivity $[k_f/k_m]$
k_m	matrix thermal conductivity
L	height of basic building block
q_0	uniform heat flux
T	temperature distribution
$\overline{T}$	average temperature at a horizontal plane
T_0	temperature distribution at the bottom of the building block
T_L	temperature distribution at the top of the building
V	blockfiber volume fraction $[A_f/A_c]$
λ_i	the coefficients of quadratic equation, i=1,2,3

REFERENCES

Hasselman, D. P. H. and Johnson, L. F., Journal of Composite Materials, Vol. 21, June 1987.

Benveniste, Y., Journal of Applied Physics, Vol. 61, No. 8, 15 April 1987.

Chiew, Y. C. and Glandt, E. D., Chemical Engineering Science, Vol. 43, No. 11, 2678-2685, 1987.

Beck, J. V. and Arnold, K. J., "Parameter Estimation in Engineering and Science", Wilen & Ames, NY, NY (1977).

Beck, J. V. and Osman, A. M., Proceeding of Sixth International Conference on Composite materials/Second European Conference on Composite Materials, Vol 4, Edited by Matthews, F. L., Buskell, N. C. R. and Hodgkinson, J. M., pp4.287-4.296.

Spencer, A., Composite Science and Technology, Vol. 27, 1986.

METALLIC GLASS REINFORCEMENT OF GLASS-CERAMICS

Rajendra Vaidya, K. N. Subramanian
Met., Mech., & Mat. Science
Michigan State University
East Lansing, MI 48824-1226 USA

ABSTRACT
The potential of using metallic-glass ribbons as reinforcements for brittle glass-ceramic matrices was investigated. Conventional pressing and sintering techniques were found to be suitable for manufacturing such composites. Small volume fractions of metallic-glass reinforcements were found to increase the mechanical properties of the brittle glass-ceramic matrices quite significantly. The strengthening and toughening mechanisms in such composites were also studied. The tests carried out so far indicate the potential of manufacturing such composites for structural applications using low cost techniques.

A large number of modern structural applications, especially in the field of aerospace and aeronautics, demand high temperature capabilities. Ceramics are potential materials for such applications due to their relatively high melting points. As compared to metals, they retain their mechanical properties to higher temperatures, have lower densities, and are more resistant to oxidation and corrosion.

The major drawback of ceramics which has been inhibiting their widespread use is their brittle nature and tendency to fail catastrophically. Various means such as crack deflection and branching [1-3], micro-crack toughening [4], and transformation toughening [5-7] have been found to improve the fracture toughness of ceramic matrices quite significantly. Of the various techniques investigated so far, fiber reinforcement has been proven to be the most effective means to increase the fracture toughness and to prevent catastrophic failure of such ceramic matrices [8-14].

Among the various ceramic matrices studied so far, glass-ceramics offers some unique advantages. They are fabricated in the glassy state possessing low softening temperature. By suitable heat treatment, they can be converted to an almost 100% crystalline structure. The resulting crystalline material does not soften till the regions near the melting point of the crystalline phase, and possess high temperature capabilities. In addition, they can be almost devoid of porosity, unlike conventional crystalline ceramics which usually contain some porosity as a result of the fabrication processes used. Appropriate selection of the glass system, can result in a glass-ceramic during the fabrication process itself, without requiring any additional heat treatment to crystallize the matrix. Glass-ceramic possess superior mechanical properties as compared to their parent glass and crystalline ceramics obtained by conventional processing techniques. The feasibility of using metallic-glass ribbons as reinforcements for ceramic matrices has not been investigated so far. Metallic-glass reinforcement of polymer matrices has been investigated by Hornbogen et al. [15-17]. They have demonstrated that even a very small volume fraction of metallic-glass reinforcements dramatically improved the strength and toughness of brittle polymer matrices. Unidirectionally reinforced metallic-glass/polymer matrix composites, were also shown to have transverse strengths almost 50% of their longitudinal values, and transverse Youngs moduli of almost 90% of their longitudinal values, as compared to conventional graphite-epoxy composites which have transverse strength and Youngs moduli only about 10% of their longitudinal values. Metallic glasses possess some unique advantages. They possess relatively higher fracture strengths and fracture strains, and are more resistant to oxidation and corrosion, as compared to their metallic counterparts. Their unique geometry also provides a large surface area to bond with the matrix. The ribbon shape of the metallic glasses make them not suitable for structural applications in their original form. Hence, metallic glass ribbons have to be incorporated into suitable matrices in order to exploit their structural potential. Although polymer matrices have been found to be quite successful, they have limited high temperature capabilities. The crystallization temperature of the metallic glasses also limit the potential to metallic matrices with low melting points. The main aim of the present study was to develop metallic-glass ribbon reinforced glass-ceramic matrix composites for structural applications.

Based on the crystallization temperatures of the metallic-glasses, Corning Glass Code 8463 was selected as the matrix material. Corning code 8463 is a $PbO-SiO_2-B_2O_3-ZnO$ based glass containing about 84% PbO. The major crystalline phase is $2PbO \cdot ZnO$. The coefficient of thermal expansion of this material is comparable with that of the metallic-glass ribbons used in this study.

The specimens were prepared by compacting the powders in a steel die under a pressure of 3000psi. Amyl acetate was used as the binder. After initially heating to $200^\circ C$ to drive off the organic binder, the specimens were sintered at $400^\circ C$ for one hour. After sintering, the specimens were devitrified at $450^\circ C$ for 20 minutes, and furnace cooled down to room temperature.

Rectangular bar-shaped specimens were used for determining the mechanical properties of the composite system. The elastic properties of the system were measured by using Sonic Resonance Technique [18]. Strength measurements were made using a three-point bend test (in accordance with ASTM specification C-203/85), and the fracture toughness was measured using single-edge notched-beam specimens in three-point bending.

EXPERIMENTAL PROCEDURES

The metallic-glasses used in the present study were obtained from Metglas Products, a business unit of Allied Signal Inc. Two metallic glasses were used as reinforcements in the present study: one was an iron-based metallic glass METGLAS 2605S-2 alloy, and the other a nickel-based metallic glass METGLAS MBF-75 alloy. Both have recrystallization temperatures in the range of $550^\circ C$.

Initial studies on the composite fabrication were carried out using borosilicate slide glass and Corning Glass Code 0080. Various processing techniques including hot-pressing were attempted. However, these techniques were found to be unsuitable for the current system, which incorporated metallic-glass having recrystallization temperature around $550^\circ C$. Both, the borosilicate slide glass and Corning Glass Code 0080, had softening temperatures well above $600^\circ C$. Sol-gel technique was also found to be unsuitable, due to the high softening temperature of the nearly pure silica involved in this technique. It was also difficult to incorporate metallic-glass ribbon reinforcements into the gel.

RESULTS AND DISCUSSION

This study has demonstrated the feasibility of fabricating glass-ceramic matrix composites reinforced with multiple metallic-glass ribbons. Typical micrograph of one such specimen is shown in Figure 1.

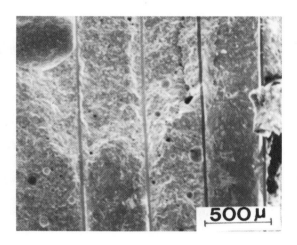

FIGURE 1 Multiple METGLAS 2605S-2 ribbon reinforced Corning Glass Code 7572 matrix composite produced by pressing and sintering. The matrix is almost 100% crystalline as a result of crystallization treatment at $450^\circ C$.

In such composites the matrix and the ribbons bond extremely well as illustrated in Figure 2.

METGLAS is a registered trademark of Allied-Signal Inc. for amorphous metallic alloys and brazing alloys.

FIGURE 2 Strong (void-free) bonding between METGLAS MBF-75 and glass-ceramic obtained from Corning Glass Code 8463.

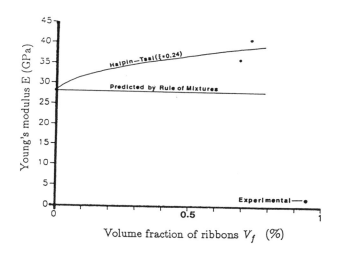

FIGURE 3 Plot of Young's modulus of METGLAS MBF-75 reinforced glass-ceramic produced from Corning Glass Code 8463 as a function of volume fraction of metallic glass reinforcement.

The nature of such strong bonding is evidenced by the presence of the matrix material adhering to the ribbon surface after specimen failure.

The results of the experimentally obtained Youngs modulus and fracture strength of Corning Glass Code 8463 (glass-ceramic) reinforced with METGLAS MBF-75 are presented in Figures 3 and 4. From the data it is evident that even small volume fractions of metallic-glass reinforcements significantly improve the mechanical properties of the glass-ceramic matrix. Recent fracture toughness measurements made on another $PbO-ZnO-B_2O_3-SiO_2$ system (Corning Glass Code 7572 matrix) have indicated that even small volume fractions (1.24%) of metallic-glass reinforcements improve the fracture toughness (K_{IC}) of the matrix (0.4 Mpa m) significantly (by a factor of about 3.5).

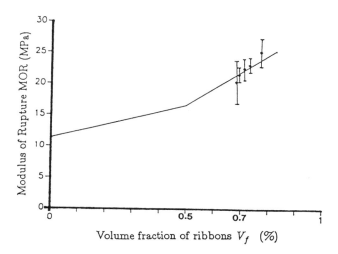

FIGURE 4 Plot of fracture strength of METGLAS MBF-75 reinforced glass-ceramic produced from Corning Glass Code 8463 as a function of volume fraction of metallic-glass reinforcement.

High strength and large elongation to fracture of metallic glass ribbons contribute to the over-all composite behavior. In addition, crack arrest and deflection near ribbon-matrix interface plays significant role. Such features can be observed in the micrograph presented in Figure 5.

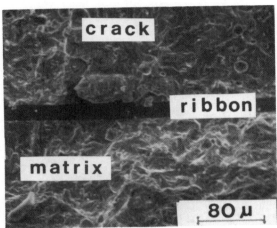

FIGURE 5 Crack arrest and deflection in metallic-glass ribbon/matrix interface in METGLAS 2605S-2 reinforced glass-ceramic produced from Corning Glass Code 7572.

In the composites studied, the extent of crack deflection along the interface is limited by the strong interfacial bond that exists between the metallic-glass ribbon and the glass-ceramic matrix.

The improvement in the fracture toughness arises because of the contribution of the ribbons to energy absorption during crack propagation. Fracture energy contributions due to ribbons by themselves consists of four components, namely energy aborbed due to (i) ribbon debonding, (ii) ribbon pull-out, (iii) bending of the ribbons, and (iv) ribbon fracture. For the systems under investigation, the main contributions to the fracture energy are due to the bending and fracture components. The ribbon debonding and pull-out components are relatively small. This fact was also confirmed by the absence of ribbon pull-out in pull-out tests.

The ribbon orientation also has a significant effect on the fracture toughness of these composites. Composites containing ribbons oriented with their broad surfaces parallel and perpendicular to the crack front were prepared. The fracture toughness values for the specimens containing ribbons oriented perpendicular to the crack front were found to be four to five times that for specimens containing ribbons in the parallel orientation. This improvement can be explained on the basis of difference in the energy absorbed by the ribbons during bending. It is also believed that the ribbon orientation changes the crack velocity in the matrix, which in turn affects the energy absorption by the matrix.

Although continuously reinforced composites containing multiple ribbons in a layered configuration (as shown in Figure 1) can be produced, fabrication of discontinuously reinforced composites with high volume fraction of ribbons will cause less difficulties. Most discontinuous fiber reinforced composites are characterized by a critical fiber length, which is the minimum length of the fiber required for efficient load transfer between the matrix and fiber. For the systems under consideration, it was found that the ribbon length and ribbon width were instrumental in affecting the load transfer between the matrix and the ribbons.

The capabilities of these composites for high temperature structural applications are also being investigated. Initial results have shown that the high temperature properties of such composites are limited mainly by the crystallization temperature of the metallic-glass reinforcements. Mechanisms such as debonding and frictional sliding of the ribbons enhance the strain to fracture and prevent catastrophic failure at elevated temperatures.

CONCLUSIONS

The studies carried out so far have shown metallic-glass ceramic matrix composites can be manufactured by relatively low cost techniques such as cold pressing and sintering. Even very low volume fractions of such reinforcements lead to significant improvements in mechanical properties. Although most of the studies carried out so far have utilized very low volume fractions of continuous ribbons, the possibility of developing discontinuous ribbon reinforced composites with relatively higher volume fractions of reinforcements exists. Such composites should exhibit excellent mechanical properties suitable for structural applications.

ACKNOWLEDGEMENTS

The authors would like to acknowledge The Composite Materials and Structures Center at Michigan State University for supporting and funding this project.

REFERENCES

1. K. Faber and A. Evans, Acta Metall., 31 (1983) 565.
2. K. Faber and A. Evans, Acta Metall., 31 (1983) 577.
3. K. Faber and A. Evans, J. Am. Ceram. Soc., 66 (1983) 94.
4. K. Faber and A. Evans, J. Am. Ceram. Soc., 67 (1983) 255.
5. A. Evans and A. Heuer, J. Am. Ceram. Soc., 63 (1980) 241.
6. A. Evans, Advances in Ceramics, Vol. 12, Ed. N. Claussen M. Ruhle and A. Heuer, Am. Ceram. Soc., Ohio, 1984, p. 193.
7. R. McMeeking and A. Evans, J. Am. Ceram. Soc., 65 (1982) 242.
8. K. Gadkaree and K. Chyung, Am. Ceram. Soc. Bull., 65 (1986) 370.
9. P. Shalek, J. Petrovic and F. Gac, Am. Ceram. Soc. Bull. 65 (1986) 351.
10. S. Rishbud and M. Herron, Am. Ceram. Soc. Bull., 65 (1986) 342
11. J. Homeny, W. Vaughn and M. Ferber, Am. Ceram. Soc. Bull., 67 (1987) 333.
12. T. Tiegs and P. Becher, Am. Ceram. Soc. Bull., 66 (1987) 339.
13. S. Bulijian, G. Baldoni and M. Huckabee, Am. Ceram. Soc. Bull., 66 (1987) 347.
14. A. Caputo, D. Stinton and T. Besmann, Am. Ceram. Soc. Bull., 66 (1987) 368.
15. A. Fels, K. Friedrich and E. Hornbogen, J. Mat. Sci. Letters, 3 (1984) 569.
16. A. Fels, E. Hornbogen and K. Friedrich, J. Mat. Sci. Letters, 3 (1984) 569.
17. T. Tio, K. Friedrich, E. Hornbogen, U. Koster and A. Fels, J. Mat. Sci. Letters 3 (1984) 415.
18. S. Screiber, D. Anderson and N. Soga, Elastic Constants and their Measurements, McGraw Hill, New York, 1974, P. 82.

INTERNAL FRICTION CHARACTERISTICS OF A KEVLAR/EPOXY SYSTEM

Kirk S. Burson
Aerojet Solid Propulsion
Sacramento, CA USA

William N. Weins
University of Nebraska-Lincoln
Lincoln, NE USA

ABSTRACT

The study discusses the changes in composite internal friction due to increases in Kevlar content and proposes mechanisms for these changes. Damping characteristic of Kevlar/epoxy and neat anhydride epoxy systems were surveyed over a temperature range of 90° K to 363° K at frequencies ranging from 0.93 Hz to 2.70 Hz. The investigation used an anhydride cured epoxy matrix based on the LRF-092 formulation for each specimen in the study. The Kevlar/epoxy samples varied from 0.641 volume percent (v/o) to 9.61 v/o Kevlar. Control specimens of neat epoxy were molded in conjunction with each Kevlar specimen. The study revealed a large internal friction peak to occur at approximately 185° K (-88° C) at a frequency of 1.70 Hz. For each specimen, the profile of the internal friction peak was large and unsymmetric, never returning to background levels. This phenomenon suggests the possibility of multiple relaxations occurring in the material. The activation energy on neat epoxy specimens measured 12.7 cal /Mol °K while the 1.60 v/o and 9.61 v/o Kevlar specimens were found to be 6.43 cal/Mol °K and 5.05 kcal/Mol °K respectively. Inversely, the relaxation time contents (τ) increased with increasing Kevlar content. Time constants varied from 2.12×10^{-16} seconds to 2.47×10^{-7} seconds for the neat epoxy and 9.61 v/o samples respectively. The effects of increasing Kevlar content on epoxy in addition to possible mechanisms associated with the process are discussed further.

THE DEMAND FOR LIGHT WEIGHT, high strength materials has lead to the development of composite materials. Engineering plastics, reinforced with continuous fibers made of glass, graphite, or aramid materials make up the vast majority of composites used today. The advantage of composite materials lies in the extreme versatility at which the fibers can be positioned. Optimum laminate orientation produces a remarkably light weight, high strength part which has equivalent or greater stiffness than conventional engineering materials.

The strength and stiffness of composite materials is dependent upon the direction of the reinforcing fiber. Because of this, the laminates are extremely anisotropic. Due to the high degree of anisotropy, the engineering plastic (matrix) plays a very important role in distributing loads between the reinforcement fibers. The composite material under investigation in this study consist of a man-made aramid fiber known under the DuPont trade mark as "Kevlar" with an anhydride cured epoxy matrix used as the engineering plastic.

Internal friction is an energy loss or dissipation in a stressed material system not due to external processes. Internal friction arises from a material's deviation from Hooke's Law, such as anelastic behavior. This deviation may be due to grain boundaries or solute atoms in metals or fiber/matrix mechanical slippage or molecular motion in composite materials. Measurements are usually taken by driving a sample at various frequencies and measuring the phase lag angle (ϕ) between the stress and strain. Another method is to allow a specimen to free decay at various temperatures and measure the decay in amplitude or log decrement of the vibrational decay. By taking these measurements at various frequencies or temperatures, it is possible to determine parameters such as the activation energy and relaxation time constant associated with the process, as well as information about the symmetry and strength of the defect.

This work was initiated to investigate how internal friction measurements could be used to study the Kevlar/epoxy structure. In particular, the processes occurring at the Kevlar/epoxy interface were of interest in addition to the neat epoxy structure itself.

THEORY

The dynamic response of an anelastic solid subjected to a cyclic stress is shown in Eq. (1). Because the solid is not perfectly elastic, the strain response will be out of phase with the applied stress by an amount ϕ, shown in Eq. (2).

$$\sigma = \sigma_0 \, e^{[i\omega t]} \qquad (1)$$

where σ_0 = applied stress
ω = circular frequency ($2\pi f$)
f = vibration frequency.

$$\varepsilon = \varepsilon_0 \, e^{[i(\omega t - \phi)]} \qquad (2)$$

where ε_0 = strain amplitude
ϕ = angle the strain lags behind the stress

The phase lag, ϕ, is known as the internal friction and defined by the Debye expression shown in Eq. (3).

$$\tan \phi = \Delta \, \omega\tau/(1 + \omega^2\tau^2) = \phi \text{ if } \phi \ll 1 \qquad (3)$$

where τ = relaxation time
Δ = relaxation strength of the defect, typically related to the moduli

The relaxation time τ is given by the following expression:

$$\tau = \tau_0 \, e^{[Q_a/RT]} \qquad (4)$$

where τ_0 = relaxation time constant
Q_a = activation energy

The internal friction characteristics of a material can be measured by several procedures. The Debye expression, Eq. (3), suggests that the internal friction can be measure at a constant temperature over a wide range of frequencies. Conversely, since the relaxation time (τ) is a function of temperature, as shown in Eq. (4), the phase lag can also be measured at a constant frequency over a wide temperature range. For the purposes of this investigation, the specimens were allowed to free decay in torsion at a constant frequency. The logarithmic decrement, or strain decay (δ), was measured over a broad temperature range spanning from liquid nitrogen temperature to just below the glass transition temperature. The internal friction (ϕ or Q^{-1}) is related to the logarithmic decrement δ by Eq. (5), displayed below.

$$\phi = Q^{-1} = \delta/\pi = \ln (A_0/A_n)/n\pi \qquad (5)$$

where A_0 = the initial amplitude,
A_n = the amplitude at the nth cycle
n = the number of cycles.

By taking measurements at constant frequency with varying temperatures, the maximum temperature at which the peak occurs, T_p, can be identified. By changing the frequency of oscillation several time while continuously measuring the materials strain decay over a broad temperature range, the internal friction characteristics of the material are determined. Simplifying Eq. (4) by multiplying both sides by ω and using the condition that $T=T_p$ and therefore $\omega\tau = 1$, the expression becomes

$$\ln \omega + \ln \tau_0 + (Q_a/R)(1/T_p) = 0 \qquad (6)$$

From Eq. (6), it is possible to calculate the activation energy (Q_a) and relaxation time constant (τ_0) from the plot of $\ln \omega$ vs. $1/T_p$.

The circular frequency of vibration (ω) in free decay is proportional to the mass (m) of the system and the stiffness (K_1) of the sample for small strains applied in the elastic range according to the following equation.

$$\omega_0^2 = K_1/m \qquad (7)$$

Therefore, with a constant mass system, changes in frequency, especially with temperature, are directly proportional to changes in the modulus.

The theoretical considerations associated with anelastic behavior up to this point have strictly been concerned with a single relaxation response. Materials, however don't always respond with a single relaxation peak, but rather, can have multiple relaxations occurring over a broad frequency range. Multiple relaxation responses are usually identified in two ways.

First, several completely different and separate relaxation effects may occur, each characterized by a more or less well-defined relaxation time τ. This presents the internal friction curve as a series of damping maxima spread out over the frequency range. Hence, the solid is said to represent an internal friction "spectrum" measured as a function of frequency or as a function of temperature [1]. Secondly, a single relaxation peak may vary appreciably from the theoretical expressions derived for the relaxation time. This deviation can be associated with fluctuations in the relaxation time at different positions within the crystal (or molecular makeup). These variations can be caused by differences in the state of internal stresses or by differences in the composition of the solid (not evenly distributed or impurities in the material). This results in the superposition of a very large number of small relaxation peaks. The overall measured I.F. peak corresponds to a relaxation time distributed about some average value. Several excellent references are available [1,2] for a more complete explanation of the theory of anelasticity and internal friction.

EXPERIMENTAL

The internal friction experiments were performed on samples consisting of Kevlar 29 yarn in an epoxy matrix. The Kevlar content of each specimen varied by the number of strands used in each sample, as seen in Table 1.

TABLE 1: SAMPLE PROPERTIES

Number of Strands	% Fiber Volume
2	0.641
3	0.961
4	1.28
5	1.60
6	1.92
15	4.81
30	9.61

The resin matrix used is this investigation consisted of a three part anhydride cure epoxy derived from a variation of the LRF-092 formulation. The epoxy was formulated from Shell's Epon 825, Nadic Methyl Anhydride (NMA), and a Benzyldimethylamine catalyst (BDMA Sherwin-Williams) in the following ratios by weight:

Epon 825..........100 parts
NMA................80 parts
BDMA...............2 parts

The dimensions of the specimens used in this investigation were 0.125 inches square in cross-section and 2.25 inches in length. The aramid fibers were positioned along the axis of the sample, completely suspended in the epoxy matrix. Each of the specimens were cast in cured RTV Silicone Rubber molds, 8 inches in length. Due to the superior surface finish of the as-received samples, no surface polishing was necessary. In order to maintain a control for each of the specimens tested, a neat epoxy sample was cast in conjunction with every Kevlar/epoxy sample. The procedure helped to eliminate potential problems associated with resin formulation and cure cycle. The specimens were cured for 4 hours at 212° F with a ramp rate of approximately 5° F per minute.

The cured specimens were finally mounted in an inverted pendulum type, low frequency internal friction machine similar to the unit described by Schwartz [3]. The machine consisted of a fixed weight pendulum system and was capable of operating at frequencies ranging from 0.1 to 5 Hz. The pendulum was operated in free decay from an initial amplitude of approximately 1×10^{-4} radians. The strain decay of the specimen was recorded on a computerized data acquisition system in addition to an analog strip chart used as a redundant backup. Measurements were taken over a temperature range of approximately 90° K to 363° K in a partial pressure atmosphere of helium. The dynamic modulus and internal friction were measured at approximately 0.94, 1.70, and 2.65 Hz. The logarithmic decrement was calculated according to Eq. (5) while the activation energy and relaxation time constant were determined from Eq. (6)

RESULTS

The full investigation into the internal friction characteristics of Kevlar/epoxy were performed on specimens ranging from 0.641v/o to 9.61 v/o Kevlar content. Due to space restrictions, a detailed explanation for each test can not be provided. Table 2, however, summarizes the I.F. peak temperature, peak magnitude, and frequency of ocsillation of each specimen.

TABLE 2. INTERNAL FRICTION PEAK MEASUREMENTS OF KEVLAR/EPOXY SPECIMENS

Kevlar Content	Peak Temp (°K)	Peak Mag. (x10^{-4})	Frequency (cyc/sec)
0.641	158	226	1.75
epoxy	160	215	1.74
0.961	188	238	1.60
epoxy	193	204	1.68
1.28	155	232	1.68
epoxy	158	211	1.73
1.60	184	289	1.70
epoxy	189	277	1.58
1.92	186	323	1.66
epoxy	192	296	1.67
4.81	163	242	1.69
epoxy	189	277	1.58
9.61	194	272	1.48
epoxy	192	296	1.67

I. F. RESPONSE OF NEAT EPOXY - The internal friction response for the neat epoxy specimen as measured over a temperature range from 92° K to 364°K is shown in Figure 1. The high temperature end (left side) of the curve indicates that the logarithmic decrement exponentially decreases in magnitude until just below room temperature. Because of the smooth exponential decline, i.e. no abnormal deviations, this region is known as high temperature background, and is not truly an anelastic region. The rise in the logarithmic decrement at 286° K is the beginning of the internal friction peak for the neat epoxy sample. In this investigation, the peak region was isolated from the high temperature background making the visualization of the peak easier to detect. The frequency of oscillation for the I.F. scan, displayed in Figure 1, varied from 1.17 Hz at 364° K to 1.90 Hz at 92° K. Examination of the I.F. peak revealed an unsymmetric shape, skewed on the cold temperature side, decreasing at a slower rate and never returning to background levels. This phenomenon implies that multiple relaxations are occurring within the epoxy in the cold temperature region. Similar procedures were repeated for two additional frequency settings. Figure 2 displays the internal friction peaks for three neat epoxy scans performed at 0.93 Hz, 1.67 Hz, and 2.66 Hz respectively. The I.F. peak for the low frequency response was found to occur at 186° K and 0.95 Hz. The medium frequency response, displayed in full in Figure 1, peaked at 192° K and 1.67 Hz, while the high frequency scan produced a peak at 192° K and 2.58 Hz. The activation energy and relaxation time constant were determined from the shift in peak temperature with frequency to be 12.68 +/- 5.16 cal/Mol °K and 2.12 x 10^{-16} seconds respectively.

I.F. RESPONSE OF KEVLAR/EPOXY - The internal friction characteristics of epoxy with the addition of 1.60 v/o and 9.61 v/o Kevlar are displayed in Figures 3 through 6. The initial scan performed on the 1.60 v/o (5 strand) Kevlar sample was conducted at temperatures ranging of 90° K to 364° K at a frequency of 1.52 Hz at room temperature, shown in Figure 3. Figure 4 displays the temperature (T_p) and frequency of oscillation of the internal friction peaks for the 1.60 v/o Kevlar sample performed at three frequencies. The I.F. peak for the low frequency scan occurred at 181°K with a frequency of oscillation of 0.94 Hz. The medium frequency peak occurred at 185 °K and 1.66 Hz while the high frequency scan peaked at 189°K with a frequency of 2.70 Hz. The linear least squares fit of the natural logarithm of the frequency vs. 1000/Tp revealed an activation energy equal to 6.43 +/- 2.05 cal/ Mol °K with a relaxation time constant of 2.83 x 10^{-9} seconds.

The initial scan of the 9.61 v.o (30 strand) Kevlar sample was performed at 1.35 Hz over a temperature range of 95°K to 340°K, shown in Figure 5. The I.F. peak for the low frequency response occurred at 191°K with a frequency of oscillation of 1.00 Hz. The medium and high frequency scans produced peaks at 192°K at 1.13 Hz and 195°K and 1.44 Hz respectively. The activation energy for the 9.61 v/o Kevlar specimen was calculated at 5.05 +/- 2.58 cal /Mol °K. The relaxation time constant was determined to be 2.47 x 10^{-7} seconds.

MULTIPLE RELAXATIONS - The skewed, un-symmetric shape of the internal friction peaks for the neat epoxy and 9.61 v/o Kevlar specimens lead to the investigation of multiple relaxations. Figure 7 shows the first four relaxations associated with the neat epoxy sample while Figure 8 displays the relaxations found in the 9.61 v/o Kevlar article. Table 3 summarizes the peak temperature and magnitudes of the four relaxations for both of the above mentioned samples. As demonstrated in Figures 7 and 8 as well as Table 3, the Kevlar I.F. peaks occur at slightly higher temperatures with lower magnitudes than the neat epoxy specimen. The relative difference of the multiple relaxation peaks between the two samples is almost insignificant. The curves do show, however, that the epoxy material contains a spectrum of three closely spaced relaxations as opposed to multiple

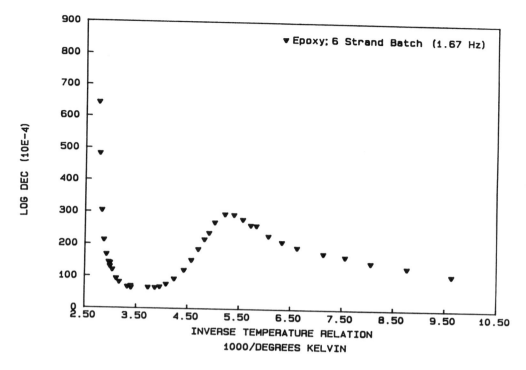

Figure 1 - Internal Friction of Pure Epoxy Versus 1000/T at Medium Frequency

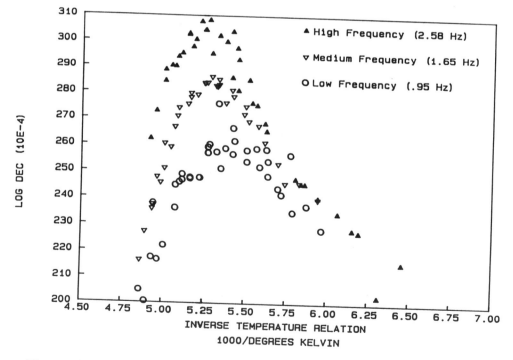

Figure 2 - Superposition of Net Internal Friction Peaks for Low, Medium, and High Frequency Scans for Neat Epoxy

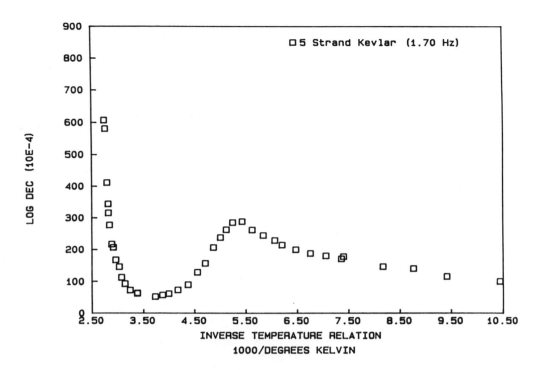

Figure 3 - Internal Friction of 1.60 v/o (5 Strand) Kevlar Versus 1000/T at medium Frequency

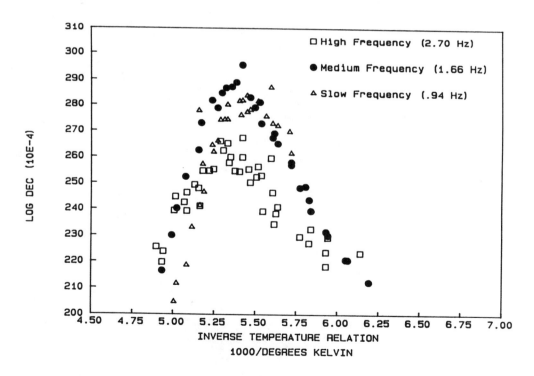

Figure 4 - Superposition of Net Internal Friction Peaks for Low, Medium and High Frequency Scans for 1.60 v/o Kevlar

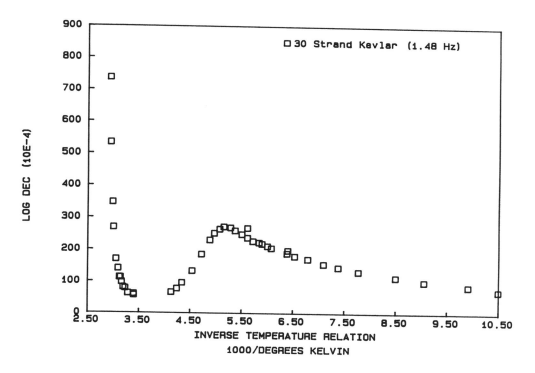

Figure 5 - Internal Friction of 9.61 v/o (30 strand) Kevlar Versus 1000/T at Medium Frequency

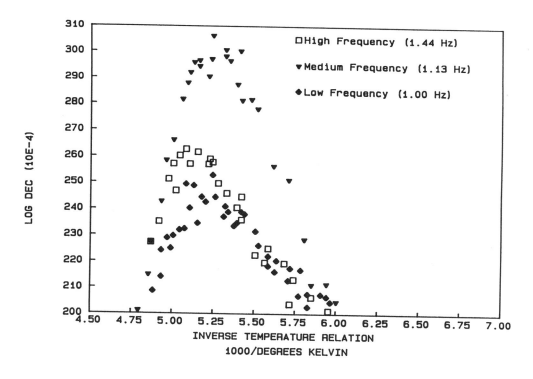

Figure 6 - Superposition of Net Internal Friction Peaks for Low, Medium, and High Frequency Scans for 9.61 v/o Kevlar

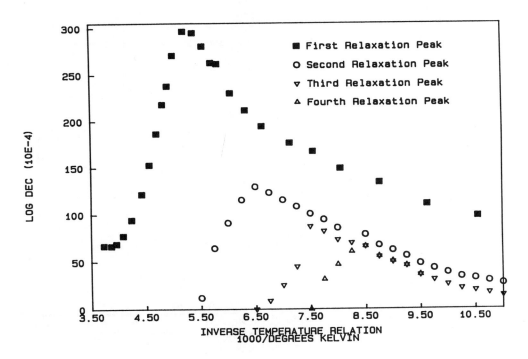

Figure 7 - Superposition of Multiple Relaxations for Neat Epoxy

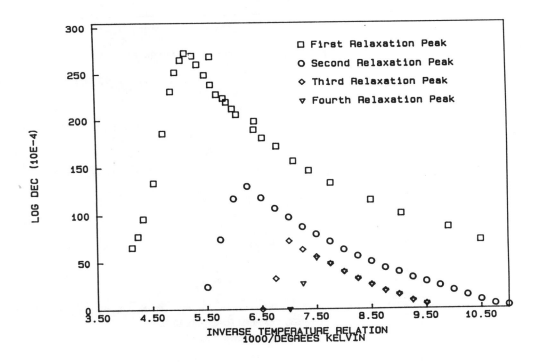

Figure 8 - Superposition of Multiple Relaxations for 9.61 v/o Kevlar

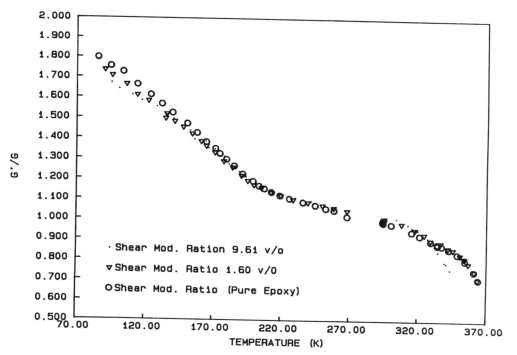

Figure 9 - Effect of Temperature on Shear Modulus for Neat Epoxy, 1.60 v/o, and 9.61 v/o Kevlar Specimens, where G is the Relative Modulus at 298° K

isolated peaks. The second relaxation was calculated by superimposing the high temperature rise of the I.F. peak onto the cold temperature side. The difference between the as-measured curve and the superimposed background noise produced the second relaxation curve. The third and fourth relaxations were calculated in a similar manner using the second and third relaxation curves respectively.

TABLE 3. MULTIPLE RELAXATIONS

Material	Relaxation Curve	Peak Temp (°K)	Peak Magnitude ($\times 10^{-4}$)
Kevlar (9.61 v/o)	1	194	227
	2	167	126
	3	143	72
	4	133	55
Neat Epoxy	1	192	296
	2	154	126
	3	133	88
	4	118	67

DISCUSSION

Examination of the internal friction characteristics for the neat epoxy, 1.60 v/o and 9.61 v/o Kevlar samples reveals that the activation energy decreases while the relaxation time constant increases with increasing Kevlar content. Originally, the decreasing activation energy with increasing Kevlar content was thought to be a measurement of the internal friction characteristics of the Kevlar fiber. Figure 9, however, reveals a consistency in the ratio of the shear modulus at varying temperatures with the shear modulus at room temperature (G'/G). The close similarities in shear modulus of the three samples suggests the Kevlar has an insignificant effect on the shear modulus of the sample, and therefore, the internal friction characteristics measured are those of the epoxy and not the Kevlar.

The magnitudes of the I.F. peaks for the neat epoxy specimens taken from the 0.641, 0.961, and 1.28 v/o Kevlar batches range from 204×10^{-4} to 215×10^{-4} with an average of 210×10^{-4}. The peak magnitudes for the neat epoxy samples formulated in the 1.60 - 4.81 v/o and 1.92 - 9.61 v/o batches were measured at 277 $\times 10^{-4}$ and 296 $\times 10^{-4}$ respectively. The relative difference in peak magnitudes is thought to be associated to the amount of constituents added during formulation. Table 4 displays the actual ratios of constituents added to each batch. The deviation in internal friction peak magnitude between the first three batches and the last two is thought to be caused from the additional NMA added to the epoxy.

The epoxy matrix is comprised of a network of Epon 825 - NMA chains. Complete polymerization would produce a continuous crosslinking chain of Epon - NMA molecules. Within this network, the NMA molecule exhibits higher elastic properties than the epoxy molecule [4,5,6,7]. The phenyl link within the NMA molecule is capable of rotating, producing movement within the network. It is the rotation of the NMA phenyl which is believed to cause the major internal friction peak observed within the pure epoxy sample. The introduction of Kevlar to the epoxy is thought to interrupt or quench the polymerization of the epoxy at or near the Kevlar interface. If the matrix is not fully polymerized, there exists a greater capability for movement to occur within the given molecules. The ease of molecular reorientation would therfore be seen in the form of a lower activation energy. Mechanical slip at the Kevlar/epoxy interface was considered as a possible contributor to the lower activation energy but this theory was ruled out. If this assumption was correct, a plot of fiber surface area vs. activation energy should produce a linear relation based upon previous studies concerning internal friction characteristics in two phase materials [8].

CONCLUSION

The evaluation of the internal friction curves for the 0.641 v/o through 1.92 v/o Kevlar and corresponding neat epoxy samples reveals the Kevlar to exhibit a logarithmic decrement magnitude of 21 $\times 10^{-4}$ higher than the pure epoxy specimens on the average. The I.F. peak temperature of the Kevlar samples was 4 °K lower on average than that of the epoxy. The log decrement magnitude for the 4.81 v/o and 9.61 v/o (15 and 30 strand) Kevlar specimens were lower than their corresponding neat epoxy counterparts by 35 $\times 10^{-4}$ and 24 $\times 10^{-4}$ respectively. This trend change is thought to be attributed to the excess amount of NMA in the 4.81 v/o and 9.61 v/o batches.

The internal friction peaks for each of the tested specimens were unsymmetric and skewed toward the low side of the temperature spectrum. Using a subtraction technique assuming a theoretical peak profile, the 9.61 v/o Kevlar and corresponding pure epoxy sample revealed three relaxation peaks. The exact nature of the three peaks is still unknown at this time, however. Since similar relaxations were observed for both samples, the effect is thought to be primarily associated with the epoxy rather than the Kevlar.

The consistency in shear modulus at varying temperatures per shear modulus at room temp. (G'/G) of the pure epoxy, 1.60 v/o, and 9.61 v/o Kevlar samples suggests the Kevlar has very little effect in the measurements. Therefore, it is considered that the major I.F. peak which occurs at approximately 192°K is primarily influenced by the epoxy. The activation energies of the samples were found to decrease with increasing Kevlar content. The activation energy of the neat epoxy, 1.60 v/o, and 9.61 v/o Kevlar samples were measured at 12.7, 6.43, and 5.05 cal/Mol °K respectively. The relaxation time constant for the same specimens were found to increase with increasing Kevlar content. The time constants ranged form 2.12 $\times 10^{-16}$ seconds for the neat epoxy to 2.47 $\times 10^{-7}$ seconds for the 9.61 v/o Kevlar. There was no linear relationship associated with activation energy or relaxation time constant and volume percent Kevlar. The change in activation energy and time constant with the addition of Kevlar fibers is thought to be attributed to a quench in the polymerization of the epoxy. The crosslink network is thought to be interrupted at or near the Kevlar interface. The primary internal friction peak, however is considered to be caused by the movement of the phenyl link within the NMA molecule.

REFERENCES

1. R. DBaptiste, <u>Internal Friction of Structural Defects in Crystalline Solids</u>, (North-Holland Publ.: Amsterdam, 1972), p.73

2. A.S. Nowick and B.S. Berry, <u>Anelastic Relaxation in Crystalline Solids</u>, (Academic Press: New York, 1972)

3. J.C. Schwartz, <u>Review of Scientific Instruments</u>, Vol 32, (1961) p. 335

4. D. Timm, A. Ayrinde, F. Huber, C. Lee, "Molecular Characterization of Network Structure by Gel Permeation Chromatography and Dynamic Mechanical Spectroscopy", <u>Proc. International Rubber Conference</u>, Moscow, U.S.S.R., Vol A2, (1984) p. 66

5. A. Ayrinde, C. Lee, D. Timm, W. Humphrey, "Determination of Thermoset Resin Crosslink Architecture by Gel Permeation Chromatography", <u>American Chemical Society, ACS Symposium Series No. 245,</u> (1984), p.329,330

6. D. Timm, A. Ayrinde, R. Foral, "Epoxy Mechanical Properties: Function of Crosslink Architecture", <u>British Polymer Journal</u>, Vol 17, No. 2, (1985), p. 230.

7. D. Timm, A. Ayrinde, et. al, "Kevlar 49 Composite Performance: Dependence on Thermoset Resin Microstructure", <u>Polymer Engineering and Science</u>, Vol 24, No. (1984), p. 932.

8. K.E. Vidal, "Internal Friction in Aluminum-Iron-Nickle High-Strength Powder Metallurgy Alloys", <u>M.S. Thesis, University of Nebraska</u>, (1986), p.67

ANALYTICAL CHEMISTRY AS APPLIED TO RECYCLED PLASTICS

William H. Greive
Monarch Analytical Laboratories, Inc.
Toledo, OH USA

Analytical Chemistry as Applied to Recycled Plastics

by

W. H. Greive

Monarch Analytical Laboratories, Inc.

Toledo, OH 43607

Abstract

While the use of recycled plastics has increased progressively during the past several years, the need for analytical data has just been realized. Today most chemical analyses are performed on recycled polyethylene and polyethylene terephthalate. Often these are performed for the identification and quantitation of the impurities. Many factors are involved in these analyses which include: 1) a need to know the end use so that the analysis is tailored to that need, 2) what sample sizes are appropriate, 3) what types of analyses can be performed within the margin of costs and 4) what degree of speed is needed for the analyses.

Discussion

During the last five years the number of industries involved with recycled plastics has increased each year. These industries include the collection, distribution, reclamation, sales and usage of those plastics. A technical person should immediately note that little emphasis has been placed on the testing of the recycled plastics. To that, a need to address the testing problems will be discussed in this paper. Most analytical services today are performed on recycled polyethylene (PE) and polyethylene terephthalate (PET).

I. The following tests are often times valuable for the evaluation of the recycled plastics:

1. Visual Examination

   A moderately trained individual will be able to examine the plastics for unwanted foreign components such as paper or label stock.

2. Pressing of a Thin Film

   The presence of non-compatible materials, lumps or blisters will assist the manufacturer in knowing whether this plastic can be used in a particular application.

3. Infrared Spectroscopy

   While this technique requires a trained technician it is useful to know the identity of a substance. The problems with the technique is the specialized equipment and personnel needed for the proper application.

4. Thermal Analysis

   Thermal analytical equipment under the direction of a skilled analyst can provide such useful information as glass transition and melting points (ranges). This information obtained can be useful to determine processing conditions and other information about the sample.

5. Size Exclusion Chromatography (SEC)

    Sometimes referred to as GPC, SEC under the direction of a skilled chemist can provide such information as processability and possibly foreign components.

6. Metals Analysis

    Equipment such as emission spectroscopy, energy dispersive x-ray (EDX), and x-ray fluoresence (XRF) can aid in the identification of the metals present in the samples. In addition, EDX and/or XRF can enable the analyst to determine the presence of some non-metals such as chlorine which may be an indication of polyvinyl chloride in the recycled products.

II. Current Tests Being Applied to PET

    Most of the current work as to the analytical chemistry of recycled plastics is addressed to PET.

    1. A simple visual examination for the presence of paper or label stock can be carried out at the plant level.

    2. The forming of a sheet or plague for the examination of streaks and other visual defects can be done with the use of a simple heating device and press.

    3. The measurement of the amount of hot melt adhesives remaining in the PET requires the use of a extraction solvent followed by infrared spectroscopic measurement. The equipment and technical personnel requires some investment and a laboratory facility. A technician can be trained to do the extraction and perform the infrared analysis. The equipment necessary to perform these tests should be in an area isolated from the plant manufacturing.

    4. A number of tests can be conducted on the PET materials such as inherent viscosity, diethylene glycol measurements and ashing. These tests require equipment and trained personnel. Other tests can be conducted on the PET and the requirements of the tests are likely to be more expensive than the end use of the recycled product. It should be noted that tests requiring specialized equipment and services often take a longer time period for the analysis. Those are not likely to be used in a marginal cost operation or in an operation requiring immediate answers.

III. Plastics Such As Polyethylene

    Some tests are performed on recycled polyethylene (PE) which are as follows:

    1. A relatively simple test that can be conducted at the manufacturing site is the making of a thin film in a heated press. A visual examination of the film for lumps and/or non-compatible streaks generally suffices.

    2. Measurement of the melt index (MI) or melt viscosity will assist in giving clues as to processibility. MI requires a minimal amount of technician training and equipment.

    3. Infrared spectroscopic analysis of a film sample may be valuable to identify any foreign components.

    4. Ashing of the sample for metals can be easily carried out in a laboratory

with a minimal amount of training. The identification of the metals requires additional equipment and technical training.

Additional analyses may be performed on other types of plastics such as acrylates, vinyls, etc. These tests are very dependent on the end use of the recycled plastics. For material used as drain pipes, park benches and others the tests would not be as complete or expensive as those needed for manufacture of non-food grade bottles such as detergent containers. Essentially the greater the need for manufacturing control the greater the need for good analytical data.

Often about 100 g (4 oz) of sample will suffice for most analytical needs. The degree of impurities present will control the amount of sample required. If the sample is not homogenous then the sample size required could approach 5 Kg (10 lbs). If the requirement is to know small amounts of impurities then the sample size needs increase.

The cost of the analysis increases with the complexity of the tests. The more a skilled person is used the higher the cost.

Obviously, the degree of speed many times will control the amount and types of tests performed. The visual and film examinations are often times easily and rapidly performed within the manufacturing facility. A foreman in facility may be able to use the results from these tests to adequately predict the usefulness of the recycled plastic.

An attempt has been made to address some of the problems encountered with recycled plastics. This presentation was not meant to be all encompassing but to view some of the problems found with the analysis of recycled thermoplastic materials.

PLASTICS RECYCLING-MARKETS AND APPLICATIONS

Robert A. Bennett
College of Engineering
The University of Toledo
Toledo, OH 43606 USA

ABSTRACT

A systematic approach to plastics recycling requires information in three areas. These areas are 1) collection of plastics, 2) recycling or reprocessing technology, and 3) markets for recycled products. The third area is where this research was performed.

Product development and market demand for recycled plastic products is a long term force which will help to make the plastics recycling industry successful. Market demand will assure that post consumer polymer scrap has an economic value as do the competing materials glass and metal. Additionally, an economically viable market for recycled plastic products has the social benefit of reducing the flow of plastics into the solid waste stream. Increased plastics recycling will assist the U.S. Environmental Protection Agency in achieving the 1992 goal of either reducing or recycling solid waste by 25%. Companies were contacted to determine interest in utilizing recycled plastics in manufacturing products. Questionnaires were sent to recyclers to determine quantity, products and types of plastics being recycled. Information was entered into a personal computer database which allows efficient and convenient data access.

Supply, demand and pricing considerations currently being experienced were analyzed. Market potential estimates for PET and HDPE show that market penetration is still relatively small. Six markets have been targeted for mixed commingled recycled plastics. Product testing, which is used to determine recycled plastic product applications, will be described. Specifically the holding ability of nails in plastics lumber versus wood were performed. Temperature sensitivity tests of maximum nail holding strength have been expanded to include performance at 32°F freezing in addition to the room temperature and above room temperature measurements. Various compositions of mixed commingled plastics were tested for performance variations.

Emphasis has been placed on utilizing mixed plastics for potential high volume applications such as plastic pallets. Market estimates are that 370 million pounds of recycled plastics could be consumed in this market if only a 1% penetration was achieved. A survey of plastic pallet manufacturers indicate interest in utilization of recycled plastics to reduce material costs. Possible opportunities for the utilization of recycled thermoset plastics as inert fillers for lumber substitute fillers and hard boards for construction are explored.

Supply

OVER 65 BILLION POUNDS OF PLASTICS were sold in the U.S during 1988. (Exhibit 1). Sales data were obtained from <u>Modern Plastics,</u> Jan. 1989 and the <u>Textile Organon</u>, Jan. 1989. Polyethylene (high and low density) remained the dominant resin with a total of over 18 billion pounds sold which represents about 30% of the total nontextile plastics resin sales. Polyester resin for nontextile products accounted for 2 billion pounds or 3.5% of total plastics sales. Polyester used in textiles accounted for another 3.6 billion pounds bringing both uses of PET to total over 8% of total plastic sales. Polyethylene and polyester are the dominate plastics currently used in recycling post consumer plastics. Approximately 85% of plastic bottles are made from these plastics. Post consumer scrap generated from these billions of pounds of thermoplastics offer recycling opportunities. Reuse of thermoplastics will reduce raw material costs to manufacturers and reduce the burden caused by plastics on the solid waste stream. Recent price increases for plastic resin will provide further incentives for recycling plastics.

Exhibit 1 - U.S. PLASTIC SALES 1988

Material	1988 Million lbs.
ABS	1238
Acrylic	697
Alkyd	320
Cellulosics	90
Epoxy	470
Nylon	558
Phenolic	3032
Polyacetal	128
Polycarbonate	430
Polyester, thermoplastic(PBT,PET)	2007
Polyester, unsaturated	1373
Polyethylene, high density	8244
Polyethylene, low density*	9865
Polyphenylene-based alloys	180
Polypropylene and copolymers	7304
Polystyrene	5131
Other styrenics	1220
Polyurethane	2905
Polyvinyl chloride and copolymers	8323
Other vinyls	958
Styrene acrylonitrile(SAN)	137
Thermoplastic elastomers	495
Urea and melamine	1515
Others	288
Sub Total	**56,908**
Textile Fibers	
Yarn and Monofilaments	
Nylon - Industrial	375.2
- Carpet	976.2
-Textile	386.9
Polyester - Industrial	358.7
-Textile	871.9
Olefin	1191.3
Total Filament Yarn	**4,160.2**
Staple and Tow & Fiberfill	
Nylon	963.9
Acrylic - Modacrylic	590.2
Polyester	2449.1
Olefin	363.7
Total Staple, Tow & Fiberfill	**4,366.9**
Total Textile Fibers*	**8,527.1**
Total U.S. Sales	**65,435.1**

*Textile totals include domestic and export shipments.
Source of data:
Modern Plastics, Jan., 1989; Textile Organon, Jan., 1989

Rising Resin Pricing will Increase Recycling

Raw material cost saving will continue to be a major factor in the growing demand for recycled plastics. The increased value of plastics resins are making the utilization of recycled plastics a low cost alternative raw material resulting in increased profits margins. Recent changes in resin prices for PET and HDPE are shown in Exhibit 2.

Exhibit 2 Resin Pricing Increasing

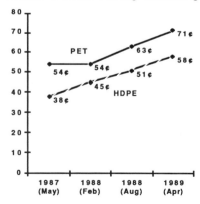

Modern Plastics - market prices

Demand for Recycled PET and HDPE is Greater than Supply

The national survey conducted as part of this research has shown that established PET and HDPE recyclers are experiencing a supply shortage of post consumer plastic scrap. These results are corroborated by a report in Chemical Engineering Progress, February 1989 where it was stated that "Recyclers find Plastics in Short Supply...". Many recyclers are advertising in trade journals for additional scrap. Strategic alliances have been formed to assist in increasing collection. Wellman, Inc. the largest plastics recycler has announced a joint venture with Browning Ferris, Inc. (BFI), a national waste hauler. Cooperative ventures will open channels to capture post consumer plastics before these materials become part of the solid waste stream. Currently the major source of collection is mandatory deposits in certain states. By utilizing the solid waste stream many more types of containers and varieties of plastics can be reclaimed for recycling.

Another cooperative effort has been announced by E.I. duPont de Nemours & Co. and Waste Management, Inc. (Wall Street Journal, April 1989) They have signed a letter of intent to form a joint venture that will sort and recycle plastics from municipal solid waste. It is estimated by duPont that a new PET resin plant would cost between $1.25 and $1.50 per annual pound to build while a recycled resin plant is about 15 cents per pound to build. These economic incentives make recycling plastics an attractive investment.

Other Plastics Recycling Market Alliances

Procter & Gamble Co. and Plasitpak Packaging, Inc. (Chemical Marketing Reporter, October 31, 1988) have entered into a joint venture development to make products from PET. The Spic & Span Pine cleaner bottle has been chosen for test marketing. Chemical & Engineering News, March 20,1989 reports that Dow Chemical and wTe Corp of Bedford, Mass. plan to develop a demonstration plastics recycling program in Akron, Ohio. Dow announced in October of 1988, Chemical & Engineering News, March 20,1989, that they entered into a joint venture with Domtar, Inc. of Montreal to form a North American plastics recycling company. Mobil Chemical Co. and Genpak Corp, N.Y. are now recycling polystyrene foam products such as food trays, cups and containers. Lunch room scrap made of foamed polystyrene will be recycled into insulation and industrialized packaging foam. CR Inc., a Massachusetts recycler, will collect used polystyrene for the recycling plant. The entrance of major companies into plastics recycling will accelerate the growth of this industry.

Growth of Recycling Post Consumer Plastics - PET and HDPE

The rapid growth of PET container recycling is continuing. As mentioned in the preceding section, additional companies have entered into the plastics recycling market and new ways of securing additional post consumer plastic scrap from the waste stream are being developed. Cost advantages and reduced investment realized by using reclaimed plastics rather than using or manufacturing virgin plastics will be a driving force in developing the volume of recycled plastics. Current research estimates that approximately 165 million pounds of PET soft drink containers were recycled in 1988. Data from this year's national survey is still being received and analyzed on 1988 recycling activity. Deposit legislation on soft drink containers in 9 states influenced this rapid growth. Recycling legislation aimed at reducing solid waste is expected to continue the growth of plastics recycling along with other materials. Exhibit 3 shows the rapid increase in recycling plastic beverage containers.

Exhibit 3

PET BEVERAGE CONTAINER RECYCLING

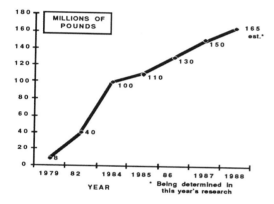

Markets for recycled PET continue to be in carpeting, fiberfill, unsaturated polyester, polyols for rigid urethane foam, strapping, engineering plastics and extruded products. New applications, such as thermoformed products and textiles/geotextiles offer additional opportunities. In 1988, there were 735 million pounds of PET used in manufacturing soft drink containers. Estimates showed that there exists a potential market for over 530 million pounds of this material in non food applications, see Exhibit 4, for PET market projections. Since an estimated 165 million pounds are only being recycled, this market is far from being saturated.

Exhibit 4
PET Potential Markets

Major Markets [million lbs.]	1988 Actual Sales	Potential % Recycled	Potential Volume Recycled
Polyester Thermoplastic (PET)			
Blow molding			
Soft-drink Bottles	735	0%	0
Custom Bottles	200	10%	20
Extrusion			
Film	510	10%	51
Magnetic Recording Film	80	0%	0
Ovenable Trays	30	0%	0
Coating for ovenable board	10	0%	0
Sheeting (for blisters,etc)	15	10%	2
Strapping	30	40%	12
Exports	235	10%	24
Total PET	**1,845**		**108**
Polyester, Unsaturated			
Reinforced Polyester			
Molded	834	2%	17
Sheet	180	2%	4
Surface Coating	19	0%	0
Export	16	0%	0
Other	324	0%	0
Total Unsaturated	**1,373**		**20**
Reinforced Polyester; Unsaturated			
Aircraft/aerospace	34	0%	0
Appliance/business	93	2%	2
Construction	414	2%	8
Consumer	135	0%	0
Corrosion	338	0%	0
Electrical	53	0%	0
Marine	375	2%	8
Transportation	202	2%	4
Other	51	0%	0
Total Reinforced Unsat.	**1,695**		**22**
Polyurethane - Rigid Foams			
Building Insulation	540	2%	11
Refrigeration	160	2%	3
Industrial Insulation	95	2%	2
Packaging	64	2%	1
Transporation	40	2%	1
Other	45	0%	0
Total Polyurethane	**944**		**18**
Textile			
Filiment Yarn	1,231	10%	123
Staple and Tow	2,449	10%	245
Total Textile	**3,680**		**368**
GRAND TOTAL	**9,537**		**536**

HDPE Recycling

Total sales of HDPE in 1988 was in excess of 8.2 billion pounds. About 72 million pounds per year of HDPE have been identified as being recycled in 1987. This is an increase of 24% from the approximately 58 million pounds in 1986. This volume represents less than 20% market potential identified for recycled HDPE for non food applications, see Exhibit 5. A potential market of an estimated 460 million pounds could be developed to utilize recycled HDPE. Individual major potential products for HDPE are soft drink basecups, plastics pipes, lumber, and various containers.

Exhibit 5
HDPE Potential Markets

Major Markets [million lbs.]	1988 Actual Sales	Potential % Recycled	Volume Recycled
Blow Molding			
Bottles			
Milk	750	0%	0
Other Food	321	0%	0
Household Chemicals	1,038	10%	104
Pharmaceuticals	204	0%	0
Drums (>15 gal.)	126	5%	6
Fuel Tanks	74	0%	0
Tight-Head Pails	92	10%	9
Toys	78	5%	4
Housewares	54	0%	0
Other Blow Molding	288	0%	0
Total Blow Molding	**3,025**		**123**
Extrusion			
Coating	51	0%	0
Film (< 12 mil.)			
Merchandise Bags	216	0%	0
Tee-shirt Sacks	168	0%	0
Trash Bags	102	0%	0
Food Packaging	96	0%	0
Deli Paper	16	0%	0
Multiwall Sack Liners	54	0%	0
Other	122	0%	0
Pipe			0
Corrugated	126	25%	32
Water	64	0%	0
Oil & Gas production	76	0%	0
Industrial/Mining	54	0%	0
Gas	122	0%	0
Irrigation	42	50%	21
Other	57	0%	0
Sheet (> 12 mil)	276	10%	28
Wire & Cable	116	0%	0
Other Extrusion	39	10%	4
Total Extrusion	**1,797**		**84**
Injection Molding			
Industrial Containers			
Dairy Crates	64	10%	6
Other Crates, Cases, Pallet	131	10%	13
Pails	407	10%	41
Consumer Packaging			0
Milk-bottle Caps	26	0%	0
Other Caps	63	0%	0
Dairy Tubs	145	0%	0
Ice-cream Containers	88	0%	0
Beverage-bottle Bases	130	50%	65
Other Food Containers	52	0%	0
Paint Cans	38	10%	4
Housewares	260	0%	0
Toys	92	5%	5
Other Injection	230	10%	23
Total Injection Molding	**1,726**		**157**
Rotomolding	140	10%	14
Export	704	0%	0
Other	852	10%	85
GRAND TOTAL	**8,244**		**463**

Mixed Plastics - Six Potential High Volume Markets

Collection of plastics from the solid waste stream results in a mixture of many types of plastics. Separation of this mixture initially is usually limited to gleaning out the easily recognized PET soft drink and milk containers. The remaining mixed commingled plastics can be manufactured into noncritical product applications. Six potential high volume products to be manufactured from mixed recycled plastics were identified in conjunction with the End-Use Market Committee of the Plastics Recycling Foundation. These products are:

1. Treated Lumber
2. Landscape Timbers (1/2 Billion pound market)
3. Horse Fencing
4. Farm Pens for Poultry, Pigs, and Calfs
5. Roadside Posts
6. Pallets (over 400 million wooden pallets used annually)

Survey Recycled Plastics for Pallets

Questionnaires were developed for use in surveying plastics recycling activity in the plastic pallet industry. These questionnaires were sent to 70 plastic pallet companies across the country. This survey's objective is to determine if recycled plastics are currently being used or are being considered for use in pallet manufacturing. Shown in Exhibit 6 is a graph of the board feet of wood being utilized in pallets and other shipping material. Also shown, are estimates of equivalent amounts of mixed plastics based on the volume of wood used in pallet manufacturing.

Exhibit 6

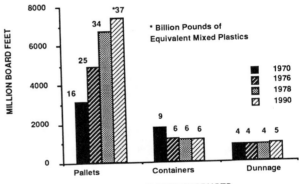

POTENTIAL PLASTICS CONSUMPTION

Source of board feet consumption: U.S. Department of Agriculture, Forest Service.

Plastics poundage estimates based on approximately 5 pounds of mixed plastics per board foot.

If only 1% penetration of recycled plastics was made into this pallet industry, then approximately 370 million pounds of plastics could be utilized, see calculation in Exhibit 7.

Exhibit 7 Pallets - Mixed Plastics Potential

Market Size (1990)	7.4 billion bf	37 billion lbs
Estimated Penetration		1 %
Estimated Markets		370 million lbs

The National Wooden Pallet and Container Association (NWPCA) in Washington, D.C. estimates new pallet production at over 465 million. The growth in unit production is displayed in the following Exhibit 8. This large material market offers considerable opportunities for utilizing recycled plastics. Superwood International Ltd., of Wicklow, Ireland identified the manufacture of pallets from recycled material as a major niche market for its production technology Plastics Technology, March 1989. Their process is best suited for recycling industrial scrap. Feed material could include HDPE, LDPE, PP and even wood chips as filler.

Exhibit 8 Pallet Production

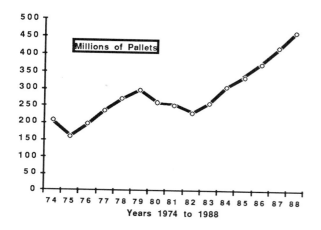

Thermoset Recycling Potential Markets

Products similar to those presented for thermoplastics must be developed for thermoset plastics in order to demonstrate the potential for recycling. Prototype products may assist in encouraging entrepreneurs in establishing businesses utilizing recycled thermoset materials. Two potential product areas are explored; a filled plastic lumber and a particle board type product for construction.

Recycled thermoset material could be used as a filler in a thermoplastic binder. The inert properties and strength of the recycled thermoset material should improve the strength and rigidity of these recycled products. Residual glass fiber content of SMC could allow superior screw holding characteristics, especially at higher temperatures. Shown in Exhibit 9 is a schematic of lumber like post made from mixed thermoplastic scrap plastics with non melted thermoset filler.

Exhibit 9 Recycled Plastic Post

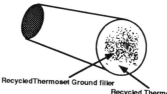

Hard board type products could be manufactured from recycled thermoset materials. Recycled thermoset plastics could be ground and formed into sheets An appropriate adhesive system would be utilized to bind the ground thermoset material into a rigid board. This type of sheet could potentially be utilized in construction applications such as walls and flooring. Toughness and water resistance are positive properties for this material. Shown in Exhibit 10 is a schematic of this product concept. This type of product could find applications in subflooring and walls as well as in other construction applications.

Exhibit 10 Recycled Plastics Board

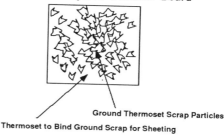

Product Testing

A series of nail and screw pull-out experiments shows plastic lumber to behave quite differently than wooden lumber. This year's tests were expanded to include performance at temperatures below and above room temperature and with various compositions of mixed commingled recycled plastics. Nail pullout tests were performed on wood and recycled plastic samples to compare their nail holding strength.

Nail pullout tests performed at freezing temperature was achieved by soaking the samples in ice water at 0°C (32°F) for six hours. The nails were driven in the samples prior to soaking. Above room temperature measurements were made by soaking in a water bath at 60°C. The nails used were smooth with 0.15" diameter. Nail penetration was 1-1/2" for all the tests. Four recycled plastic samples had the following compositions:

Sample 1. 100% milk bottles
Sample 2. mostly PE wire cables insulation
Sample 3. 50% HDPE, 25% miscellaneous containers, 20% LDPE, 5% base cup resin
Sample 4. commingled residue stock

Exhibit 11 shows that soaking the recycled plastic samples in ice water for 6 hours had an inconsistent effect on the nail holding strength of the samples. Both positive and negative variations in nail holding ability were observed. Possibly most of the variation was produced by the inhomogeneities of the materials rather than temperature. Additional testing must be performed to increase the statistical significance of the results. Experiments at an elevated temperature of 60°C exhibited consistently a loss of nail holding strength of about half of that at room temperature. For comparison, wood retained much of its strength while being heated in a water bath. Consequently, utilization of lumber made from recycled plastics must be carefully evaluated regarding the environment in which it will be used in order to avoid inappropriate applications.

Exhibit 11 Temperature Effects

Recycled Plastic Sample	Max pullout force-lbs (6 hrs. in ice water) 0°C	Max pullout force-lbs (room temp)	Max pullout force-lbs (in heated water-60°C)
Sample 1 (100% milk bottles)	212.5 -10.6%*	237.7	117.7 -50.5%*
Sample 2 (PE wire cables insulation)	190 +20.8%*	157.3	61.3 -61.0%*
Sample 3 (50% HDPE, 25% misc. 20% LDPE 5% resin)	251 +6.7%*	235.3	119.7 -49.1%*
Sample 4 (commingled residue stock)	165 -29.6%*	234.3	142.7 -39.1%*

* Change compared to room temperature

Plastics Recycling Future

Plastics recycling is facing increasing pressure due to dwindling supply of landfill space. Another force is that recyclers are experiencing a shortage of scrap material supply. The entrance of new companies and new businesses into the marketplace will generate even more demand for scrap plastics. Increasing the amount of recycled plastics in the future will require improved high speed sorting and novel collection techniques. Further research in product and market development will be required to maximize the value and usage of recycled plastics.

Acknowledgements

Research was performed in conjunction with other plastics recycling research projects funded through The Plastics Recycling Foundation/Center for Plastics Recycling Research. Additional funding was provided by the Ohio Department of Natural Resources, Division of Litter Prevention & Recycling and The University of Toledo.

RECYCLED POST-CONSUMER HDPE: PROPERTIES AND USE AS A MATRIX FOR WOOD-FIBER COMPOSITES

Susan E. Selke
Michigan State University
East Lansing, MI USA

Abstract

Recycled HDPE from dairy bottles has properties little changed from virgin resin. Composites of this material and wood fibers can be prepared easily utilizing a twin-screw extruder. The resultant material has improved stiffness and creep resistance compared to HDPE alone, and offers improved ease of forming compared to traditional wood composites. Additive systems can improve the tensile strength of the material. Current efforts at MSU are directed towards exploring the use of additives for this purpose, as well as investigation of other recycled plastics as the matrix.

THE GROWING SOLID WASTE PROBLEM has lead to increased efforts to utilize recycling as a means of diverting material from dwindling landfills. However little information has been published about the properties of recycled plastics produced from post-consumer waste. It is known that exposure to heat and to ultraviolet light can cause various degradative reactions in polymers, resulting in oxidative, chain-scission, and crosslinking reactions in varying amounts depending on the specific environmental influences and the polymer in question. Potential users of recycled plastic have been reluctant to depend on this material because of concerns about reliability of sources in terms of both quantity and quality. In addition to concerns about deterioration of properties due to degradative reactions, contamination is a major concern.

At Michigan State University we have investigated the properties of post-consumer recycled high density polyethylene (HDPE) from dairy bottles. Dairy bottles are a very significant use of HDPE, consuming 341 million kg of resin in 1988. An additional 146 million kg of HDPE bottles were used for other food products, and an additional 565 million used for bottles for household chemicals, pharmaceuticals and cosmetics (1). Thus this material represents a large potential source of recycled plastic.

PROPERTIES OF RECYCLED HDPE

Standard ASTM procedures were used to compare properties of virgin HDPE resin intended for use in milk bottles, regrind HDPE from bottles which were blown but never filled, and recycled HDPE from post-consumer bottles, all prepared from the same resin batch. Results showed no significant change in melt flow index, and no significant change in tensile strength, modulus of elasticity, or elongation at yield for the three materials. Elongation at break was significantly reduced, as was Izod impact strength (statistically significant at the .01 level of confidence). Reduction in elongation at break was about 50%, and reduction in impact strength about 60% (2).

Properties were also determined for a mixed batch of post-consumer milk bottles, originating from a variety of sources. Table 1 contains values obtained for a sample of virgin resin and samples of mixed post-consumer milk bottles. It should be noted that in this case the differences seen cannot be interpreted as representing effects of processing, since the virgin and recycled materials were not identical. They do serve to indicate the range of properties which can be expected from a

Table 1 - Properties of Virgin and Recycled HDPE (2)

Property	Virgin HDPE	Recycled HDPE
Melt Index (g/10 min)	0.727	0.715
Tensile Strength (nt/m^2)	3.37×10^7	3.42×10^7
Tensile Modulus (nt/m^2)	5.96×10^8	6.40×10^8
Elongation at Yield (%)	17.0	16.2
Elongation at Break (%)	69.7	36.9
Izod Impact Strength (nt m/cm notch)	1.34	1.18

recycled stream.

Further details regarding testing methods and results are available in reference (2).

USE OF HDPE WITH WOOD FIBERS IN A COMPOSITE MATERIAL

Plastics are limited in their usefulness in a number of applications by their low stiffness and susceptibility to deformation due to high temperatures or heavy loads. These limitations can be overcome in some cases by reinforcing the plastic with fibers of various types. Both glass and carbon fibers are often used in these applications. However glass and carbon fibers are relatively expensive. Processing of these materials is also complex and expensive, especially when the fibers are used in continuous form. Short fibers have less reinforcing potential than long fibers, but are easier to process since extrusion can be used. However the abrasive nature of glass and carbon fibers is a drawback in extrusion. Wood fibers are nonabrasive, inexpensive, nonhazardous, and abundant, and offer stiffness and strength characteristics that can make them excellent reinforcers or extenders for plastics (3).

As mentioned above, contamination of recycled plastic streams presents serious problems in their reuse. Plastic bottles are frequently contaminated with label materials, which are often paper. Separation technologies including air classification and washing can be employed to remove these unwanted materials. An alternative approach is to develop applications where the presence of small amounts of contaminants is of less concern, as the potential for interfering with needed properties is less. The technologies developed for utilizing commingled plastics, such as the ET-1, are obvious examples (4). Wood-fiber HDPE composites are expected to be inherently relatively insensitive to small amounts of contamination, especially from paper fibers. For all of these reasons, we decided to investigate the use of wood fibers to reinforce recycled HDPE.

Wood fiber/HDPE composites were formed by blending wood fibers and HDPE in a corotating twin-screw extruder with two feed ports. Recycled HDPE was fed into the first feed zone and melted before the fibers were introduced. This method results in less fiber damage than if both materials were introduced together. Composites were prepared for a range of fiber loadings up to 60% by weight. The extrudate was then compression molded to produce samples for testing. Fiber length was also examined by dissolving the HDPE from the extruded material in xylene and measuring samples of fibers. Fiber length decreased as a result of processing, and was sensitive to the processing conditions used, with more severe processing causing increased shortening of fibers.

The wood fiber composites were significantly stiffer than the recycled HDPE alone. Stiffness, as measured by both tensile and flexural modulus, increased with increasing fiber content, and was as much as 250% higher at 60 weight % fiber. Tensile strength of the composites was slightly less than that of the unreinforced HDPE, decreasing with increasing fiber content. Flexural strength showed a slight increase at fiber loadings of 25-50%. Izod impact strength was significantly less in the composites, showing a precipitous drop at even low fiber loadings. Water sorption was examined using a total immersion method, and was found to increase with increasing fiber content. Fiber loadings of less than 40 weight % sorbed less than 10% water by weight after 1000 hours of immersion, but composites containing 59% fiber by weight sorbed nearly 20 weight % water in 600 hours (with no significant increase at 1000 hours). Samples made to compare the effectiveness of recycled HDPE and virgin HDPE as the matrix showed no decrease in properties using recycled HDPE. Table 2 shows properties of composites with 30% wood fiber in comparison to recycled HDPE alone. (Note: The recycled HDPE values in this table are not identical to those in Table 1 because this was a different batch of material.) Additional information on methodology and results can be found in references (5-7).

EFFECTS OF ADDITIVES

One of the difficulties encountered in securing the reinforcing potential of wood fibers in a thermoplastic matrix is the hydrophilic nature of the fibers and the hydrophobic nature of most commodity thermoplastics, including high density polyethylene. This difference acts to impede both the even dispersion of the fibers in the matrix and the adhesion between the fibers and the thermoplastic. A general principle of adhesion is that chemically similar materials are likely to adhere to each other and chemically dissimilar materials are unlikely to do so. Adhesive formulations, including resins for tie layers in coextruded structures, often operate by combining functional groups with chemical similarities to the substrates to be joined within a single molecule. The adhesive can then, by adhering to both substrates, bond the whole material together. Similar approaches have been taken to providing fiber-matrix bonding. The problem of adequate mixing of the fibers within the matrix can sometimes be alleviated by the use of dispersing agents.

In an effort to improve the mechanical properties of recycled HDPE/wood fiber composites, we investigated the use of several additives chosen for potential in improving either adhesion of the fibers to the matrix or dispersion of the fibers within the matrix. The additives investigated included ionomer, maleic anhydride modified polypropylene, low density polyethylene, chlorinated polyethylene, and stearic acid (7). An initial investigation suggests that both ionomer and maleic anhydride modified polyethylene may have potential for improving adhesion and therefore improving mechanical properties, as determined by measurements of mechanical properties and water sorption. Maleic anhydride modified polypropylene appears especially promising. Use of these additives at a 5% concentration in composites with 30% wood fiber suggested gains in tensile strength and in elongation at break. Tensile modulus increased with 5% ionomer and with 2% maleic anhydride modified polypropylene, though apparently not with 5%. Impact strength, however, showed a further decrease with addition of these materials. While the results are too few for reliable conclusions to be drawn at this point, they are encouraging, and additional studies are currently under way.

SUMMARY AND CONCLUSIONS

As our nation's solid waste disposal problems grow more acute, the need for recycling materials increases. One common material which can be viably recycled is high density polyethylene bottles, especially milk bottles. We have demonstrated that recycled resin from milk bottles has properties very similar in many respects to virgin resin, and thus is suitable for a wide variety of uses.

A potential new use for this recycled material is in composite materials containing wood fibers. This material, made from two inexpensive substrates, is extrudable and easily formable by compression molding. Its stiffness is improved considerably over HDPE alone, and its tensile strength, though somewhat lower, is still acceptable for many applications. It appears to be possible to improve the properties of this composite by the use of additives to improve the adhesion between the wood fibers and the HDPE matrix.

Table 2
Properties of 30% Wood Fiber Composites (7)

Property	Recycled HDPE	30% Wood Fiber and Recycled HDPE
Tensile Strength (nt/m^2)	3.43×10^7	2.70×10^7
Tensile Modulus (nt/m^2)	7.7×10^8	12.7×10^8
Elongation at Break	270%	1.38%
Izod Impact Strength (nt)	34.2	14.2

REFERENCES

1. Modern Plastics, pp. 69-119, Jan. (1989)

2. Pattanakul, C., S. Selke, C. Lai, E. Grulke, J. Miltz, ANTEC '88, Society of Plastics Engineers, pp. 1802-1804 (1988)

3. Cruz-Ramos, C. A., in "Mechanical Properties of Reinforced Thermoplastics" D. W. Clegg & A. A. Collyer, eds, pp. 65-81, Elsevier Applied Science Pub., New York (1986)

4. Selke, S., Packaging Tech. Sci., 1, 93-98 (1988)

5. Yam, K., V. Kalyankar, S. Selke, C. Lai, ANTEC '88, Society of Plastics Engineers, pp. 1809-1811 (1988)

6. Yam, K., b. Gogoi, C. Lai, S. Selke, "Composites from Compounding Wood Fibers with Recycled High Density Polyethylene" presented at Third Chemical Congress of North America, Toronto, Ontario, June 5-10, 1988, submitted to Polymer Composites (1989)

7. Selke, S., K. Yam, K. Nieman, ANTEC '89, Society of Plastics Engineers, pp. 1813-1815 (1989)

PREFORMING FOR LIQUID COMPOSITE MOLDING

Earl P. Carley, John F. Dockum, Jr., Philip L. Schell
PPG Industries, Inc.,
Pittsburgh, PA USA

ABSTRACT

The process of Liquid Composite Molding (LCM) to produce structural composites has gained considerable attention over the last several years. One barrier to the process gaining further acceptance is the lack of knowledge and process expertise in preform reinforcement production. If the LCM process (which claims to have the advantage of low cost) is to remain economically viable, equally low cost methods of preform production must be coupled with the molding.

The two forms of glass fiber reinforcement and preforming methods generally available to process three-dimensional preforms for liquid composite molding are thermoforming continuous strand mat and fiber directed preforming with a multi-end roving. The processes and forms of glass reinforcement will be compared, and examples of fiber directed preforming of the reinforcement for a spare tire cover and commercial bumper beam will be presented.

LIQUID COMPOSITE MOLDING, which includes both Resin Transfer Molding (RTM) and Structural Reaction Injection Molding (SRIM), is poised for explosive growth in the next five years. A recent report[1] projects that RTM and SRIM will grow at annual rates of 34% and 42%, respectively. If this double-digit growth rate is to be realized and carried through the 1990s, cost-effective and production-worthy preforming processes will be required. The process of molding a composite part from preplaced (dry, in-the-mold) glass reinforcement and a wet thermoset resin is at least 35 years old. Newer Structural Liquid Composite Molding (SLCM) resins are stronger and lower in viscosity; however, they still require in-the-mold placement of a fiber glass preform to yield a structural part. This study will present the basic processes of preform production, compare mechanical property data for various types of reinforcements, and present two examples of composite parts fabricated from preforms and LCM.

According to an automotive end user in a paper presented to the 1988 Advanced Composites Conference, the key issue in order for SLCM to be widely accepted is preforming.[2] The desired attributes of preforms and the process to produce them are low cost, near net shape, adequate structural performance, durability, suitability for volume production, compatibility with resins, and preform-to-preform reproducibility.

PREFORMS: BASIC TYPES

There are two basic input forms of fiber glass available to the LCM molder for fabrication of a stiff three-dimensional preform. These are a thermoformable continuous strand mat and a multi-end roving. In addition, either of the above reinforcements can be combined with unidirectional or bidirectional forms of reinforcement in selected areas during preform fabrication. The added specialty reinforcement will provide additional strength in areas of the composite that may encounter high stresses.

Processing thermoformable mat into a three-dimensional preform requires an oven to heat the mat, a frame to hold it while it is stretched into shape, and a tool to form the mat into a preform. To preform a typical 0.125 inch part at 40% glass by weight, four plies of the 1.5 oz/ft^2 mat would be cut to the approximate rectangular shape of the molded part, allowing extra material to be held in a frame. The frame containing the material is placed in an oven for a period of approximately two minutes while it is heated to 350°F. After removal from the oven, the frame and hot glass mat must be transferred to the forming tool within 10 seconds. The tool is closed, forming and cooling the mat for a period of 30 seconds. Once the frame is removed and the waste glass clamped in the frame is trimmed, the preform is ready for molding. With an assembly line type of thermoforming machine proposed by the author,[3] it should be possible to produce a preform approximately every 90 seconds. Parts that have deep vertical draws where the mat must be stretched may result in resin-rich radii and/or a nonuniform glass content on the vertical walls. Inconsistent stretching of the mat can lead to nonreproducible preforms.

Continuous roving fabricated into a chopped fiber directed preform employs a tried and proven process developed over 35 years ago. Photographs of the 4-station rotary preform machine at PPG's Fiber Glass Research Center are shown in Figure 1. Briefly, the process utilizes an air-assisted chopper/binder gun which conveys glass and binder to a perforated metal screen shaped identical to the part to be molded. The chopped glass is held in place on the screen by a large industrial blower drawing air through the screen. After all of the glass and binder have been deposited, the ferris wheel mechanism containing the preform screens indexes 90 degrees, thereby rotating the preform into the oven. In the oven, hot air (250°F) is drawn through the glass to dry and set the binder. The ferris wheel is indexed another 90 degrees for cooling and removal of the chopped fiber preform from the screen. With this process, there is no need to draw and stretch the glass. The glass is deposited on the screen in the exact location where it is needed and remains there. Uniform glass deposition on the screen is the rate-limiting step in fiber directed preform production. This, in turn, is regulated by the complexity of the preform to be constructed and the size of the chopper/binder system. PPG's directed fiber/chopper has the capability of chopping and depositing glass at a maximum rate of 4.5 lbs/minute. For higher preform build rates, either additional choppers can be employed, or a chopper to handle larger volumes can be constructed.

PREFORM REINFORCEMENT: COMPOSITE PROPERTIES

Current LCM resin systems require some type of reinforcement to produce a functional composite part. In most applications, the cost of that reinforcement is of primary importance, and fiber glass is the reinforcement of choice. Load bearing or structural parts will often require glass contents of 35% to 50% by weight in order to provide adequate mechanical properties. PPG has developed a series of roving products (types 5540 and 5542) that will process during fiber directed preforming and yield excellent mechanical properties when molded with LCM resins. Mechanical properties were determined from chopped fiber directed preforms made on a flat test plaque preform screen and from both thermoformable and nonthermoformable continuous strand mats. The flat test plaques were molded with either an SRIM urethane resin system at 50% glass by weight or a Structural RTM vinyl ester resin system at both 35% and 50% glass by weight. Two glass contents were chosen for the RTM system to provide mechanical property information at structural glass contents (50%) and at cosmetic or appearance levels (35%).

The test panels were cut and tested for tensile strength, flexural strength and modulus, notched Izod impact, and a falling dart (Dynatup) instrumented impact. Half of the tensile and flexural samples were subjected to 24 hours in boiling water and then tested to determine the retention of their initial strengths. Boiling water is an aggressive aging technique. The strength retention after water boil gives a relative indication of the adhesion at the interface between the resin and the glass reinforcement and how well the structural laminate will hold up with age.

Table 1 lists mechanical property data of the various glass reinforcements in a vinyl ester resin system. Two types of chopped fiber directed preforms were evaluated and compared to thermoformable and nonthermoformable continuous strand mat. The fiber directed preforms contained either 5540 or 5542 preform roving, which are fine and coarse strand geometry rovings, respectively. Table 2 lists mechanical property data for the same resin system and glass reinforcements but at a 35% glass content.

Two points are initially obvious from the vinyl ester mechanical property data. First, in contrast to previous publications,[4,5] the continuous strand mat reinforcement does not provide the composite with superior mechanical properties compared to the fiber directed preform reinforcement. Secondly, in almost all cases, the chopped fiber directed preform retains more of its initial strength after a 24-hour water boil compared to either continuous strand mat product. This demonstrates that vinyl ester LCM composites reinforced with chopped fiber directed preforms will retain more of their initial strength as they age. At the 50% glass content, the chopped fiber preforms had an average property retention of 85%, while the continuous strand mat average property retention was only 71%. At 35% glass, the difference was even greater, with the chopped fiber preform composites retaining an average of 94% of their initial strengths and the continuous strand mat only 73%. Clearly, in regard to mechanical property retention in an RTM resin, the chopped fiber preform is a superior reinforcement.

The two methods of impact testing produced contrasting results. At both glass contents, the 5542 chopped fiber preform reinforcement produced the highest notched Izod impact value (see Tables 1 and 2). The falling dart instrumented impact test (Dynatup) measures a different type of impact and produces three values of interest. These are: (1) the maximum load, which is the highest point on the load versus impact/time curve; (2) the maximum energy, which is the energy absorbed by the test specimen up to the point of maximum load and is also the amount of energy a specimen can absorb before failure starts; and (3) total energy, which is the amount of energy absorbed during the complete test, including failure of the laminate.

At the 50% glass content, the 5542 reinforced laminate produced the highest maximum energy and would be the most resistant to impact without sustaining damage. However, for the total amount of energy absorbed during the test, the 5542 was lowest with 198.3 ft-lbs/inch versus a high of 219.1 ft-lbs/inch for the 5540 reinforced laminate. Surprisingly, the maximum energy values of the laminates at 35% and 50% glass were very close, which indicates that the 15% change in glass content does not influence the impact resistance of the composite before it incurs damage. The difference in glass content does show up in both the notched Izod impact and in the total energy absorbed during the falling dart impact test which includes penetration through and failure of the laminate. On average (for all of the plaques tested), the maximum energy absorbed without sustaining damage decreased only 0.8% (71.8 to 71.2 ft-lbs/inch), while the notched Izod impact decreased 21% (19.3 to 15.2 ft-lbs/inch), and the total energy decreased 20% (208.3 to 166.1 ft-lbs/inch) in going from 50% to 35% glass content.

Two types of chopped fiber preform reinforcement were evaluated in the RTM resin system. These were 5540 and 5542 which differ in strand geometry or strand coarseness, with 5542 having approximately twice the strand size of 5540. Previous mechanical property data in a urethane SRIM resin indicated that, with the exception of the notched Izod impact, mechanical properties were essentially constant with changes in strand geometry. The results from Tables 1 and 2 are less clear. At the 50% glass content, tensile strengths for the two chopped fiber preform composites were similar, but flexural strengths and moduli were significantly different. In contrast, at 35% glass content, tensile strengths were significantly different, while flexural strengths and moduli were similar. At both glass contents, notched Izod impact was greater for the coarser strand geometry reinforcement 5542.

In general, there appear to be higher mechanical property values for smaller strand size reinforcement (i.e., 5540 properties are greater than 5542 properties and the 5540 strand size is less than the 5542 strand size). This would be expected since a smaller strand size would allow for more surface area of the reinforcement to be available for the resin. However, a coarser strand tends to produce a more porous, open preform which will allow the resin an easier flow path, especially in filling out large parts.

Listed in Table 3 are mechanical property values for both fiber directed preform and continuous strand mat reinforcement at 40% glass content in an SRIM resin. As observed previously in the RTM resin, the chopped fiber preform yields equivalent or superior mechanical properties compared to the continuous strand mat. Retention of the initial properties after water boil was comparable to the vinyl ester resin with the chopped fiber preform retaining an average of 86%, while the nonthermoformable and thermoformable continuous strand mat retained 74% and 76%, respectively.

Table 4 lists mechanical property data for a novel type of fiber directed preform reinforcement that is in the early stages of development. This reinforcement utilizes a large filament diameter single end roving. After processing on a fiber directed preform machine, the resulting preform is highly filamentized. Glass contents of 35% to 40% in an SRIM resin have been achieved with the current processing variables. With the exception of notched Izod impact and total energy on the instrumented impact, other mechanical properties are comparable to 5540/5542 roving or continuous strand mat reinforcements. A major advantage that this product could bring to the preform market is a significant cost savings over a multi-end roving.

PREFORM REINFORCEMENT: PROCESS

The PPG fiber directed preform machine, which is a 4-station rotary machine, was designed and built by the I. G. Brenner Co. It is capable of preforming any part that will fit within a 40-inch diameter circle, with a vertical draw not exceeding 24 inches. A resin-compatible water emulsion binder is deposited along with the glass on the preform screen. The binder, which is present at 6% to 10% by weight, is used to hold the preform together. This enables the preform to retain its shape and integrity during removal from the preform screen, conveying and positioning in the mold, and resin injection or impregnation. Binder costs may range from $0.80 to $1.50 per dry pound.

Two major concerns with the fiber directed preform process, since it is currently under operator control, are preform-to-preform weight variation and glass weight consistency (distribution) within a preform. Since the amounts of glass and binder applied to the preform screen are controlled by microprocessor timers to the tenth of a second, the variability from preform to preform on total weight is less than 2%. Within a preform intended for uniform glass content, the variation is nominal ± 3%. As automated deposition methods are developed, placement of the glass at precise areas on the preform screen will be controlled even more closely.

The productivity issue relative to fiber directed preforming is often linked to the preform binder in terms of having to remove moisture from the wet preform. While investigating preform binders, we found that certain thermoplastic emulsions will work very well and can be dried at only 250°F. With a 7% solids preform binder applied to a flat plaque preform at 7% total Loss-On-Ignition (LOI) by weight, the moisture content after preforming and before indexing into the oven is approximately 27%. At 250°F, a preform weight of 6 oz/ft^2 requires only 30 seconds in the oven to remove the moisture. The amount of offal, moisture before drying, preform-to-preform weight variation, and weight variation within a preform are listed in Table 5. These values were obtained for 2, 6, and 8 oz/ft^2 flat plaque preforms. As expected, when higher glass loadings are added to the screen, the amount of offal increases slightly. Although approximately equal weights of binder (moisture included) and glass are deposited on the preform screen during fabrication, there is only 27% by weight moisture present as the preform is indexed into the oven. The majority of the water is removed through the forming blower.

Reinforcement cost will be a primary concern for the LCM molder as it directly impacts the final composite part cost. Table 6 lists approximate selling prices of reinforcements that may be used in structural composites. When this information is factored in with the amount of scrap glass or offal generated in fabricating preform reinforcements, the actual cost of the reinforcement in the part can be calculated.

With the current PPG fiber directed preform machine, glass and binder can be applied controllably to the screen, depending on the part, at a maximum rate of 72 oz/minute with one chopper. The productivity limiting step for most preforms is application of the glass to the screen, not drying the preform in the oven. At the above glass deposition rate, a 5-lb. preform could be produced at a rate of at least 44 preforms/hour (assuming an 80% job efficiency) or an overall preform production time of less than 82 seconds/preform.

Screen design and construction are obviously very important in fiber directed preforming. To be as close as possible, the screen is made from an actual part or a "splash" from the model. The preform screen is generally fabricated from the male or plug side of the mold so that the finished preform will fit snugly on the plug. In addition to the direction that the operator gives the glass as it is sprayed onto the screen, the air flow generated by building the screen out of different degrees of open perforated metal sheet will also help direct the glass. Typical perforated metal sheet used for preform screen construction is 53% open. However, in areas where less glass is desired, a 40% or less open screen can be used. Depending on size and complexity, screen cost can range from a few hundred to several thousand dollars.

An obvious goal of the LCM molder is to be able to shoot a net part with little, if any, trim. PPG's preform screen design and development work make this goal closer to a reality. Typical preform screens are flat in the transition area from the perforated metal sheet where the glass is deposited to the solid metal sheet frame that holds the screen. This results in glass laying at all orientations around the edge of the preform which must be trimmed to prevent fibers from projecting into the land area of the mold. By building the screen with an edge wall perpendicular, or near perpendicular, to the flat screen and approximately 3/4 inch high (Figure 2), the chopped roving tends to orient parallel with the edge of the preform. This provides preform reinforcement with a smooth edge and potentially a net preform that does not have to be trimmed. Additional benefits of this new screen design are the greater preform edge strength, reduced offal, and better housekeeping. Orientation of the reinforcement parallel to the edge of the part will prevent inward crack propagation.

Current rotary 4-station fiber directed preform production machines vary in size from as small as 36 inches up to 120 inches (turntable size). Any part to be preformed must fit within that diameter of turntable. With only limited industry interest in preforming prior to the availability of the new LCM resin systems, sources of such machines are limited. However, there are currently several companies in the process of building fiber directed preform machines. An approximate cost from one manufacturer of a 60-inch diameter 4-station preform machine is approximately $150,000, fully instrumented but excluding installation and environmental controls, if required.

PREFORMS FOR LCM EXAMPLES: GM SPARE TIRE COVER, COMMERCIAL BUMPER BEAM

One of the LCM parts that is commercial in a significant volume is the GM spare tire cover molded by Ardyne Inc. of Grand Haven,

MI, at an annual volume exceeding 750,000 units. Although the spare tire cover is not a true structural part, it has been an ideal part for the LCM process with in-the-mold placement of reinforcement. The reinforcement for the part has been a nonthermoformable continuous strand mat present in the composite at approximately 25% by weight. The reinforcement was cut from multiple plies of mat by a steel rule die. Waste from cutting out the spare tire cover was on the order of 20%. A preform screen for the spare tire cover was built for the PPG fiber directed preform machine. For laboratory production of the spare tire cover preforms, glass was deposited from two 205-yield roving packages at the rate of 1.5 lbs/minute. Total time to produce a preform was 30 seconds. The offal generated when the spare tire cover is fabricated by the fiber directed preform process is less than 2% by weight. Cost savings in the reinforcement from both raw material and waste reduction is approximately 50% (Table 7).

Another structural composite application area which represents complex shapes is bumper beams.[6] A commercial bumper beam was molded by Dow Chemical Co. from preforms produced on the PPG fiber directed preform machine. Since the length of the bumper beam exceeded the diameter of our turntable, the preform was produced in two pieces with a tapered overlap of some 5 inches in the middle. The bumper beam design called for glass loadings on the order of 40% by weight. Due to the thickness of the beam, this was accomplished by using an inside and outside preform so that one nested inside the other when placed in the mold. The total time to produce a complete set of four preforms weighing approximately 8 lbs. was slightly less than 4 minutes. The fiber directed preforms and continuous strand mat preforms were successfully molded. Impact test results are listed in Table 8.

At glass contents in the mid to upper 30% range, bumpers reinforced with either continuous strand mat or chopped fiber directed preform passed both the center and corner impact tests. In both impact areas, the chopped fiber directed preform had a slightly larger deflection. When slightly lower glass content bumpers were impacted, the two bumpers containing continuous strand mat cracked. The fiber directed preform bumper at the lower glass content passed both the center and corner impact tests without any damage. This example demonstrates the utility of fiber directed preform reinforcement for true structural applications such as bumper beams.

CONCLUSIONS

Applications for Structural Liquid Composite Molding (SLCM) are projected to grow at a faster rate than most other composite production processes. If parts produced by SLCM are going to compete economically not only with other composites, but ultimately with steel, the most economical components and processes must be put together.

We have presented data which demonstrates that composites produced via LCM with chopped fiber directed preform reinforcement yield excellent mechanical properties when compared to both thermoformable and nonthermoformable continuous strand mat reinforcement. In addition, when a fiber directed preform reinforcement and process are used, these properties coupled with near net shape can be obtained at a much lower cost. A liberal example of this is the GM spare tire cover, where a reinforcement cost savings of 50% was realized by changing from continuous strand mat to a fiber directed preform. The successful testing of a commercial bumper beam provides evidence that fiber directed preforms can be utilized in a true structural part.

REFERENCES

1. Butler, R. Business Communications Co., Norwalk, CT.

2. Johnson, C. F. and N. G. Chavka, "Preform Development For A Structural Composite Crossmember," in Proceedings of the Fourth Annual Conference on Advance Composites, ASM International (September 1988).

3. Mazzoni, B. C., "Equipment for Processing Structural Rim," in Proceedings of the Fourth Annual Conference on Advance Composites, ASM International (September 1988) and in Proceedings of the 43rd Annual Conference SPI Composites Institute, Session 11-F (February 1988).

4. Dunbar, S. G., "Preformable Continuous Strand Mat," in Proceedings of the Fourth Annual Conference on Advance Composites, ASM International (September 1988) and in Proceedings of the 43rd Annual Conference SPI Composites Institute, Session 4-D (February 1988).

5. Cloud, M. J., "Structural RIM for Automotive Applications," in Proceedings of the Fourth Annual Conference on Advance Composites, ASM International (September 1988).

6. Ellerbe, G., "Economic Evaluation of a New Structural Composite," SPE RETEC (1987).

ADDITIONAL REFERENCES

1. Carley, E. P., J. F. Dockum, Jr., and P. L. Schell, "Preforming for Liquid Composite Molding," in Proceedings of the 44th Annual SPI Conference, Session 10-B (February 1989).

2. Miller, B., "Preforms Pave the Way For Lower Cost Structurals," *Plastics World* (January 1988).

3. Kallaur, M., "New Materials for RTM Processing," in Proceedings of the 43rd Annual Conference SPI Composites Institute, Session 22-A (February 1988).

4. Scrivo, J., "Structural RIM - Manufacturing Technology for High Volume Production of Composites," in Proceedings of the 43rd Annual Conference SPI Composites Institute, Session 22-B (February 1988).

5. Voeks, S. L., et al., "Advances in the Art of Preforming," in Proceedings of the 43rd Annual Conference SPI Composites Institute, Session 4-E (February 1988).

6. Shirrell, C. D., "Liquid Composite Molding-The Coming Revolution," AutoCom '88 Conference, Society of Manufacturing Engineers, Paper No. EM88-228 (May 1988).

7. Emrich, P., "Fiber Glass Preforming," in Proceedings from the Third Annual Conference on Advanced Composites, ASM International (September 1987).

TABLE 1 MECHANICAL PROPERTIES OF FIBER DIRECTED PREFORMS AND
CONTINUOUS STRAND MAT IN A VINYL ESTER RESIN AT 50% GLASS[1]

Mechanical Property	Chopped Fiber Preforms		Continuous Strand Mat	
	PPG 5540	PPG 5542	Nonthermoformable	Thermoformable
Tensile Strength[2]	27,140	25,230	26,560	32,050
% Retention[3]	71.9	92.4	65.5	70.0
Flexural Strength[2]	51,810	42,520	45,310	48,470
% Retention[3]	81.3	82.8	50.1	71.9
Flexural Modulus[2]	1.75×10^6	1.54×10^6	1.67×10^6	1.76×10^6
% Retention[3]	93.9	87.5	69.7	96.9
Notched Izod Impact[4]	17.6	24.5	16.9	18.1
Instrumented Impact				
Maximum Load[5]	1149	1130	915	976
Maximum Energy[4]	69.2	75.5	70.6	64.8
Total Energy[4]	219.1	198.3	202.9	212.7

[1] Dow Derakane® 411-50
[2] psi
[3] property retention after a 24-hour water boil
[4] ft-lbs/inch
[5] lbs.

TABLE 2 MECHANICAL PROPERTIES OF FIBER DIRECTED PREFORMS AND
CONTINUOUS STRAND MAT IN A VINYL ESTER RESIN AT 35% GLASS[1]

Mechanical Property	Chopped Fiber Preforms		Continuous Strand Mat	
	PPG 5540	PPG 5542	Nonthermoformable	Thermoformable
Tensile Strength[2]	20,400	14,900	19,220	20,490
% Retention[3]	98.0	91.0	66.3	68.3
Flexural Strength[2]	35,490	33,100	34,720	35,580
% Retention[3]	87.0	90.5	62.9	66.3
Flexural Modulus[2]	1.27×10^6	1.28×10^6	1.21×10^6	1.27×10^6
% Retention[3]	95.5	99.2	81.3	91.5
Notched Izod Impact[4]	15.4	18.8	12.4	14.3
Instrumented Impact				
Maximum Load[5]	878.0	774.0	689.3	772.0
Maximum Energy[4]	76.1	58.6	72.2	77.9
Total Energy[4]	178.9	151.9	155.8	177.7

[1] Dow Derakane® 411-50
[2] psi
[3] property retention after a 24-hour water boil
[4] ft-lbs/inch
[5] lbs.

TABLE 3 MECHANICAL PROPERTIES OF A FIBER DIRECTED PREFORM AND CONTINUOUS STRAND MAT IN AN SRIM RESIN AT 40% GLASS[1]

	5540 Chopped Fiber Preform	Nonthermoformable Continuous Mat	Thermoformable Continuous Mat
Tensile Strength[2]	24,500	22,600	22,400
% Retention[3]	78.8	69.5	75.9
Flexural Strength[2]	36,400	33,500	36,000
% Retention[3]	87.6	67.5	68.3
Flexural Modulus[2]	1.30×10^6	1.20×10^6	1.23×10^6
% Retention[3]	92.3	85.0	83.7
Notched Izod Impact[4]	10.2	12.0	11.8

[1] Mobay STR-400
[2] psi
[3] After 24-hour water boil
[4] ft-lbs/inch

TABLE 4 MECHANICAL PROPERTIES OF <u>NOVEL</u> FIBER DIRECTED PREFORM REINFORCEMENT IN SRIM RESIN SYSTEM AT 38% GLASS[1]

Mechanical Property	PPG Novel Preform
Tensile Strength[2]	19,670
Flexural Strength[2]	32,900
Flexural Modulus[2]	1.47×10^6
Notched Izod Impact[3]	3.9
Instrumented Impact	
Maximum Load[4]	622
Maximum Energy[3]	67.1
Total Energy[3]	99.8

[1] Mobay STR-400
[2] psi
[3] ft-lbs/inch
[4] lbs.

TABLE 5   PRODUCTION CHARACTERISTICS OF A FIBER DIRECTED PREFORM

Oz/ft^2	% Glass Offal	% Moisture Before Drying	% COV* Preform to Preform	% COV* Within a Preform
2	1.14	24.0	0.49	4.89
6	1.60	27.0	0.23	5.74
8	6.25**	29.2	0.85	4.03

*COV = standard deviation/mean
**High % offal due to limiting air flow in machine

TABLE 6   APPROXIMATE SELLING PRICES OF FIBER GLASS PREFORM MATERIALS*

Basic Forms		Price/Pound
Continuous Strand Preform Roving		$0.80
Continuous Strand Mat		1.35
Continuous Strand Mat - Thermoformable		1.40

Additional Reinforcements	(oz/yd^2)	
Biaxial Woven Roving	24	$1.00
Woven Roving Backed Chopped Strand Mat	37.5[a]	1.21
Uniaxial Nonwoven Roving (Stitched or "Knit")	26	1.60
Biaxial (0°/90°) "Knit" with Chopped Strand Mat	36.5[b]	1.65
Biaxial (0°/90°) "Knit"	23	2.00
Biaxial Biased "Knit" (±45°) with Chopped Strand Mat	37.5[b]	2.00
Biaxial Biased "Knit" (±45°)	24	2.20
Triaxial "Knit" (0/±45°, or 90°/±45°) with Chopped Strand Mat	36.5[b]	2.70
Triaxial "Knit" (0/±45°, or 90°/±45°)	23	3.00
Biaxial Braided Tapes (roving based)		4.00
Biaxial Braided Tapes (yarn based)		5.00
Triaxial Braided Tapes (roving based)		5.00
Triaxial Braided Tapes (yarn based)		6.00

Specialty Preforms (textile yarn based)		
3D Weft Knitted		$3.40+
3D Braided		10.00

*2nd Quarter 1989
[a]WR (24 oz/yd^2) + Mat (1.5 oz/ft^2)
[b]"Knit" (23 or 24 oz/yd^2) + Mat (1.5 oz/ft^2 given in examples, although same price per pound applies when 1 oz/ft^2 mat is used)

TABLE 7 GM SPARE TIRE COVER

REINFORCEMENT MATERIAL COST
Continuous Strand Mat – $1.30/Lb – 20% Trim Waste – $1.63/Lb in Mold Reinforcement – (0.6 Lbs/Part) ($1.63/Lb) = <u>$0.98/Preform</u>
Roving/Chopped Fiber Preform – $0.80/Lb – 3% Waste – $0.83/Lb in Mold Reinforcement – (0.6 lbs/Part) ($0.83/Lb) = <u>$0.50/preform</u>

TABLE 8   1987 T-BIRD/COUGAR BUMPER BEAM PERFORMANCE TEST RESULTS

		5 MPH Impact			3 MPH Impact	
Glass Preform	% Glass	Center Medium Lbs. Force	Deflection Inches	Comments	Corner Medium Lbs. Force	Deflection Inches
Thermoformable Continuous Strand Mat	33 – 35	10180	2.44	No Damage	4920	1.73
Chopped Strand Fiber Directed	38 – 39	9020	2.52	No Damage	4270	1.80
Thermoformable Continuous Strand Mat	31 – 34	8480	3.31	8" Crack	–	–
Thermoformable Continuous Strand Mat	27 – 30	9020	3.20	8" cRACK	–	–
Chopped Strand Fiber Directed	28 – 30	8940	2.66	nO dAMAGE	4330	1.70

FIGURE 1a. Overall view of 4-station rotary preform and chopper mechanism.

FIGURE 1b. Initial stage of preform fabrication cycle with application of chopped fiber glass.

FIGURE 1c. Conclusion of preform fabrication cycle with application of chopped fiber glass and binder.

FIGURE 1d. Removal of finished preform.

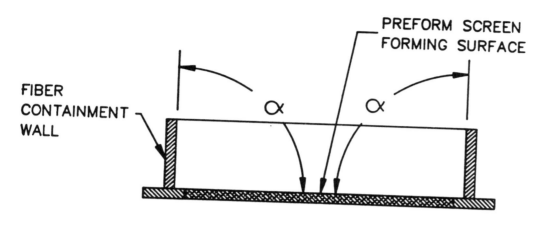
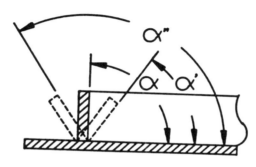

VARIOUS ANGLES OF
FIBER CONTAINMENT WALLS

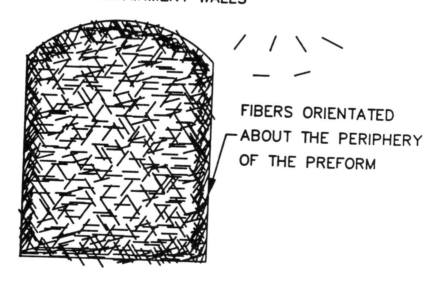

FIGURE 2

1990 CORVETTE REAR UNDERBODY— THE CASE FOR PREFORM

Thomas P. Schroeter, R. Keith Leavitt
Molded Fiber Glass Co.
Ashtabula, OH USA

Preformed glass reinforced plastic parts are returning to the Corvette after a quarter century of absence. Along with some other 1990 Corvette parts, the rear underbody will be made from a glass directed fiber preform liquid molding. The liquid molding or wet mix formulation will be a low profile filled polyester system.

First, we will look at the motivations for the conversion from Sheet Molding Compound (SMC) to preform liquid molding for structural applications. Secondly, we will show some preliminary properties of a sample of preform liquid molded rear underbodies. Thirdly, we will explain and illustrate the preforming and molding process for the rear underbody. In the process explanation we will address the various areas in which the rear underbody's complex geometry puts unusual requirements on the preforming process. Lastly, we will summarize and draw conclusions from what we have accomplished.

MOTIVATIONS FOR THE CONVERSION FROM SMC TO PREFORM LIQUID MOLDING

There have been various studies at Molded Fiber Glass Research Company (MFG Research) to study the differences in properties between SMC and preform liquid molding. The first extensive study at MFG Research was undertaken in 1974 and was designated as Project Choice. Project Choice was set up as a study of alternative fiberglass plastic molding materials. In the initial study nearly 5,000 molded test specimens were checked to evaluate differences between liquid mat molding, liquid preform molding, and SMC molding. We will not get into the details of the results of this early work, but merely mention it here as the preface for more recent studies.

For the purpose of the Corvette rear underbody, we are interested in the comparison of low profile SMC to low profile liquid preform molding. For this purpose we will focus on the results of Project Choice IV, a portion of which compares low profile SMC and low profile liquid preform molding in an updated sequence of testing with the more recent advanced low-profile formulations. Project Choice IV was performed by MFG Research in early 1984.

The molding that was used for the experimentation was an open box with round edges. The dimensions of the boxes were 15 inches long by 12 inches wide by 9 inches deep. The test specimens from each box were taken as shown in figure 1: 8 vertical side wall samples, 6 horizontal side wall samples, and 4 top wall samples. The boxes were all molded at a thickness of 0.096 inch. Six boxes of each type of molding were sampled. The results for the low profile SMC molding and the low profile preform molding are shown in Table I. The average strengths reported in Table I are based on the 6 part average of the individual part averages. The average ranges reported are based on the 6 part average of the individual ranges. Average minimum values reported are based on the 6 part average of the individual part average of the 5 lowest strength values. The purpose of this is to stay away from false conclusions that could be made by freak minimum points due to testing or testing fabrication of the test specimens. It is our feeling that this gives the average minimum strengths reported here more value. The coefficient of variation reported is defined as the standard deviation divided by the average.

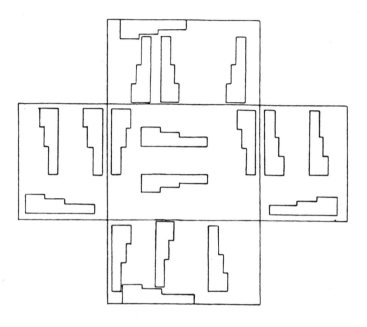

Figure 1. Sample specimen locations on the experimental box moldings.

Looking at Table I, it is notable that the coefficients of variation of the average properties for the low profile preform have a maximum of 3.9% and an average of 3.1% which is very good. The average property coefficient of variation for the low profile SMC is 7.5% with a maximum of 9.7%.

The average property coefficient of variation of the average minimums for the low profile preform is 3.9% with a maximum of 5.6%. This indicates good reproducibility part to part. The low profile SMC showed an average property coefficient of variation of the average minimum of 12.7% with a maximum coefficient of variation of 15.5%.

The above figures show good reproducibility part to part for both systems, but the low profile preform shows a much tighter distribution around the average than the low profile SMC.

The average property coefficients of variation for the average ranges were 20.5% for the low profile preform and 13.3% for SMC.

From above and looking at the values observed for the coefficients of variations for the average range it is noticed that the figures show higher for preform than for SMC. This in some ways can appear misleading. Since there is an extremely large difference in the average of average ranges in favor of preform (±20.9% for preform and ±44.8%), the total distribution of the average of the average ranges is much closer to the desired average properties for preform. To illustrate this, the approximate distribution for the average of the average ranges based on the average plus or minus three average range coefficents of variation is ±7.7% to ±34.1% for preform and ±26.8% to ±62.7% for SMC. This clearly shows the ability of preform to keep a smaller overall average range.

In summary of the physical properties comparison, the most important advantage of preform over SMC is the tighter distribution around the average minimum strengths. This allows preform to target a lower average strength but yet still meet a higher minimum average strength than SMC. This indicates an advantage to utilizing preform liquid molding where it is feasible for structural applications.

With preforms, we know where the glass is at and how it is distributed because we can see this in the preform before it is molded. This contributes to a more uniform glass distribution along with a random fiber direction orientation. With SMC molding we have to consider knit lines and glass orientations due to flow and how these elements will affect the strength properties in certain areas of the structural part. Keep in mind that it is not our intention to discredit SMC for its accomplishments. SMC has advantages in highly detailed parts with complex ribbing, large numbers of bosses, or deep bosses. Here we only intend to compare SMC to preform liquid molding where structural applications are feasible to preform

Table I - Part to Part Variation of
Various 0.096" Thick Fiber
Glass Reinforced Moldings

Formulation Type	Low Profile Preform (60% Carbonate Filled Wetmix)	Low Profile SMC (60% Carbonate Filled Paste)
Tensile Strength		
Average PSI	9,008	6,574
Coef. of Variation (%)	3.9	8.9
Average Range (%) ±	29.4	± 55.3
Coef. of Variation (%)	18.9	15.5
Avg. Min. (%)	71.3	54.1
Coef. of Variation (%)	3.6	13.2
Flexural Strength		
Average PSI	21,736	19,936
Coef. of Variation (%)	2.1	9.7
Avg. Range (%) ±	18.4	± 41.2
Coef. of Variation (%)	24.6	15.9
Avg. Min. (%)	81.8	57.9
Coef. of Variation (%)	5.6	15.5
IZOD		
Average FT-LB/IN	19.3	17.2
Coef. of Variation (%)	3.2	4.0
Avg. Range (%) ±	14.9	± 37.9
Coef. of Variation (%)	18.0	8.6
Avg. Min. (%)	86.6	65.5
Coef. of Variation (%)	2.4	9.5
AVG. GLASS CONTENT (%)	22.5	26.8

liquid molding. This places a limit on the extent of the comparison.

We feel we can also make accurate and unbiased statements on cost comparisons between SMC and preform liquid molding. At the Ashtabula Division of Molded Fiber Glass Company, over 2.4 million pounds of preforming glass roving and over 1.6 million pounds of SMC glass roving was consumed in 1988. From these quantities, one can have an appreciation for our ability to develop adequate cost comparisons between the two types of molding.

On an equal part thickness comparison, preform liquid molding is generally slightly lower cost than SMC molding. Preforms cost more to prepare for molding than SMC, but at the press area preform liquid molding more than overcomes the additional pre-processing costs by its ability to be molded in a much lower tonnage press than its SMC counterpart. Preform liquid molding also obtains another cost advantage at the press area since preformed parts can be molded faster than SMC parts.

When the enhanced physical properties of the preformed parts are put to its advantage, the part walls can be made thinner than that for SMC and still exhibit higher minimum average strengths. This can substantially lower total part cost utilizing preform over SMC since less materials can be used to achieve superior physical properties.

In summary, the cost comparison here shows that when preforming can be applied to structural applications it is generally more economical to utilize preform liquid molding over SMC molding. This is exactly the case with the Corvette rear underbody.

PRELIMINARY STRENGTH PROPERTIES OF A SAMPLE OF PREFORMED LIQUID MOLDED REAR UNDERBODIES

A sample of 5 preproduction preliminary moldings were made at identical molding conditions. Each underbody part had 26 specimen areas tested. The average thickness and average glass percentage of all the

Table II - Part to Part Variation
of Preliminary Preformed
Underbody Parts

Tensile Strength

Average PSI	14,148
Coef. of Variation (%)	6.4
Avg. Range (%)	± 44.9
Coef. of Variation (%)	20.5
Avg. Min. (%)	69.3
Coef. of Variation (%)	6.5

Flexural

Average PSI	29,757
Coef. of Variation (%)	1.2
Avg. Range (%)	± 28.2
Coef. of Variation (%)	13.9
Avg. Min. (%)	79.3
Coef. of Variation (%)	6.4

IZOD

Average FT-LB/IN	21.8
Coef. of Variation (%)	1.9
Avg. Range (%)	± 35.8
Coef. of Variation (%)	25.6
Avg. Min (%)	78.4
Coef. of Variation (%)	4.7
Avg. Glass Content (%)	35.8
Avg. Sample Thickness In.	0.092

sample specimens are reported. The same strength properties as illustrated in Table I are also shown for the sample of 5 preliminary underbodies in Table II. Comparing the values from both tables, the preliminary preformed underbody sample illustrates property strength values that were anticipated. However, of particular interest the average range values are slightly higher than we would expect although still appearing to be better than that of SMC. We feel the slightly higher than anticipated average range values are due to the inherent nature of preliminary part sampling. Molding adjustments still need to be made before production begins. Once production begins we feel a better average range will be observed. Overall, at this time we believe that the preformed underbody is meeting its predetermined expectations.

AN OVERVIEW OF THE PREFORMING AND MOLDING PROCESS FOR THE 1990 CORVETTE REAR UNDERBODY

We will now explain and illustrate the preforming and molding stages of the rear underbody section. Other production stages will not be addressed here since they would be similar for either preform liquid molding or SMC molding. Along with the explanation of the preforming and molding stages, we will also discuss the various aspects of the underbody that represented difficult challenges to the preforming and molding processes.

First, we start with the rotating perforated screen in the shape of the rear underbody as shown in figure 2. The rotating screen is under suction so that the chopped glass and binder are attracted to the

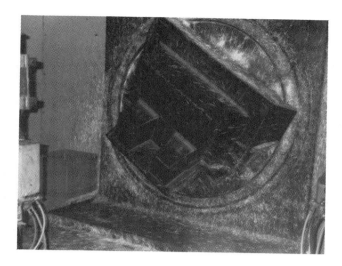

Figure 2. Underbody preform screen on a preform machine

screen when they are applied. Screens require baffling to insure that the chopped glass fibers do not accumulate in certain areas and appear sparse in other areas. This required a substantial challenge because of the shape and complexity of the outer winged sections of the underbody. This is important so that uniform glass distribution is achieved.

Next, chopped glass and binder are applied to the screen using a combined chopper and spray gun as shown in figure 3. Areas that are of greater thickness or low glass affinity can be reinforced by stopping the application at given times and inserting additional reinforcement or mat by hand.

The use of a single X-Y axis robot is used to help apply glass fibers. This helps to minimize the labor costs of preforming and increases part to part reproducibility.

After all the chopped glass and binder have been applied to the screen, the preform covered screen is rotated into the oven. There the binder cures while material is applied to another preform screen. Once the screen has rotated again the finished preform is exposed to the stripping area. Here the preform is cooled and removed as shown in figure 4.

Figure 4. Removal of a finished rear underbody preform

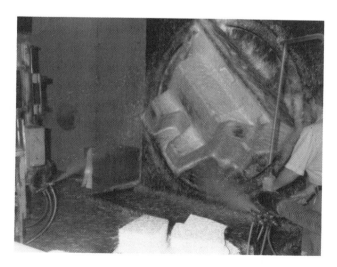

Figure 3. Glass and binder being applied to the rotating preform screen

Figure 5. The finished rear underbody preform

Here we have the finished preform as shown in figure 5.

The preforms are quite sturdy and have sufficient body to permit ease of handling and storage. The preforms are stacked and nested together as shown in figure 6. The stack-nested preforms can then either be stored or sent directly to the press area depending on scheduling needs.

Figure 7. Wet mix poured onto the rear underbody preform

Figure 6. Underbody preforms awaiting further processing

Now, we can proceed to the press area to mold the preforms using a wet mix or liquid molding technique. A preform is removed from a stack and surface mat is attached to the bottom of the preform and placed in a pour fixture. At this time the designated amount of the low profile wet mix formulation is poured onto the preform in a predetermined pattern throughout the preform as shown in figure 7. Additional surface mat is placed on the top surface of the poured preform which completes the charge. A special technique not described here is used to mold bosses onto the bottom of the preform during the molding process. The charge is then loaded into the die as shown in figure 8. After the charge is loaded, the press is cycled.

Figure 8. Loading the completed charge into the die

When the press opens the preform liquid molded part is removed as shown in figure 9. After the flash is removed and the part is sanded, we see the sanded molded part ready for secondary operations such as drilling, etc. as shown in figure 10. After the molding stage and seondary operations we can see the finsihed part in the Corvette chassis in figure 11.

Figure 9. Removing the molded rear underbody part

Figure 10. A deflashed and sanded molded rear underbody part ready for secondary operations

Figure 11. The 1990 Corvette rear underbody in its chassis location

not be ignored. For most applications preformed parts can reduce wall thickness over SMC parts and still have higher minimum average strengths. As composites advance, there will be more and more opportunities to utilize preform liquid molding advantages to enhance structural part quality and lower overall per part costs.

CONCLUSION

As presented, there are many reasons why the Corvette rear underbody went to preform liquid molding from SMC molding. The major reasons we see are improved physical property data, particularly in the range of strength values throughout the parts, and reduced per part production cost. Preform liquid molding has much to offer for structural applications and should

PRODUCTION OF AN AUTOMOTIVE BUMPER BAR USING LCM

Dan A. Kleymeer
Ardyne Incorporated
Grand Haven, MI USA

James R. Stimpson
CPC Group—GMC
Warren, MI USA

ABSTRACT

Production of an impact bumper bar was developed for General Motors' Corvette using the Liquid Composite Molding (LCM) process which will be described. The activities leading to the design, analysis, prototyping, and testing will be presented as well. The Liquid Composite Molding process has many material choices which yield various properties. These properties and the selection rationale used for this bumper application will be presented. The steps for production implementation as they occurred will be given as related to the LCM materials used. Advantages of Liquid Composite Molding will become evident throughout the project's discussion - advantages that can be extended to other applications in automotive and industrial markets. This is the first production use of structural RIM resins processed by LCM as applied to a truly structural automotive part.

THE ACTIVITIES FOR PRODUCTION of an automotive impact bumper bar made with the Liquid Composite Molding process will be presented herein. The companies involved in the project targeted bumpers as the best candidate to demonstrate the structural capability of composites using LCM. Having modest production volumes and high visibility, General Motors' Chevrolet Corvette was chosen as the best candidate to receive an LCM bumper.

The project required studying the front bumper system which had been in place since 1984. The results of the study and load data given by GM, provided a basis against which new designs could be compared. The iteration determined by analysis to give the best performance was drawn on paper with respect to package constraints and manufacturing concerns. A model and prototype mold were fabricated which yielded parts that could be tested both statically and dynamically. The performance and economics were both favorable enough for GM to continue support of a production phase program.

Revisions were required at some places where the bumper assembly interfaced with other body parts before production tooling could be started. The final outcome from the production tools was tested for certification to GM performance standards. Pilot parts were delivered to the Corvette assembly plant so input could be incorporated prior to actual production. Techniques and methods developed during prototyping were carried into production. Once all the details were handled, production was approved.

DESIGN

The scope of the initial project was to design, prototype, and validate a front impact bumper bar demonstrating the advantages of Liquid Composite Molding. The scope was further defined by requiring the LCM bumper bar to function as a one-for-one replacement for the then existing bumper. Since the project was initially conceived by people from Dow Chemical, General Motors, and Ardyne, it was structured so that each member could contribute jointly to the program. The stage was then set to aggressively complete the project in as short a time as possible. Design efforts began in May, 1987.

BASELINE SYSTEM - Very little performance data were available as the baseline bumper bar design was not CAD generated and load path data were unavailable. Results of independently conducted performance testing were given as follows:

Table 1
Dynamic Performance Of Baseline Bumper System

Vehicle Weight: 3,300 lbs.(1,275 kg.)

Load Location	Load, lbs. (N)
Center	11,700 (52,000)
Corner High	6,000 (26,700)
Corner Low	3,100 (13,800)
Barrier	16,000 (71,200)

Table 2
Baseline Bumper System FEA Summary

Stiffness	16,210 lbs./in. (2.839 E+6 N/M)
Specific Strain Energy	764 in-lbs./lb.
Impact Load (Given)	16,000 lbs. (71,200 N)
Centerline Deflection	0.987 in. (25.1 mm)

For a proper finite element analysis to be conducted, each component in the existing system had to be modeled. The information produced would serve as a baseline against which proposed designs could be compared. The exact details of the analysis will not be presented because they appear elsewhere in print. The existing double-shell reinforced polypropylene system was modeled using the same impact loads found in testing. The model included the bumper bar, aluminum support brackets, shim plates, a steel cross-member, and chassis rails. The results of this analysis are shown in Table 2. These results were obtained by using the property data given in Table 3.

DESIGN PACKAGING - The baseline system that was the target for replacement included the 0.102 inch (4.0 mm) thick, bonded, double-skin impact bar, three fascia reinforcement supports, two fascia reinforcement end rods, and a fascia reinforcement channel rail. The assembly of these components required approximately twenty fasteners. Figure 1 shows a drawing of the system. It was decided to carry over the rail mounting screw retainers, the aluminum support plates, and the egg-crate energy absorber. This design package allowed attaching the bumper bar to the rails and attaching the following to the bumper in the existing manner: energy absorber, skid bars, air dam, vapor canister bracket, turn signal/fog lamp assembly, wiring harness, and fascia. The reason for retaining so many of the current components was to make it easier to implement as a running change at assembly. With the design packaging determined, proposed LCM bumpers could be modeled and analyzed.

PROPOSED NEW DESIGNS - It was quickly determined that single-shell construction that integrated fascia attachment components, would maximize economic advantages. Three geometries were generated for review relative to manufacturing. One was eliminated because it called for sharp corners in the rail mounting areas. Such corners are difficult to form in the glass reinforcement and still maintain uniform stress distribution. Another was disouraged because it added the complexity of a foam core. The remaining geometry was analyzed relative to constant and variable thickness configurations. The resulting design called for a centerline thickness of 0.114 in. (4.5 mm) and a thickness of 0.152 in. (6.0 mm) from the rail mounting area outboard. The vertical fascia reinforcement strip along the top of the bumper was 0.051 in. (2 mm) thick. A center-to-outboard-end sketch of this design is shown in Figure 2.

The materials and properties assumed for the analysis are shown in Table 4 and the results of the initial analysis are summarized in Table 5.

The goal of the design effort was to generate a proposed construction that would match the results of the existing baseline system. As can be seen by comparing the two FEA summaries, this goal was achieved. Similar comparative analysis was performed for the other load cases, i.e., corner high, corner low, outboard high, etc. A final optimal geometry was developed that could lead the project into a prototype phase.

PROTOTYPE

The next step in the project was using the packaging envelope previously described, and the single-shell bar was integrated onto GM's layout drawing of the baseline system. This ensured that all background components would fit properly. Section, plan, and top views were drawn with modifications made to ease manufacturing. The completed drawing was given to a model maker in September, 1987. Three weeks later, a wood model of the forward surface was delivered to the Ardyne Composite Development Center. An epoxy mold was constructed from this model for prototyping LCM bumpers.

It was determined during the final

Table 3
Baseline System Properties

Flexural Modulus	800 ksi (5,500 MPa)
Poisson's Ratio	0.30
Tensile Stress	14 (97)
Flexural Strength	24 (166)
Reinforcement Type	Random Fiberglass
Reinforcement Fraction	40 Wt. %

Table 4
Proposed LCM Materials and Properties

Resin Matrix	Polyisocyanurate (Dow Spectrim MM354)
Reinforcement Type	Continuous Strand Mat
Reinforcement Fraction	57 Wt. %
Flexural Modulus	1,820 ksi (12,550 MPa)
Poisson's Ratio	0.30
Tensile Stress	29 (199)
Flexural Strength	55 (381)

FEA iterations that production of the bumper bar would not be practical using all continuous strand mat; especially at the weight fraction specified. The tighter bundling of directional mat offered a solution to the preform manufacturing problem. Some random fibers were kept to handle shear and compressive forces exerted on the bumper system through corner impacts and attachment loads. Based on fraction calculations and property curves, the bracketing technique was used to select three reinforcement packages. Preforms were made using heat and pressure. After trimming the preforms to size, parts were shot using a proprietary designed resin injection machine. The second part shot was completely filled and deemed acceptable for testing. The initial shot time was a generous seven seconds and total cycle time was about two minutes, button-to-button. Several parts of each reinforcement package were finished by drilling holes, sawing cut-outs, and attaching supports and retainers. The prototype parts were then ready for testing.

TESTING

In November, 1987, static and dynamic tests were conducted at an in-house testing facility. Static tests were performed first so initial baselines could be established for all bumpers. A load cell and linear displacement transducer were mounted in a hydraulic frame press. A mounting fixture was made to simulate bumper bar attachment to the chassis rails. The fixture was designed so centerline and corner load cases could be conducted. In every test, identical conditions were maintained from bumper to bumper. Three LCM bumpers with different reinforcement packages were tested as well as the baseline bar. Static testing showed some areas of the LCM bumpers that were stronger than the baseline, and some that were weaker. Table 6 lists a few of the averaged results of static loading at failure.

The static test indicated weaker performance in the lower corners. A horizontal flange had been added during the design phase, but its exact geometry was not well defined. Modifications to the shape and thickness of that flange were made prior to dynamic testing. More bumpers were molded.

General Motors provided Corvette vehicles for dynamic testing using a pendulum having mass equal to that of the vehicle. A total of fifteen tests were run on each of two bumper systems; current and proposed. Each system, complete with energy absorber and lamp assemblies, were mounted to the Corvette according to assembly instructions. Since this test would not be used for validation, full vehicle inspection was omitted. Some of the typical results appear in Table 7.

After each impact, the bumper and all attachments were examined for deformation and disfiguration. Highly stressed areas were easily identified by their characteristic "whitening" that occurs as the resin-reinforcement bond begins to fail. An example of impact load plotted against time is shown in Figure 3. Note how closely the two systems compare. Neither system suffered catastrophic failure but some small surface cracks in the proposed system suggested the use of a resin with higher elongation.

Testing was completed in late November, 1987. The results satisfied the original scope of the program. A presentation was made to General Motors that led the program into a production development phase.

PRODUCTION

DESIGN CHANGES - Before production tooling could be started, all geometry and interface details had to be addressed. The people who could best provide input were those at the C-P-C

Table 5
Proposed LCM Bumper System FEA Summary

Stiffness	16,249 lbs./in. (2.846 E+6 N/M)
Specific Strain Energy	808 in-lbs./lb.
Impact Load (Given)	16,000 lbs. (71,200 N)
Centerline Deflection	1.008 in. (25.6 mm)

Table 6
Static Test Data Comparison At Failure

Load Location	Bumper	Load, lbs. (N)	Deflection, in. (mm)
Center	Baseline	4,500 (20,020)	0.92 (23.4)
	Proposed	4,600 (20,470)	0.85 (21.6)
Corner High	Baseline	2,960 (13,170)	1.43 (36.3)
	Proposed	3,140 (13,970)	1.45 (36.8)
Corner Low	Baseline	2,700 (12,015)	1.85 (47.0)
	Proposed	2,200 (9,800)	1.85 (47.0)

Group, Bowling Green assembly plant. Many quality improvement suggestions were fielded by the authors and incorporated into the system design. One suggestion was to have the design accommodate easier fascia-to-hood gap adjustment. The vertical flange along the top of the LCM bumper bar was folded rearward so it was nearly horizontal. The flange supported adjustable metal reinforcements and provided additional section modulus along the upper portion of the bar. The reinforcements enhanced the ability to fine tune the placement of the fascia.

Bowling Green was satisfied that desired adjustability and improved quality could be achieved by the LCM bumper system. The design simplified work on the assembly line by reducing the number of parts and eliminating one assembly station. Line workers were supportive of the proposed changes and offered suggestions.

Other details included vapor canister bracket mounting and screw torques. Mounting an existing vapor canister bracket was not a simple task, although the final solution would make it appear so. The bracket was designed to mount to a double-shell bumper bar and its mounting tabs became points in space when held against the LCM bumper. After several attempts to devise a new, twisty metal stamping, the final solution came in the form of a resin-rich horizontal tab protruding rearward from the bumper's backside, which matched the tab of the bracket. Generous radii and comfortable thicknesses were employed to minimize the risk of breaking the tab during handling. Careful development with all the parties involved ultimately provided a simple, low-cost solution.

Screw torques for holding the fascia reinforcement strips in place on the top flange were calculated. Three fasteners for each of the two strips were recommended since the strips contained slots for mounting. Various screw thread configurations, typical for composites, were tested and none could consistently reach the desired torque. Additional thickness was added to the underside of the flange in the mounting locations only. Even though the new six millimeter thickness did not contain additional glass, torques 25% in excess of the specification were consistently achieved using standard sheet metal screws.

Other design modifications included recessed pocket areas on the backside of the bumper so rivet heads would be flush with the surface. Local thinner sections provide a tolerance for rivet misalignment during energy absorber attachment. The baseline design called for a large square opening to be pierced in the molding for lamp socket removal. This opening was enlarged and made round to simplify manufacturing. It allowed easy removal of a lamp shroud as well. Part thickness was increased locally behind some attachment screw locations to accommodate usage of a universal screw throughout the bumper system. Three different types of screws used in the baseline system were reduced to one. Bowling Green was supportive of the inventory reduction and simplified assembly. All design changes were documented to drawings that GM could use directly.

MATERIALS - During the Spring of 1988, Dow Chemical was readying the commercialization of Spectrim MM310, which

Table 7
Dynamic Test Data Using A Weighted Pendulum

Vehicle Weight = Pendulum Weight = 3,375 lbs. (1,532 kg.)

Impact Location	Bumper	Load, lbs. (N)	Deflection in. (mm)	Velocity mph	Duration msec
Center	Baseline	10,600 (47,200)	0.71 (18)	5.2	106
	Proposed	10,100 (45,000)	0.63 (16)	5.5	112
Corner High	Baseline	3,600 (16,000)	1.50 (38)	3.0	176
	Proposed	3,775 (16,800)	1.67 (42)	3.0	186
Corner Low	Baseline	3,800 (16,900)	1.50 (38)	3.1	186
	Proposed	3,800 (16,900)	1.36 (35)	3.0	183

is a polycarbamate system. Bumpers incorporating most of the above design changes were prototyped using the new resin. The results of static testing showed similar performance to the bumpers made with MM354, polyisocyanurate. This was encouraging since polycarbamate has higher elongation and better impact properties. For an energy absorbing application such as this bumper, it seemed very appropriate to use the new resin system.

Bumper system weights were monitored throughout the program. With a production version nearly finalized, it was time to make a serious comparison. Looking at each bumper bar trimmed but without hardware, the LCM bar weighed slightly more than the baseline one. With all the hardware attached, thereby making each bumper a complete system assembly, the LCM system weighed slightly less than the baseline. It was known, however, that production mold efficiency would reduce the part weight further because better resin flow would occur. A steel production mold was fabricated and parts made. The system weight decreased by 0.4 lb. (0.2 kg.). The only remaining step before a purchase order would be released was validating the performance of the LCM bumper system to GM corporate standards.

VALIDATION TESTING - A Corvette was outfitted with a proposed bumper system and delivered to GM's test facility. It was then learned that the vehicle mass had increased. The reoriented top flange would help handle this extra load, it was decided. When pendulum testing occurred, slight hood scuffing was noticed during high corner impacts. Careful study revealed that the fascia reinforcement strip was mounted with a screw located outboard of the bumper bar's natural hinge point. The screw was relocated inboard of the hinge point and the reinforcement strip was redesigned to provide additional cantilevered stiffness. The modifications were implemented and more bumpers produced for testing. This time no hood scuff was detected and the LCM bumper system received full GM-United States validation approval. The test results were similar to those presented at the onset of the project. With validation completed, a purchase order was released in September, 1988, for Ardyne to manufacture front impact bumper bars for the current model Corvette.

MANUFACTURING - Production of bumpers for the volume expected at Bowling Green began in January, 1989. The process steps can be listed as follows: cut mat; stack and form mat; trim mat; mold part; deflash part; drill holes; attach brackets; inspect and pack for shipment. Two shifts began production until learning curves could be stabilized.

The most difficult challenge has been to adequately match the reinforcement preform to the mold. It was decided early in the program to make a net part; that is, one which has no reinforcement extending beyond the part's edges. The difficulty comes in making the preform fill the mold cavity exactly. When it does not, the resin flows where it wants, which is out the nearest vent. Careful preform trimming and proper mold sealing are necessary to make a full part.

Secondary operations to drill holes and slots are done with conventional methods to correspond to the modest annual volumes. Holes are singularly drilled with stationary drill-and-bushing units driven by a manually-positioned drill motor. Large holes are made with conventional hole saws. Slots are made with hand-operated routers. Even with such manual intensiveness, a bumper having over forty holes can be finished in less than five minutes. The chassis bolt retainers are attached to the bumper with rivets. The replaceable support plates are attached with sheet metal screws.

Each bumper is inspected for holes and brackets. It is also reviewed for surface integrity and visible reinforcement fibers. As part of the quality assurance plan, bumpers are dimensionally checked on a GM-supplied gage. Bumpers are statically tested on a random basis to ensure performance does not erode. In addition, plaque samples are tested for physical properties of each lot of resin. Such measures are taken to grow the data base for extension to future programs.

SUMMARY

There are few truly structural applications of Liquid Composite Molding using high speed polymer systems. To help advance the technology, an automotive bumper was targeted as a piece of business that would have far-reaching effects. A program of low risk was presented and supported by management behind the Corvette. As long as the proposed LCM bumper performed equally to the incumbent system and demonstrated other advantages, there was corporate support for production. Inspired by this opportunity, Dow Chemical and Ardyne engaged in a project to make it happen. An analysis of the incumbent bumper system served as a baseline against which proposed designs could be compared. The best design was drawn with respect to making a one-for-one change-out with the baseline bumper. A model and prototype mold were fabricated. Shortly, thereafter, parts were made and tested. The results were very encouraging; and when combined with the LCM bumper's advantages, the program attained critical mass and became a production reality.

Designing with LCM properties and processing features brought about several advantages in the proposed bumper system besides lower cost. The LCM bumper is simpler yet tuned to perform specifically as intended, no more and no less. The number of attachment parts and fasteners were reduced. Added adjustment was incorporated where desired. Mass was reduced. Service access openings were enlarged. Only one screw size is used throughout the impact bumper system. The assembly process at the

Bowling Green plant was simplified. As production continues, a data base of performance history is being collected. With these and other advantages, the LCM bumper system is welcomed by Corvette Engineering. Leadership in technology such as this is essential to leadership in the marketplace.

ACKNOWLEDGEMENTS

Dr. Luis Lorenzo
Engineering Analysis Service, Inc.
Auburn Hills, Michigan

Dr. King Him Lo
Shell Development Company
Houston, Texas

Mr. Tom Lohmann
Bowling Green Plant
Bowling Green, Kentucky

Mr. David Fischer
Dow Chemical U.S.A.
Southfield, Michigan

Mr. Ron Pruitt
Dow Chemical U.S.A.
Midland, Michigan

APPENDICES

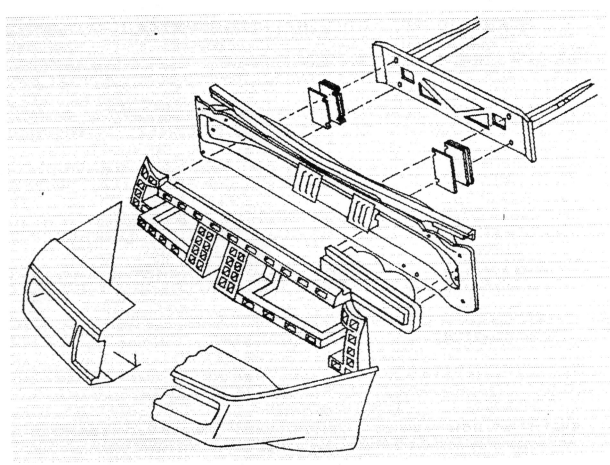

Figure 1: Baseline bumper system showing the fascia, energy absorber, one of two lamp assemblies, impact bumper bar with fascia reinforcement, shim plates, and chassis rails with cross-member.

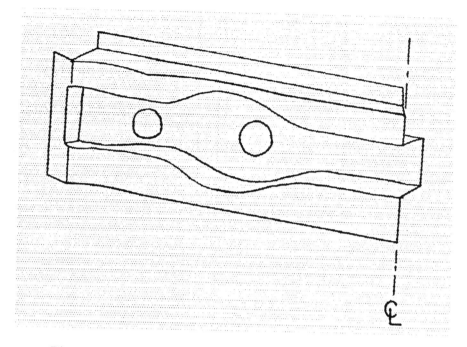

Figure 2:   Proposed LCM impact bumper bar.

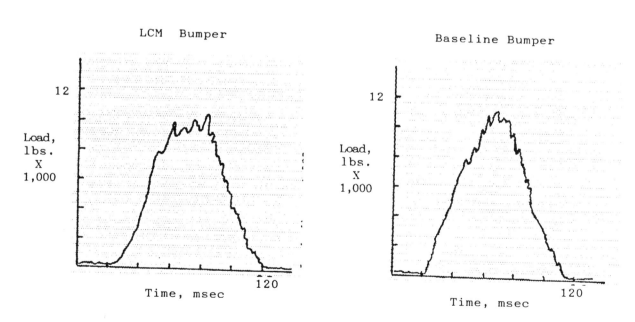

Figure 3:   Dynamic impact test results of LCM and Baseline bumpers.

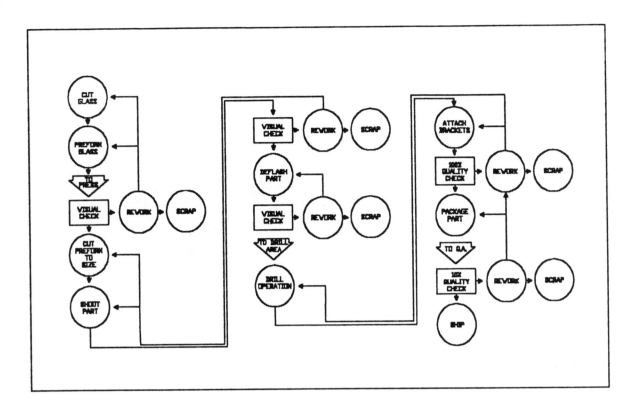

Figure 4: Process Flow Diagram for Bumper Bar Production.

EVALUATION OF FABRIC PREFORMS FOR HIGH VOLUME MANUFACTURE OF AUTOMOTIVE COMPOSITES

John J. Kutz, Frank K. Ko
Fibrous Materials Research Center
Drexel University
Philadelphia, PA USA

ABSTRACT

In order to facilitate the selection of optimum preform architecture for high volume stamping manufacturing processes, several commercially available multidirectional fabrics were characterized by measuring the mechanical properties under shear, bending, and omnidirectional tensile loadings. Special methods were developed to quantify the response of directional fabric in the stamping process. The results of this testing and their implications on fabric preforming are discussed in this paper.

1.0 Introduction

The driving force in the automotive industry is the ability to provide unique design flexibility at the low monetary investment levels required for low volume economic feasibility.[1] It is anticipated that through large scale parts consolidation and judicious selection of composite fabrication processes, the systems cost of composites has competitive potential over steel through savings in tooling and assembly costs, while providing far superior specific performance.

Additional stiffness requirements for automotive structural composite applications is leading to an increasing interest in multidirectional continuous fiber preforms. There has been an increased interest in textile preforms which, because of their relatively high strength translation efficiencies, lead to an overall reduction of system weight and cost and the potential for process simplification.

The key to successful implementation of composite fabrication processes is an understanding of the dynamics of materials-process-structure interaction. While fibers and matrices are of basic importance in the performance of the composite, the fiber architecture of the preform plays a crucial role in the translation of fiber properties to the composite and dictates the ease of fabrication of the composite.

In order for structural automotive composites to become a production reality, several open issues must be addressed. Demonstration of rapid process cycle times and improved preforming of reinforcement materials are necessary elements for structural automotive composites to be acceptable for mass production.

Stamping is a mature and highly automated production technique used in producing high volume parts for the auto industry. The automotive industry would ideally like to see a fiberglass preforming process which gives equivalent production efficiency to that of the sheet metal forming process.

Automated fiberglass preform fabrication processing involves the stamping of the fiberglass preform to achieve a near net shape for rapid insertion into a resin transfer molding tool. The current commercially viable preforming process for structural composite automotive applications involves the combination of a thermoformable random continuous fiberglass mat [1,2] and a directional oriented fiberglass fabric. The random and continuous preform materials are combined in layers and heated to melt the thermoplastic binder, then quickly transferred to the matched tool and formed. Little difficulty is experienced in the forming of random continuous material alone but forming difficulty arises with the addition of the directional oriented material.

In order to further the development of structural composites in the automotive industry, progress must be made on optimizing the formability of directional fabric while maintaining necessary strength translation of directional oriented fabrics to fulfill stiffness requirements.[3] The question is then how must fabrics be analyzed in order to select the most appropriate for a given application.

The database on preforming for structural composites is not extensive because the technology is relatively new. An extensive data base is available on analysis of sheet metal forming [4] and one is being developed in the analysis of forming thermoplastic composite sheet materials. [5,6,7] Prior work in the field concerning fabric formability is limited to more traditional textile applications [8,9] and is not necessarily concerned with composite strength translation efficiency.

Similar problems to thinning and wrinkling in sheet metal forming are encountered in the forming of fabrics. Analysis work done in the sheet metal forming area concerns itself with measurement of strains in the major and minor axes of the sheet in the forming process and correlations of those strains are used to predict forming limits. Beneficial outcomes of sheet metal forming analysis have included insights into tool and part design, predictions of final part thickness and process developments such as optimizing forming pressure cycles, thus reducing the number of prototypes or forming trials required to produce an acceptable part.

Although there are some similarities in the characteristics of the forming of metal sheets and forming of fabrics, there is a vast difference in the analysis stemming from the fact that in general metals behave isotropically and fabrics anisotropically. Unlike metals, the formability of fabric does not depend on the plasticity of the material but rather on yarn to yarn interaction within the fabric. The mechanical and physical response of a fabric depend on the structural arrangement (fiber architecture) of the component yarns. Fiber architecture plays a primary role in dictating the processing and mechanical behavior of the fabric.

Analysis work concerning fabric formability done in the field has concerned itself primarily with apparel type fabrics. A very precise understanding of the dynamics of materials-process-structure interaction. While fibers and matrices are of basic importance in the performance of the composite, the fiber architecture of the preform plays a crucial role in the translation of fiber properties to the composite and dictates the ease of fabrication of the composite.

In order for structural automotive composites to become a production reality, several open issues must be addressed. Demonstration of rapid process cycle times and improved preforming of reinforcement materials are necessary elements for structural automotive composites to be acceptable for mass production.

Stamping is a mature and highly automated production technique used in producing high volume parts for the auto industry. The automotive industry would ideally like to see a fiberglass preforming process which gives equivalent production efficiency to that of the sheet metal forming process.

Automated fiberglass preform fabrication processing involves the stamping of the fiberglass preform to achieve a near net shape for rapid insertion into a resin transfer molding tool. The current commercially viable preforming process for structural composite automotive applications involves the combination of a thermoformable random continuous fiberglass mat [1,2] and a directional oriented fiberglass fabric. The random and continuous preform materials are combined in layers and heated to melt the thermoplastic binder, then quickly transferred to the matched tool and formed. Little difficulty is experienced in the forming of random continuous material alone but forming difficulty arises with the addition of the directional oriented material.

In order to further the development of structural composites in the automotive industry, progress must be made on optimizing the formability of directional fabric while maintaining necessary strength translation of directional system for analysis of apparel weight fabric has been developed by Kawabata [10]. Much of Kawabata's theory is applicable, but few of the test methods can be used on the industrial weight fabrics of the automotive industry, or are representative of the large deformations (fiber and yarn bending, yarn and fiber tension and compression, intra and inter yarn shear) encountered during the stamping process. Variables of each individual type of fabric may include method of construction, yarn orientation, fabric weight, fabric volume fraction, yarn linear density, yarn density in the warp and weft directions, and the density of stitch yarns. Variables are listed as fabric characterization in table I. Change in the variables mentioned will have effect on the a fabrics composite and manufacturing performance. It is the goal of this study to identify the critical variables involved and to develop some basis to aid in the evaluation of the formability of fabrics.

2.0 Materials and Methods

There are a large family of textile preforms available for advanced composites ranging from 3-D integrated net shapes to thin and medium gage multilayer fabrics. In addition to property translation efficiency, structural integrity, and formability, a key requirement of preform material for near term automotive structural composites is their availability at a reasonable cost. Fabrics for this study were chosen in regards to their ability to be rapidly and economical produced, commercial availability, high degree of strength translation efficiency, and initial subjective evaluation of there ability to be formed. Five E-glass fabrics were obtained from the following companies:

- Advanced Textiles, Inc., Seguin, TX
- J. B. Martin Company, Inc., Leesville, SC
- King Fiber Glass Corp., Arlington, WA
- Hexcel, Inc., Seguin, TX
- PPG Industries, Troy, MI

Since the objective of this paper is not to compare the relative performance of commercial fabrics, the physical and mechanical properties of the fabrics were not identified by manufacturer. It is the intent that the candidate fabrics for this study be representative of the fabric manufacturing method and fiber architecture.

All five fabrics are biaxial in orientation. Three of the fabrics employ a chain or a tricot stitch to hold the reinforcing yarns together thus being referred to as a multiaxial warp knit (MWK) fabric. Two of the MWK fabrics (+- 45, and +-+45) are made by impaled stitching method wherein the stitching yarns pierce through the reinforcing yarns. The (0/90) fabric was made by a method wherein stitching yarns wrap around and avoid impalement of the reinforcing yarns. Figures 8 and 9 illustrate the the MWK fabrics produced by the impaled and unimpaled methods. Two of the fabrics are considered woven fabrics, one a conventional (0/90) woven roving, the other (0/90/0) referred to as a "no crimp system" where the fiberglass reinforcing yarns remain linear and the woven polyester yarns hold the fabric together (figure seven).

The initial step in quantifying formability of a fabric consists of essentially of an in depth evaluation of the fabric sample and recording of the relevant findings. In the case of the evaluated samples, relevant data would consists any factor which would effect the preforming and composite performance of the fabric. This includes yarn size or linear density, fabric thickness, fabric weight, free volume, and yarn constraints such as number of stitch yarns and/or number of yarn intersections. Fabric thickness and corresponding volume fraction fraction was was measured at 560 grams/cm^2 (8 psi) in accordance to ASTM D-1777 "Standard method for measuring the thickness of textile materials"

				FABRIC CHARACTERIZATION				
Fabric type	Construction	Thickness (mm)	Weight (g/sq.m)	volume fraction	# of stitch yarns/m.	Yarn orientation	Yarn Linear density (tex)	yarns/m. warp and fill
1. +- 45	knit	0.508	597	50.5	354.3	50 / -50	685 / 685	354 / 354
2. +45-45+45	knit	0.736	861	52.1	708	50 / -50	1196 / 420 (ea.)	276 / 276 (x2)
3. 0/90	woven	0.724	746	53.8	-	0 / 90	2380 / 2380	59 / 46
4. 0/90	knit	0.635	528	45.1	276	0 / 90	1133	276 / 276
5. 90/0/90	woven	0.914	936	44.8	158	0 / 90	820 / 730 (ea.)	310 / 393 (x2)

Table I. Characterization of fabrics

It was the intention of the testing process described to simulate the dynamic action of the fabric experienced in the forming process. This experimental process combines existing techniques with newly developed techniques. An attempt has been made to quantify these interactions under simulated forming conditions.

During the process of fabric forming several modes of deformation may occur:

1. Deformation due to the trellis effect
2. Yarn straightening and related uncrimping
3. Shear slip
4. Fiber stretching

Fiber stretching due to a tensile force is of minor importance during the forming of fabrics. Most of the fiber stretching will be at relatively small strains in the elastic region of the fiber and is not considered relevant as we are more concerned with plastic strains in the neighborhood of 35% in the preforming of complex shapes.

Yarn straightening from uncrimping allows extension in woven fabrics. The amount of extension is relatively small, partially elastic, and relative to the amount of nonlinearity introduced into the fabric during manufacture.

Shear slip is the deformation of a fabric resulting from traverse displacements of fibers. This traverse displacement makes the the fabric locally tighter or looser. Shear slip is an unwanted effect which occurs during slipping of the fabric over a sharp edge or rough surface.

In the deformation of a fabric due to the *trellis effect* the yarns are not stretched, they only change direction. The process needs relatively small forces and very large deformations can be obtained. The trellis effect is limited by the maximum deformation angle between the yarns, which is about 30 degrees for most fabrics.

The stress-strain curve of a fabric in the preforming process is generally nonlinear and can be broken into three portions. (figure one) The initial portion of the curve is due to small scale mechanical yarn interaction forces such as friction, shear and uncrimping; the second, more linear portion comes from larger scale interactions such as yarn slippage and loading of the fibers; the third portion is from fiber breakage. We are concerned with the first two portions of the curve in the evaluation of fabric formability. Test methods for the evaluation of directional fabric formability have been classified into these two categories depending on the amount of deformation applied to the material.

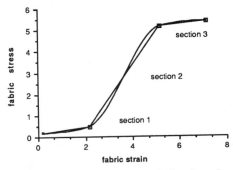

Figure 1. Typical stress/strain behavior of a fabric encountered in the preforming process

3.0 Response of the Fabric Under Small Deformation

With a small amount of deformation and low applied forces, minimal yarn interaction occurs. This relates to the subjective evaluation of "how formable a fabric feels." Tests of small deformation include the bending test and shear under normal loading. The intention is to establish a correlation between objective results and subjective evaluation.

3.1 Fabric bending stiffness:

The bending stiffness test is based on a standard textile testing method (ASTM D-1388) for measurement of the stiffness of fabrics. This cantilever type bend test measures the resistance of a fabric to bend under its own weight. Accordingly this provides an indication of conformability of a fabric in a composite formation process such as open mold lay-up where the loading on the preform is relatively small.

The bending stiffness test consists of sliding a strip of fabric off the edge of a platform until the weight of the fabric causes it to bend down to a reference angle for comparative purpose. (figure 2) The length of the strip of fabric is measured at this point, recorded and normalized to give a relative bending stiffness value. This value assists us in evaluation of stiffness by applying an objective number to what would be otherwise be just a subjective stiffness evaluation.

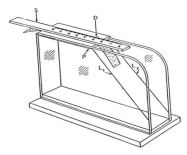

Figure 2. Fabric bending fixture

The proceedure for testing bending stiffness is as follows. Five 1" wide x 10" long fabric strips were cut at +45, -45, 0, and 90 degree directions. The fabric specimen was placed on a cantilever testing device, as shown in table II, and slid along the length direction. The length overhang "O", when the specimen is bent to 41.5 degrees as indicated in table II, is recorded. The flexural rigidity or bending stiffness, G, of the fabric was calculated based on the following equation:

$$G = W \times (O/2)^3$$

where

W = fabric weight per unit area, mg/cm^2
O = length of fabric overhang, cm

The same procedure was repeated for the opposite side of the fabrics for each of the specimens cut in the various directions. Five replications were made for each specimen. Table II provides a summary of the bending length and the average bending stiffness of the fabrics. Obviously the longer the fabric's overhang, the stiffer the fabric would be. It can be seen that the bending characteristics of the fabrics are directional and and fabric geometry dependent. It is generally true that the direction where the fibers are oriented tend to have a high resistance to bending. Fabric thickness and the freedom of fiber mobility play a significant role in bending resistance of the fabrics. It can be seen by examination of the test results that there is a relationship between the yarn linear density of the fabric and bending stiffness of the fabric. According to the results of this test the fabrics which use a relatively lower linear density yarns, appear to be the least stiff even after being normalized by weight.

STIFFNESS OF FABRICS							
material	Fabric overhang at various orientations(cm)				ave. length of overhang(cm)	fabric wt. (gm/cm)	flex. rigidity (mg-cm)
	1	2	3	4			
1. +-45 knit	9.4	9.2	7.2	3.8	7.4	65.1	3297
2. +-+45 knit	14.4	13.1	8.9	8.7	11.3	93.9	16930
3. 0/90 woven	7	7	16	15.5	11.3	74.6	13724
4. 0/90 knit	7.3	7.8	17.5	13.2	11.4	57.6	10672
5. 90/0/90 woven	6.9	6.5	14.1	10.5	9.5	101.9	10921

Table II. Results of the fabric bending test

3.2 Fabric shear under normal loading

The procedure for testing fabric shear under normal loading is described as follows. 10 x 25 cm. samples are cut at various representative orientations of a fabric. The choice of orientation represents the extremes from the highest to the lowest in resistance to shear while maintaining a balance in bias orientation. Samples were cut in 45 degree increments at 0, 45, 90, and 135 degrees, and correspondingly labeled orientations 1, 2, 3, and 4 for reference. When the yarn orientation of a sample is at an off angle, for example the +-50 degree fabric, the sample is cut as to balance the response of the fabric and not to bias yarns unevenly in either direction. The top and bottom edges of the sample are taped in the area which is clamped in the jaws of the fixture.

The specimen is placed in the fixture, clamped, and the pointer zeroed. A determined amount of dead weight is hung from the bottom clamp to simulate a normal load on the fabric. The fabric is then loaded laterally, inducing a shearing force into the fabric. The amount of force required to induce a one degree shear deflection is recorded as shown in table III. The fabrics were tested and the results documented for four different orientations. Upon examination of the force required to induce one degree of shear at a given orientation, we can note the anisotropic behavior of the fabric. Results have been plotted in polar form for various fabrics in order to illustrate the relative anisotropy between fabrics (figures 13 and 14). These plots illustrates the directional dependency of the fabrics and the magnitude of force required to deflect a fabric in a particular direction.

SHEAR RESPONSE UNDER SMALL DEFORMATION						
material	gms. wt. req'd for 1 deg. shear @ orientation				average wt. req'd (gm.)	average wt. fabric wt.
	1	2	3	4		
1. +-45 knit	142	125	500	475	270	0.452
2. +-+ 45 knit	150	125	550	525	337	0.392
3. 0/90 woven	75	75	485	475	243	0.326
4. 0/90 knit	125	100	750	600	393	0.746
5. 90/0/90 woven	175	125	1175	775	562	0.601

Table III. Results of fabric shear under normal loading test

As expected, there is a relationship between yarn interaction and the amount of force required to shear the fabric. Test results show a larger force was required to shear the fabric constructed with a high number of yarns for a given weight, and a lower load required to shear a fabric with a lower amount of yarns for a given weight. Fabric thickness may have some influence on the results of this test in that under normal loading, as one of the fabric deformations is to buckle out of plane. A thicker fabric will have more resistance to buckling and therefor require a greater force to induce a given amount of shear deformation. Examination of the normalized data shows that the results of the shear test for small deformations corresponds well to results of the fabric bending stiffness results and likewise to subjective evaluation of fabric formability.

4.0 Response of Fabric Under Large Deformations

When high loads are applied to a fabric a great deal of deformation can occur. A large deformation test more realistically simulates the current automotive process used. The tests which attempt to simulate this high loading force and high deformation include the omnidirectional tensile test, fabric shear under tensile loading in bias direction, and fabric thickness expansion test.

4.1 Omnidirectional Tensile:

In order to simulate the multidirectional nature of fabric deformation during the stamping and forming process, the fabrics were subjected to an omnidirectional tensile loading test. Originally developed for the measurement of the puncture strength of geotextiles, this method provides a means to measure the load-deformation characteristics of fabrics under multidirectional loading. With a well defined plunger geometry, one can create various loading conditions which the fabrics encounter in the molding process.

The omnidirectional tensile test simulates the forming of fabric with a constrained boundary condition, representative to if fabric is held tightly in a clamping frame during the stamping process. The fabric is rigidly clamped in a 15 cm. diameter frame and a 5 cm. diameter plunger is forced through the center. Fabric deformation comes about by uncrimping of yarns and fiber elongation. Additional localized stresses are induced by the geometry of plunger and fixture edges. This can be used advantageously in that this indicates how a fabric will perform when stamped in a high local stress situation. Results from this test relate to how a fabric will behave if rigidly clamped or constrained by the mold walls during the forming process.

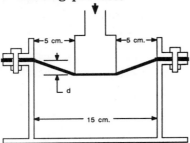

Figure 4. Omnidirectional tensile test fixture

The omnidirectional tensile test was carried out by placing a 25 by 25 cm. fabric in the test fixture. Five specimens were tested for each fabric for with a plunger speed of 50 mm/min. Taking the areal density into consideration, the load on the fabric was normalized by the following equation:

Specific stress $S = (P/A \cdot L)$ (g/tex)

where

P = load, g
A = areal density, g/m^2
L = plunger perimeter, mm

The specific stress of 1 g/tex for the glass fabric is equivalent to approximately 35.55 ksi.

The strain on the fabric is defined as follows:

$e = ((d^2 + a^2)^{1/2} - a)/a$

where
d = plunger displacement
a = gage length or distance between plunger boundary and jaw.

For discussion purposes, the ratio of the rupture strain and stress is referred to as the conformability index (C.I.). Results are calculated and summarized in table IV. The conformability index is is a measure of compliance which reflects the the conformability of the fabric and the ability of the fabric to deform without failure.

	OMNIDIRECTIONAL FABRIC TEST RESULTS				
Fabric type	Load (kg)	stress(g/tex)	deflection(cm)	strain (%)	strain/stress
1. +- 45 knit	136.6	45.7	1.67	5.3	0.116
2. +-+45 knit	156.8	36.4	3.05	16.6	0.456
3. 0/90 woven	132.6	35.9	3.5	21.4	0.596
4. 0/90 knit	93.1	35.2	1.97	7.3	0.207
5. 90/0/90 woven	159.6	34.2	2.56	12.1	0.351

Table IV. Results of the omnidirectional fabric tensile test

Under confined boundary conditions such as simulated by the omnidirectional tensile test, predominant deformation mechanisms will include yarn straightening, shear slip, and fiber stretching. Results of the omnidirectional fabric tensile test indicate that under confined boundary conditions woven fabric shows the highest degree of compliance. Factors in this test which lead to a high degree of compliance or formability of a fabric include yarn crimp, fabric thickness, a large yarn size, and fabric integrity.

4.2 Fabric Shear under Biaxial Loading:

The Automate-Yendel Fabric Testing Machine is used to characterize the shear behavior of fabrics under biaxial loading. Originally developed to investigate the response of sail fabric under biaxial loading, this test provides us with a method to determine the relationship between load and extension of fabrics under more realistic conditions. Loads are simultaneously placed in the warp and weft directions by means of a simple system of levers and hand operated screw jacks. The magnitude of the load is measured by spring scales by way which the screw jacks act. (figure 15)

Specimens are cut and tested similar to the previously described test of shear under normal loading. Hand operated screw jacks are used to biaxially load the fabric. The magnitude of the load is measured by spring weighing machines through which the screw jacks operate. A bias or shear load can be applied to the fabric by means of weight beam arrangement. The resulting deformation, measured by a simple pointer and scale, is reported as the angle of shear deformation. All samples reported here were tested under a ten kilogram biaxial force.

	SHEAR RESPONSE OF FABRIC UNDER BIAXIAL LOADING					
material	gms. wt. req'd for 1 deg. shear @ orientation				average wt. req'd (gm.)	average wt. fabric wt.
	1	2	3	4		
1. +-45 knit	1572	1387	370	370	924	1.548
2. ++ 45 knit	2590	2450	1850	1870	2190	2.544
3. 0/90 woven	2775	2660	130	130	1423	1.908
4. 0/90 knit	3052	3065	111	115	1585	3.002
5. 90/0/90 woven	1295	1307	129	127	714	0.724

Table V. Results of shear testing under biaxial loading

Results of shear testing under a biaxial force indicate quite a difference in behavior when compared to results of shear testing under normal loading. What we would consider to be the fourth most formable fabric by testing under normal loading now becomes the top ranked fabric according to testing under biaxial loading. Accordingly, the most desirable fabric according to shear testing under normal loading now drops to fourth most desirable.

Cause for this drastic change is influenced by a number of factors. A greater order of magnitude in force is required to induce a shear on a fabric under biaxial force so the initial effects of yarn interaction are not as heavily weighted. Under a biaxial force, yarn bending and buckling are more restricted. The importance of the frictional forces of yarns being forced to slide past each other is now the measure of formability. Fabric thickness no longer plays a role as out of plane fabric buckling is not a factor.

4.3 Shear under uniaxial tensile loading of fabric in the bias direction

The fabric uniaxial tensile bias shear test was developed because of observations of the bending stiffness and wrinkle analysis tests. The primary fabric deformation mechanisms encountered in this test are deformation due to the trellis effect, yarn straightening, followed by yarn shear. As evidenced by the previously discussed tests, fabric shear or more specifically the interaction between yarn bundles plays a important role in the forming of fabrics. There is a need to quantify yarn and fiber mobility under large deformation.

Samples are cut on the bias of the fabric so that yarn interaction is to be measured and not actual fiber tensile strength. The actual test area size is 15 by 20 cm.. The clamping area of the sample is taped to aid in the handling and to assure even distribution of clamping force into the sample. The sample is clamped in a wide jaw grip and pulled at a strain rate of 5 cm./min.. Load, contraction of the width, and expansion of the thickness of the fabric are recorded as a function of strain on the fabric. (see figures 5 and 6)

Factors which lead to superior performance in this test include high yarn linear density and correspondingly a minimal amount of yarn interaction. Fabrics which exhibit a minimal change in thickness and width under extension are desirable as this will indicate if a fabric is prone to wrinkling or thinning in the preforming process. The +-45 and +-+45 degree fabrics performed poorly while the 0/90 fabrics performed well. The +-45 and +-+ 45 degree samples have a inherent disadvantage as the actual measured orientation +-50 degrees. This is disadvantageous in that yarns are given less angle to rotate before jamming.

Evaluation of various fabric shear stress/strain diagrams (figure16) indicates how different fabric constructions perform relative to each other. Results of testing show that as fabric deformation increases, load on the specimen and thickness of the specimen increases. In some fabrics there was noted a point where the yarns begin to jam against each other from the force of contraction, severally restricting the formability of the fabric. Plots of percent fabric extension vs. increase in fabric thickness indicate that most of the build up or yarn thickness occurs between 10 and 40% of fabric extension. Similar results are displayed for percent extension vs. width. Deformation due to the trellis effect is limited to a maximum deformation angle of 30 degrees which corresponds to approximately 35% strain in this test.

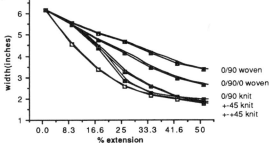

FIGURE 5. Plot of percent extension vs. width for shear under uniaxial tensile loading in the bias direction

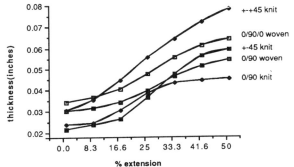

FIGURE 6. Plot of percent extension vs. thickness for shear under uniaxial tensile loading in the bias direction

4.4 Fabric Thickness Expansion:

A major problem in the current preforming process is the uneven migration of material which eventually leads to wrinkles and folds in the preform. The thickness expansion test was developed with intent to combine and measure all of the individual parameters which lead to fabric formability under one realistic test simulating the forming of a fabric. The thickness expansion test like the shear under uniaxial tensile loading in the bias direction test is concerned with the thickness expansion of a fabric as it is being formed. The thickness expansion test is performed placing the fabric between two plates and measuring the change in thickness of the fabric by the corresponding the separation of the plates. By using two plates we now have the option of variable clamping pressures. The fixture is circular in design and uses a hemispherical plunger in order to apply the load evenly to all directions of the fabric.

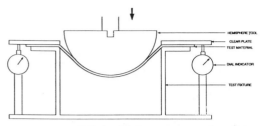

Figure 7. Fabric thickness expansion fixture

A 30 cm. diameter sample of fabric is placed between two 30 x 30 cm. acrylic plates with six inch diameter holes in their centers. Clamp pressure is variable by amount of plate surface area in contact with the fabric and amount of weight applied to the plate. The starting fabric thickness is noted and a 15 cm. hemispherical plunger is forced its depth through the holes in the plastic, forming the fabric into a hemispherical shape. The new thickness of the fabric is noted and compared to the starting thickness.

	FABRIC THICKNESS EXPANSION					
Fabric type	Construction	Thickness (mm)	volume fraction	thickness expansion @ 100 g/sq.cm.	thickness expansion @ 250 g/sq.cm.	expansion thickness (average)
1. +- 45	knit	0.508	50.5	0.024	0.009	0.032
2. +45-45+45	knit	0.736	52.1	0.096	0.049	0.099
3. 0/90	woven	0.724	53.8	0.042	0.013	0.054
4. 0/90	knit	0.635	45.1	0.045	0.024	0.054
5. 90/0/90	woven	0.914	44.8	0.024	0.012	0.021

Table VII. Results of the thickness expansion test

The primary deformation mechanism encountered in the fabric thickness expansion test is deformation due to the trellis effect. Factors leading to good performance appear to be fabric volume fraction, yarn mobility, and fabric weight. Low fabric volume fraction allows more room in the fabric for the yarns to deform and thus allows for a larger angle of deformation by the trellis effect. This is an important factor in this test and subsequently in the actual forming process. As yarns migrate during the forming process, they tend to pack together as to occupy all of the free area of the fabric until there is no where else to migrate to except out of the plane of the fabric. This migration out of the plane of the fabric or thickness expansion will eventually lead to wrinkling and folding of the fabric. Yarn mobility again plays an important role in the forming process. Restricted yarn movement in combination with a high fabric volume fraction leads to premature yarn jamming from the trellis effect and therefore out of plane fabric expansion and wrinkling. Results of the thickness expansion test indicate that with light clamping, a thin fabric appears to be better formable than a thick fabric.

5.0 Conclusions:

A series of test methods is available for evaluating the formability of industrial weight fiberglass fabrics. Some of the test methods have been adapted from existing textile test methods and some have been developed specifically for investigating the material-structure-process interaction encountered in the stamping of industrial weight fiberglass fabrics. The described method categorizes formability into low deformation and high deformation. Test results indicate that at low deformation levels, objective results correlate well with subjective evaluation. At high deformation levels which are more representative to the current stamping process, fabric is found to behave very differently. What would be the most desirable fabric for forming according to tests for small deformation of subjective evaluation, is not necessarily the most desirable fabric, according to tests for large deformations.

In addition to interpreting how fabrics should be ranked for formability within each test, there is a need to rank the relevance of each individual test to simulation of the particular forming process. The current forming process is by no means optimized and it is hoped that some of the relationships uncovered in this study will assist in the selection of a fabric for preforming and will give insight on how the forming process might be modified to give better results.

With respect to the latter, future research intends to correlate the test results with the actual forming processes.

Acknowledgements:

The authors gratefully acknowledge the financial support provided by the Ford Motor Company through the Textile Structural Composites Consortium and the Drexel Industrial Internship Program. The interest that Carl Johnson and Norm Chavka of the Ford Scientific Research Laboratory have shown in this work is greatly appreciated.

References:

[1] C. F. Johnson, and N. G. Chavka, "Preform Development for a Structural Composite Crossmember" in Proceedings of the Fourth Annual ASM/ESD Advanced Composites Conference, Dearborn, MI, September, 1988.

[2] S. G. Dunbar, " Preforming Continuous Strand Mat", in Proceedings of the Fourth Annual ASM/ESD Advanced Composites Conference, Dearborn, MI, September, 1988.

[3] F. Ko and J. Kutz, "Multiaxial Warp Knit for Advanced Composite",in Proceedings of the Fourth Annual ASM/ESD Advanced Composites Conference, Dearborn, MI, September, 1988.

[4] N. Rebelo, "FEA of Forming" Machine Design, Pages 121-124, June, 1988.

[5] R. K. Okine, "Analysis of Forming Parts From Advanced Thermoplastic Composite Sheets" SAMPE Journal Volume 25, Number 3, May/June, 1989.

[6] J. Muzzy, X. Wu, and J. Colton, "Thermoforming of High Performance Thermoplastic Composites" SPE 47th Annual Technical Conference, May, 1989.

[7] O. Bergsma and J. Huisman"Deep Drawing of Fabric Reinforced Thermoplastics" , Delft University of Technology. May 1988.

[8] "Woven Fabric Reinforced Composites for Automotive Applications" Materials Science Corporation, TFR 1605/8102,

[9] F. Ko and H. Soebroto, "Evaluation of Woven Fiberglass Fabrics Conformability" , Textile Research Lab, May 1984.

[10] S. Kawabata, "Nonlinear Mechanics of Woven and Knitted Material," Chapter 3, in *Textile Structural Composites,*T. W. Chou and F. K. Ko editors, Ensevier Publishing, 1989.

FIGURE 8. S.E.M. Micrograph of +-+45 MWK Fabric (impailed yarn)

FIGURE 10. S.E.M. Micrograph of +-+45 MWK Fabric (nonimpailed yarn)

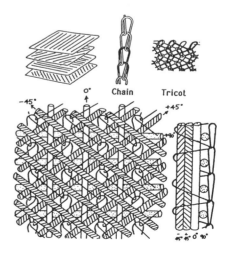

FIGURE 9. Structural geometry of MWK fabric

FIGURE 11. S.E.M. Micrograph of 0/90/0 Woven Fabric

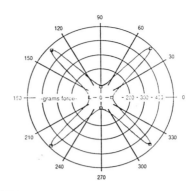

FIGURE 14. Shear behavior of 0/90 woven fabric tested at different orientations

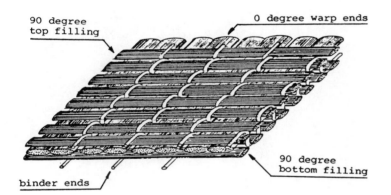

FIGURE 12. Structural geometry of 0/90/0 woven fabric

FIGURE 15. Photograph of Automate/Yendel fabric testing machine

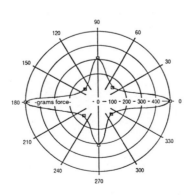

FIGURE 13. Shear behavior of +-45 woven fabric tested at different orientations

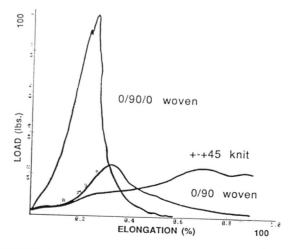

FIGURE 16. Representative load/elongation plots for shear under uniaxial tensile loading in the bias direction

DEVELOPMENT OF AN AUTOMATED CHOPPED FIBER GLASS PREFORM MANUFACTURING SYSTEM

Dean M. Perelli
General Motors Corporation
Advanced Engineering Staff
Warren, MI USA

ABSTRACT

There is a strong push in the automotive industry to utilize structural composites to achieve greater design flexibility at reduced cost and weight. Directed chopped fiber preforms offer the greatest potential to maximize the cost benefits of structural composites. A laboratory system has been developed to automatically produce complex three dimensional chopped fiber glass preforms. The system utilizes a robot which manipulates a fiber glass chopper assembly to chop and spray fiber glass onto a part-shaped, metal screen. The metal screen is contained in a closed chamber over an air exhaust platform. The system consistently produces uniform preforms in a one minute cycle time. This paper outlines the development steps followed to create this laboratory system.

Composite materials are making their way into the automotive industry in the form of structural and semi-structural body and chassis components. Typical components being considered include: bumper beams, floor pans, engine compartment panels, front apron structures, suspension cross members and pick-up boxes. Liquid composite molding (LCM) shows great promise for manufacturing these types of components. However, at present LCM advantages are negated by the high labor content and process variability associated with the manufacture of the reinforcing preform for the molded part. General Motors set out to determine the feasibility of manufacturing chopped fiber glass preforms in a one minute cycle time by building a lab scale preform manufacturing system. The system utilized a robot to reduce labor and preform variability. The purpose of this report is to present the results obtained in the development of this system.

PREVIOUS ART

Chopped fiber glass preforms have been manufactured for many years. Historically, two systems have been used; a plenum chamber system and a directed fiber preform system.

The plenum chamber system involves chopping and dropping chopped glass fibers onto a part shaped metal screen in a controlled environment. The screen rotates on a horizontal turntable and vacuum is drawn through the screen to aid in the glass deposition. Binder is sprayed onto the glass as it is deposited into the chamber. The preform is later cured and removed from the screen. Although the system appears to be automatic, it cannot provide uniform preforms for complex shapes.

The directed fiber machine is comparable to the plenum chamber machine. The screen rotates on a vertical turntable and vacuum is drawn through the screen. In this case, however, the chopped glass fibers are blown through a flexible tube and directed toward the preform screen where they are held in place by the vacuum. Binder is once again simultaneously sprayed onto the glass. Preform spray-up is followed by oven curing. Although this method can provide preforms with some degree of uniformity, it is highly labor intensive, and preform reproducibility is not guaranteed. Preform quality is usually a function of the skill level of the operator.

The concepts involved in these two systems led to a hybrid design for the single station system outlined in this report.

SYSTEM DEVELOPMENT AND DESIGN

The single station system which was designed, engineered, fabricated and tested was a hybrid of the two conventional methods mentioned above. The system employed the enclosed plenum chamber along with the concept of directly spraying the chopped fibers onto the preform screen. The hybrid manufacturing system utilized a robot to direct the flow of chopped glass fibers and binder onto a horizontal part shaped screen. A vacuum was drawn through the preform screen in the enclosed chamber. The vacuum was used to aid in directing glass/binder deposition onto the screen as well as to help contour the chopped fibers and hold them in place on the preform screen. Following spray-up, the preform was manually removed from the enclosed chamber and oven cured. The screen used in this study resembled the shape of a motor side compartment panel (figure 1). The motor side compartment panel shape was chosen because of its deep draws and complex geometry, and there was also a tool available to evaluate the preforms in the molding process. The following discussion presents and describes general features of the design, which were subsequently put into practice.

GLASS DELIVERY - System feasibility was evaluated by varying different equipment components to produce consistent preforms with uniform glass/binder content, sufficient stiffness for the typical handling requirements and integral strength for the molding process.

Major system design changes led to three systems for delivering and depositing glass fibers onto a preform screen.
1. The first system involved cutting glass roving using a chopper gun located on the wrist of a robot. The robot directs the distribution of the chopped fibers along the contour of the preform screen (figure 2).
2. The second system involved chopping glass rovings outside the enclosed chamber using a large stationary chopper and delivering the chopped glass to the preform screen through a flexible tube. The end of the tube is manipulated by the robot which directs the chopped glass onto the preform screen (figure 3).
3. The third system involved cutting glass rovings on the end of the robot wrist utilizing a fiber glass chopper assembly (figure 4).

The third system was found to perform better for complex geometries than systems one and two. Preforms made with system three contained less than 6% weight variation when measured at identical areas in different preforms. The overall glass density variation was kept to a minimum. Density variations were consistent from preform to preform, and can be controlled through accurate robot programming. The preforms were of sufficient rigidity and strength for automated transfer and material handling during the molding process.

FIBER GLASS CHOPPER ASSEMBLY - The fiber glass chopper assembly was designed to supply chopped glass from continuous glass rovings for deposition onto a preform screen by automated means. The assembly used in system three could chop fiber glass and uniformly deposit the chopped fiber glass onto a complex three dimensional preform screen. This assembly included a portable mechanism to process three rovings simultaneously, a modified inductor, binder spray guns, and a short length of rigid tubing which was connected to the discharge end of the inductor. The robot directed the assembly which discharged glass over the preform screen.

BINDER DELIVERY - The binder application was designed for liquid spray deposition. A five gallon pressure pot was utilized to feed the binder material into the spray guns. The spray system included in-line strainers and fluid regulators.

Complete coverage of the binder onto the preform was accomplished with two spray guns mounted on the chopper assembly. This arrangement allows the binder spray pattern to follow the path of glass deposition. Stationary spray guns produced neither sufficient nor uniform binder coverage.

Two automatic sprayguns were tested. Both sprayguns offered high capacity output and flexibility for atomization and spray pattern control. However, one gun was significantly lighter in weight (0.5 lbs. compared to 3 lbs.). This became extremely important when weight on the robot wrist became a control problem.

Mild steel spray fittings had to be replaced with stainless steel fittings due to corrosion caused by the water based binder. In line strainers and a regular gun maintenance schedule reduced gun fouling.

Spray atomization was found to perform best at a medium-fine setting using a medium spray pattern (5" wide at 9" from the gun tip). A wide fan (9" wide and 9" distance), and a narrow fan (2" wide at 9" distance) were both inappropriate for spraying these preforms at the operating distance. A discharge rate of 900 cc/minute/gun was used to produce preforms with optimal binder content.

FIBER GLASS FEED SYSTEM - Doffs of

fiber glass were located on the roof of the preform enclosure. Three rovings of fiber glass were fed to the chopper gun through ceramic eyelets mounted on the robot arm. Ceramic eyelets were used to optimize the roving drag to the chopper. Too much drag through the system decreased the production rate and increased wear on the chopper gun components (motor bearings, rubber cot, etc.). Too little drag caused the fiber glass rovings to become tangled around the robot arm.

SINGLE STATION BOOTH, FAN AND DUCTWORK - The booth was designed and built to maintain a continuous vacuum on the preform screen. This is necessary in both the plenum chamber and the directed fiber designs. Thus, a lower enclosed chamber was incorporated into the hybrid system and the preform screen was positioned over it.

Vacuum was maintained on the lower chamber by drawing air down through the screen. The air was drawn through an underfloor chamber and a short length of ductwork to a centrifugal fan and discharged through the roof. The fan was capable of drawing 20,000 cubic feet per minute at 9 inches water gauge pressure. Dampers were placed in the duct work on both the entrance and exit of the fan to vary the air flow rate to control the vacuum level under the screen. Wire mesh was located upstream of the fan entrance to catch any stray glass fibers which may have penetrated or bypassed the preform screen. Access doors in the ductwork allowed for cleaning of the mesh under the floor chamber.

The lower chamber and preform screen were enclosed in a booth to contain the glass fibers and binder. Panel openings were designed into the side of the booth for robot installation and freedom of robot movement. Two large doors were located in the front of the booth for transferring preforms, screens and assorted equipment in and out of the booth. Eight windows were installed along with two 300 watt explosion-proof lights to monitor the process from outside the booth.

ROBOTICS - The single-station system was designed to use robotics as opposed to hard automation. Robots were selected for their flexibility in spray up, ease of set-up, and availability. Two robot types were evaluated, a welding robot and a painting robot. The welding robot was chosen for its loading capability of up to 154 pounds. It was point to point programmable and electrically powered. The painting robot was hydraulically powered and was chosen for its flexible in movement and ease of programmability. The payload capacity for this robot was approximately 11 pounds. It outperformed the welding robot both in speed and accuracy. The robot maintained a constant spray pattern orientation at robot arm speeds up to 40 cm/sec. Its flexibility of motion and ease of programmability reduced set up time.

SUMMARY

A lab scale system has been developed which can be used to produce fiber glass preforms by automated means. Preforms produced by this system contain uniform glass and binder content, integral strength for the molding process and sufficient stiffness for handling. The results from this study can be used to develop a complete manufacturing module capable of high volume preform manufacture.

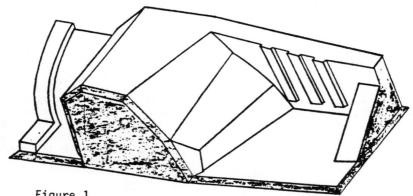

Figure 1
Motor side compartment panel preform screen

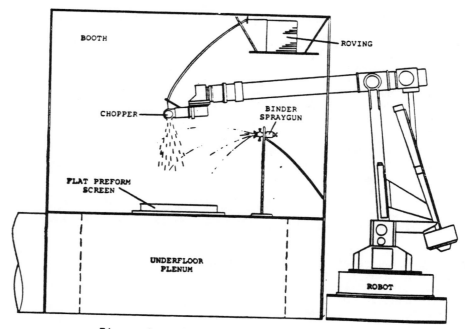

Figure 2   System 1

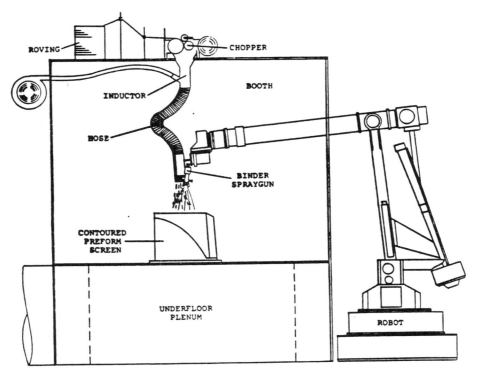

Figure 3   System 2

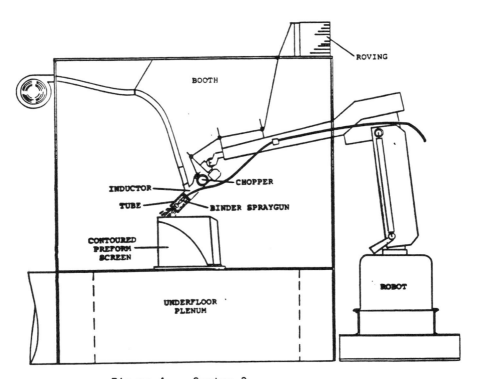

Figure 4   System 3

COMPOSITE PREFORM FABRICATION BY 2-D BRAIDING

H. Benny Soebroto, Frank K. Ko
Drexel University—FMRC
Philadelphia, PA 19104 USA

ABSTRACT

The performance of composites can be improved by optimizing the fiber architecture of the preforms. The 2-dimensional (2-d) braiding process are detailed and comparisons are made with other textile preforms especially woven fabric. The types of braiding machines are discussed and the principal components of maypole braiding machine are explained. Parameters of braiding process and several examples of braid applications are shown.

COMPOSITE STRUCTURES offer the potential of higher performance material at a lower weight due to higher mechanical properties of the reinforcing fiber and lower specific gravity. The specific gravity of composite reinforcement fibers vary from 18% for Kevlar to 32% for glass when compared to steel. The strength of reinforcement fibers vary from 400% for Kevlar to 760% for glass when compared to steel as shown in Table 1. The mechanical properties of selected engineering materials are shown in Table 1. The advantages of composite are enhanced when their mechanical properties are expressed in term of specific properties. Figure 1 shows the relationship between the specific strength and specific modulus of some engineering materials. Specific strength has a unit of inches and it is calculated by dividing the strength (psi) by the specific gravity (lb/cu in). Steel and aluminum show relatively low specific mechanical properties when compared to some reinforcement fibers in use today.

The cost of composite at present is generally higher than metal or plastic. The cost of raw materials for composite may vary from $1.00/lb for glass to thousands of dollars per pound for some graphite and ceramic reinforcement fibers and from $1.50/lb for vinyl ester to hundreds of dollars per pound for metal and ceramic matrices. Processing cost which may be in the thousands of dollars for some high temperature application composites have impeded the introduction of composites into large volume applications. The processing cost of composite can be reduced to minimal when the preform processing and composite consolidation are kept to the most efficient methods. Another method of cost reduction is to use the composite to reduce the number of parts and considering higher performance and lower weight of composites.

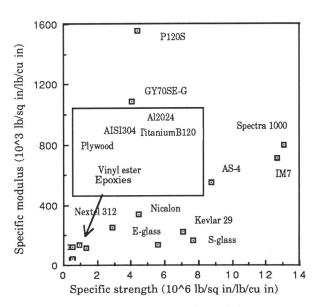

Fig. 1 - Specific strength and modulie of selected engineering materials.

Table 1 - Properties of Selected Engineering Materials*

Material	Density (g/cc)	Strength (ksi)	Specific strength(in)	Modulus (Msi)	Specific modulus (in)
Spectra 1000 ™	0.97	450	12.840	27	770
IM7 ™	1.78	800	12439	44	684
AS-4 ™	1.80	550	8457	34	523
S-Glass ™	2.48	665	7422	12	138
Kevlar 29 ™	1.44	353	6785	10	192
E-Glass ™	2.60	500	5323	10	112
Nicalon ™	2.55	390	4233	28	309
P120S ™	2.18	325	4126	120	1524
GY70SE-G ™	1.96	270	3813	75	1059
Nextel 312 ™	2.70	250	2563	22	226
Titanium B12V	4.85	200	1141	15	84
Aluminum 2024	2.77	70	699	11	106
Steel AISI304	8.03	87	300	28	97
Vinyl ester	1.12	12	284	0.5	12
Epoxies	1.21	10	229	0.4	9
Plywood(Douglas)	0.51	1.8	95	1.8	96

* From various sources

Improved composite performance can be obtained by using different materials or increasing the efficiency of the same materials. Higher efficiency is achievable through the utilization of optimal fiber architecture. Significant amount of composites had been fabricated using lay-up and filament winding processes. These processes produce structures whose properties excel in one or two direction but at the same time this arrangement allow delamination (inter laminar crack propagation) to occur as well as warping due to thermal stresses. Optimum performance may require more complex fiber architectures such as nonwoven, woven, knit and braid. These structures are classified as textile preforms.

TEXTILE PREFORMS

The classes of textile preforms are illustrated in Figure 2. Nonwoven fabric as illustrated in Figure 2a. generally contain continuous or discrete length fiber which are bonded together by secondary material, selective remelting of the fiber or fiber entanglement such as in the needle punching process. The fiber in the nonwoven fabric is generally randomly orientated in a plane but it is possible to have most of the fiber orientated at 0°, 0° / 90° or 0°/ 90° ± 45° depending on the specific method of fiber processing. Nonwoven preforms may be formed to final shape by spraying loose fibers onto a mandrel or pressing to shape from a flat sheet.

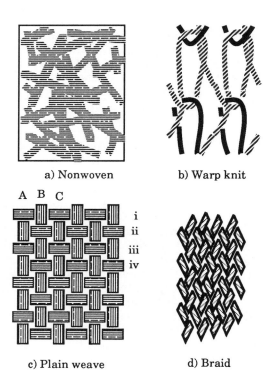

a) Nonwoven   b) Warp knit

c) Plain weave   d) Braid

Fig. 2 - Classification of textile preforms

Knitted fiber preforms usually contain 0° and/or 90° yarns in addition to the knitting fiber. The warp knit as illustrated in Figure 2b is more adaptable for fiber reinforcement than weft knit because it can accommodate laid-in yarns relatively easily. The knitting yarn is used to maintain the orientation of the 0° and 90° yarns and thus it is very small in volume compared to the 0° and 90°yarns. Some knitting machine are now capable of placing 0°, 90° as well as ±45° yarns which are referred as multiaxial warp knit (MWK).

Woven fiber preforms is one of the most commonly used textile preform for composites. Woven preforms as illustrated in Figure 2c unlike the knitted preforms does not require secondary material to form the reinforcement fabric. Woven fabric have inherently high properties in two directions, 0° (machine) and 90° (cross machine).

Braided fabric is formed by intertwining yarns and contain yarns orientated bias to the direction of manufacture. Figure 2d illustrates a braided fabric. The angle of orientation (ß) is typically from 20° to 70°. Unlike a flat woven fabric, a flat braid does not have any fiber ends along the edges of the fabric. The woven and braided fabric structure can be defined by the sequence of interlacing of the yarns.

All of the fabrics illustrated in Figure 2 are considered two dimensional. A 2-dimensional fabric is defined as a fabric whose thickness is formed using two interlacing (such as in a weave or braid) yarns or two interlooping (such as in a knit) yarns. Fabric containing additional yarns which do not interlace or interloop with other yarns such as in the MWK fabric is still considered as a 2-dimensional fabric. It should be noted that 3-dimensional fabric can be formed by weaving (ex. angle interlock) and braiding (ex. 3-D Braid). The word braid or braiding in this paper is limited to 2-dimensional braid or its process unless specified otherwise.

2-D BRAIDING MACHINE

Braiding is a process whereby yarns (any linear fiber assembly) are twisted (intertwined) around each other at a direction bias to the direction of manufacture. Braiding process has been known for centuries and its machine was invented in the mid eighteenth century.

TYPES OF BRAIDING MACHINES - There are generally two types of 2-dimensional braiding machines. The first type is The braiding machine which is discussed here and it is classified as maypole braiding machine. Another type of braiding machine is classified as high speed rotary braiding machine.

Figure 3 shows a maypole braiding machine with 32 yarn carriers. This model is floor mounted with a take up wheel located at the center and above the yarn carriers and produces fabric in a vertical direction. The rotary braiding machine is faster than the maypole type however it is less versatile in term of shapes which can be braided on the machine and the maximum number of carriers. The maypole type braiding machine is so called due to the similarity between the movement of the yarn in the machine with the action of dancers around a maypole.

MAYPOLE BRAIDING MACHINE - Figure 4 illustrates the basic process of circular 2-dimensional braiding on a maypole braiding machine. Yarn carriers, track for the carriers and the point where fabric is formed are shown. Yarn carriers are propelled along the tracks by the gear motions situated behind the tracks. The gears (also known as horn gears) are driven by a motor located outside the tracks for maximum flexibility. All of the yarn carriers move simultaneously. Half of the total number of carriers rotate clockwise and the other half rotates counter clockwise around the center of the machine. As one carrier pass a crossing between the two tracks, an overpass or an under pass with another yarn is formed.

Fig. 3 - 32 carrier maypole braiding machine

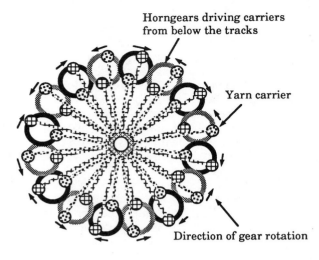

Fig. 4 - Basic process of braiding

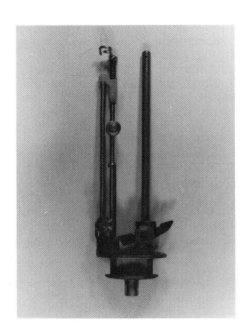

Fig. 5 - Mosspeed No. 2 yarn carrier

The fabric formed in this illustration comes out of the plane of paper. Braiding machine which produce fabric horizontally is referred as a horizontal braider.

YARN CARRIER - For a continuous production of braid, the main components of all braiding machines are yarn carriers with their spool of yarn, tracks for the yarn carriers, driving and take up mechanisms. Each of these components may vary depending on the manufacture of the machines and type of braiding machines. Maypole braiding machine can be specified by the pitch diameter of the horn gear and the number of yarn carriers. Yarn carrier as shown in Figure 5 carries a supply of yarn (not shown) and moves along one of the two tracks on the machine. There are several different sizes of yarn carrier. The yarn carrier shown in Figure 5 is identified as size no. # by it's manufacturer and it can accommodate approximately 1/2 lb of glass fiber.

Smaller yarn carriers are used for braiding small diameter structures such as suture. Larger carriers are used for rope or wire braiding. Larger carriers require larger diameter gears and enlarge the size of frame and space requirement. Since braiding require that the yarn are wound onto relatively small packages, these supply may be depleted relatively quickly if the size of the yarn is large and running at high speed.

Yarn carrier contain a replaceable spool of yarn wound onto a paper, wood or plastic tube. Figure 6 shows the maximum amount of glass fiber which can be loaded on number 2 braiding machines. This graph shows that the largest commercially available braiding machine, which is a 144 carrier machine, can be loaded with 72 lb of glass fiber for the braider carriers. Additional glass fiber can be loaded as a longitudinal lay-in component. The capacity for the longitudinal yarn is dependent on the method of supply. If the same yarn carrier as the braider is used, additional 36 lb of glass fiber can be loaded on the braiding machine.

PRODUCTION OUTPUT - The output of a braiding machine can be specified by the weight of material processed in a given time, by the speed of the braiding machine or both. Since fibrous materials have different linear density and specific gravity, comparisons should be done with the same materials or the same machine with different materials. Figure 7 shows the relationship between the number of yarn carriers and the maximum output of braided glass fabric. This is based on glass fiber with a yield of 225 yd/lb.

The maximum speed of a maypole braiding machine is limited by the rotational speed of the horn gear and its ability to propel the yarn carriers along the tracks. Braiding machines of different design have different maximum horn gear speed. Machines of the same design but with different number of carriers run at the same horn gear speed. However, due to the larger number of horn gears, the time required for a carrier to complete the circle (a track) is longer for larger braiding machine. Figure 8 shows the relationship between the number carriers in a braiding machine and the yarn speed in term of yarn carrier rotations per minute. The yarn speed is related to the number of picks per minute which the machine can produce.

Although small machines run at high yarn rpm. the circumference as represented by the number of lines of the resulting fabric is smaller than those produced in larger machines.

YARN TENSION CONTROL - The yarn in the carrier is passed through three guides. The center guide is controlled by a 6 1/2 inches tension spring which is available in several wire diameter. The function of the spring tension is to apply tension on the yarn as the yarn carrier follows the track on the machine while allowing yarn to be paid out as fabric is formed. In general finer yarns require fine diameter wire for the spring. The tension on the yarn should be such that the yarn is continuously taut without breaking the fiber or distorting the fabric.

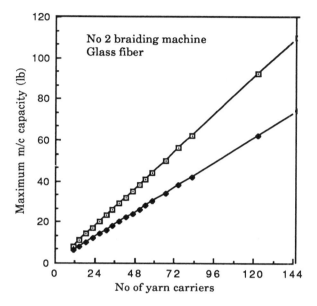

Fig. 6 - Capacity of braiding machine

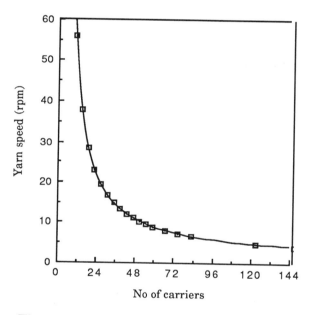

Fig. 8 - No. of carriers versus yarn rotational speed

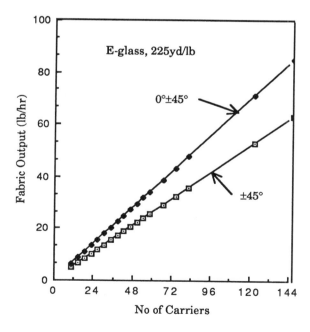

Fig. 7 - No of yarn carriers versus production ouput.

Different manufacturer of braiding machine and different style of yarn carriers use different length and diameter of spring. For the no. 2 carriers, the tension spring is 6 1/2 inches long with wire diameter ranging from 10 to 22 mil. When braiding machine is in operation, the tension on the yarn may vary from having no tension (loose) to the maximum that the spring allows. Figure 9 shows the relationship between the wire diameter of the tension spring and the maximum tension it apply on the yarn before it release new yarn. In the lower region where the wire diameter is between 10 and 12 mil, the short secondary spring comes into action during the last 1/8 in and it apply approximately 0.5 pound of tension momentarily. For industrial applications wire diameter in the medium or upper level are used.

Fig. 10 - Computer controlled horizontal braiding machine.

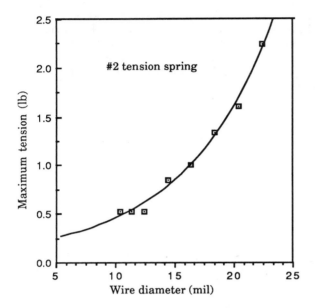

Fig. 9 - Effect of spring's wire diameter on yarn maximum tension.

COMPUTER CONTROLLED BRAIDING - A PC based hardware and computer program have been developed which can be used to run braiding machine with two independent motors.[1] The computer allows user to precisely programs the machine to lay continuously variable braid angle over a constant mandrel or identical braid angle over a continuously variable mandrel. This system also allows the user to perform automatic braiding of multiple layers. Figure 10 shows a computer controlled 24 carrier braiding machine with 12 stationary (0°) carriers located at the back and produce fabric in a horizontal manner. This braiding machine has two independent variable speed motors driving the braider carriers and the traverse movement. The traverse is the part of the braiding machine which allow the take-up of the braid. Mandrel can also be mounted on the traverse and the fabric is formed over the mandrel.

2-DIMENSIONAL BRAID

CONSTRUCTION - Braids are similar to woven fabrics in several ways. The first is the way in which the construction is specified. Both braided and woven fabrics are constructed of yarns which pass over or pass under another set of yarns. These "over" and "under" positions are illustrated in Figure 2c and 2d for a weave and a braid respectively. In the weave shown in Figure 2c, warp yarn "A" passes over and under yarns perpendicular to it (pick yarns i, ii,iii, etc). Warp yarn "A" passes over yarn i and then passes under yarn ii etcetera. The plain weave as shown in Figure 2c is the simplest construction for a weave since it only take two yarns in each direction to make one repeat (unit cell). A herring-bone pattern similar to regular braid can also be woven. Fabrics woven using a jacquard machine may have a large unit cell. A 60" wide woven fabric can have a repeat of 58" in width and unlimited in length.

The flat braid fabric shown in Figure 2d is called a diamond structure where yarn "A" passes alternately over and under the opposite yarns 1, 2, etcetera. The regular braid shown in Figure 11 have yarns which pass over and pass under two yarns in a repeat in both direction. Ajoining yarns are staggered which together give a herring-bone appearance. Most braiding machine are manufactured to produce this type of structure, however it can be converted to produce a diamond braid (1:1) as shown in Figure 2d. Some braiding machines can produce fabric which pass over and pass under three yarns in one repeat.

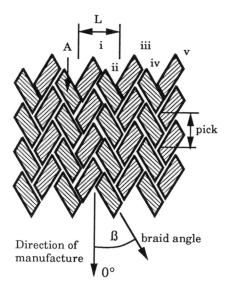

Fig. 11 - Regular braid (2:2)

Braiding machines designed for composite application are installed with hollow horn gear studs which allow axial yarns to be introduced into the fabric at each horn gear locations. The maximum number of axial yarns is equal to the number of horn gears or half the number of carriers. The longitudinal yarns do not interlace or intertwine with other yarns, instead they are trapped between the clockwise and counterclockwise yarns as illustrated in Figure 12. 2-dimensional braid which contain longitudinal yarn is also referred as a triaxial fabric. These yarns can be supplied from stationary yarn carriers as shown in Figure 10 or from a creel of large yarn packages located behind the track plates.

Fig. 12 - Triaxial braid

A weave is also specified by the number of yarns per unit length in both warp and pick directions. Yarns parallel to "A" and along the direction of manufacture are referred as warp or end yarns and those perpendicular to the direction of manufacture are referred as pick or filling. Woven glass fabrics can be obtained with weave density from 14 x 14 to 60 x 60 (warp/in x pick/in) with weights from 4 to 18 oz/sq in. Braided fabrics are more convenient to be specified by the number of unit cell in a given length and/or width. In Figure 11, the unit cell length "P" is referred as one pick or stitch and the width "L" is referred as a line. The number of pick per inch and the number of line in the circumference of a tubular braids as well as the type of intertwining (ex. 2:2) are used to specify braids. The weight is generally specified by the linear density of the yarn such as 225 yd/lb for the glass or 8000 denier for a graphite tow.
BRAID ANGLE AND COVER - A specification unique to braid is the angle by which the yarns lay from the direction of manufacture as illustrated in Figure 11. This angle is obtained from the relation between the speed of the the yarn carrier and the speed of the take up (traverse). For a given yarn carrier speed, faster traverse results in lower braid angle and vice versa. Lower braid angle produces high axial strength at the cost of low hoop strength. The mechanical properties of braided structure is dependent on the orientation of the yarns as well as the number of yarns.

The 144 carrier braiding machine is currently the largest braiding machine readily available. The maximum diameter of preform which can be braided is governed by several parameters. These parameters are: yarn size, yarn type, braid angle, machine size and type of machine. For a given machine size, the use of larger yarn size can produce preforms of larger diameter. For a given yarn size, loose fibers (untwisted) can provide more cover and thus produce larger diameter preforms. Figure 13 shows the effect of glass tow yield to the width of the tow when it is wrapped over a 2 inches diameter tube. It shows significant variation in width depending on the tension, twist and finish.

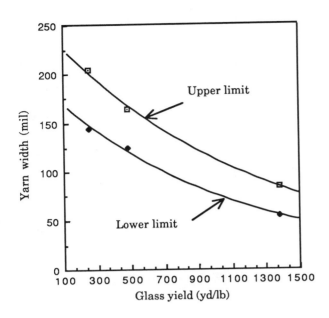

Fig. 13 - Glass yield versus width

Some braiding machines use large centralized spur gear to drive the horn gears. The location of this gear prevent any large mandrel or mold to be placed between the two traverse heads and prevent the braiding of large mandrel. Figure 14 shows a 70 inches diameter nozzle being braided using a 144 carrier braiding machine. The cover from each layer of fabric is poor in the large diameter section since the number of yarns is only limited to 144 braiding yarns and 72 longitudinal lay-in. Large preforms can be fabricated using higher braid angle to provide a full cover. Figure 15 shows the relation between braid angle the maximum diameter of fabric which can be braided at full (100%) cover. The cosine effect of the angle is very effective to produce fabric of large diameter when high braid angle is allowable. It should be pointed out that due to the flexibility of yarns, fabric of 3/4 the size suggested here can still be braided by jamming the fabric against each other. The resulting fabric is tight and has thicker wall dimension than normal.

MECHANICAL PROPERTIES - The mechanical properties of braided structures are highly dependent on the orientation of the fiber. Table 2 list the mechanical properties of graphite/PEEK composite fabricated by several braiding methods including 2-d braiding.[2] The triaxial braid tested here has a fiber orientation of 0°±45° with a fiber distribution of 74 % braider and 26% longitudinal.

Fig. 14 - Braided 70" diameter nozzle.

Table 2 - Mechanical Properties of Graphite/PEEK Composites (2)

Fiber architectue:	Tensile strength MPa (ksi):	Tensile modulus GPa (Msi):	Flexural strength Mpa (ksi):	Flexuralmodulus GPa (Msi):
3-D braid (i)	462 (67)	87 (12.6)	800 (116)	61 (8.8)
Triaxial braid(ii)	593 (86)	43 (6.2)	687 (99.5)	46 (6.6)
2-Step Braid(iii)	1070 (155)	128 (18.6)	804 (116.5)	83 (12)
Uniaxial	1518 (220)	134 (19.4)	828 (120)	97 (14.1)

(i) 100% braid
(ii) 74% braid
(iii) 10% braid

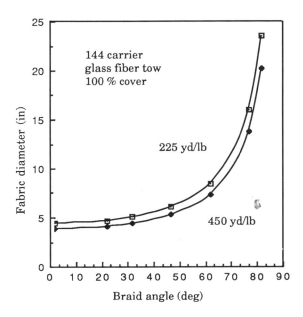

Fig. 15 - Effect of braid angle to fabric diameter

BRAID CONFORMABILITY AND APPLICATION - To measure the conformability of braid, various diameter of tubes were braided over mandrels using several sizes of glass fiber yarns. Mandrels containing convex, concave and complex surfaces were also braided. The braid was able to conform to all convex surfaces on the mandrel. The minimum diameter of curvature which the braid can conform is in the order 1/16 inch. The glass fiber is easy to process and the 4.8 % breaking strain of the fiber is sufficient to prevent breakage from sharp bending. Figure 16 and 17 show a braided drive leg and gear housing respectively. Concave surfaces could not be braided without additional accessories if the braid angle is restricted. Braiding accessories for concave surface are various wheels which force the fabric into the surface during braiding and inflated reinforced fabric or wire frame placed over the concave surface which stores additional fabric.

Braiding machine can be used to lay parallel tow of fibers when only one half of the yarn carriers are filled with yarns. For example, a 144 carrier braiding machine can be used to lay 72 yarns simultaneously over a mandrel at various helix angle. Figure 18 shows a complex shape preforms braided over a mandrel on 2-d braiding machine. "Extra" fabric were stored during braiding process which later could conform to the concave surfaces after removal of the braiding accessories. The braid angle was adjusted to minimize the number and area of the concave surfaces. Figure 19 shows an all terrain vehicle for the SAE Mini Baja competition which was built from a triaxially braided monocoque chassis with integrally braided stiffeners. Two previous mini baja bodies were also successfully built using 100% braided fabric. Other shapes which had been braided include complex drive shaft does not require universal joints, tubular struts, pressure vessels, aritifical ligament, artificial finger joint and artifical artery.

Fig. 16 - Braided drive leg

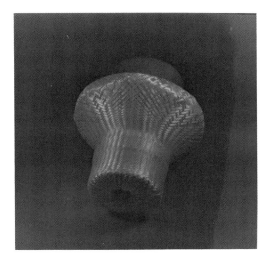

Fig. 17 - Braided coupling housing

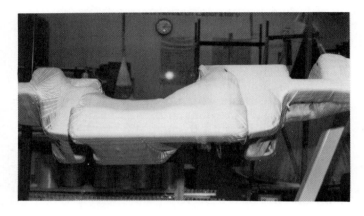

Fig. 18 - Braided complex shape mandrel

Fig. 19 - Braided monocoque all terrain vehicle with integral braided stiffeners.

ACKNOWLEDGEMENT

Much of the work presented here was supported by the Benjamin Franklin Partnership, Ford Motor Company and Wardwell Braiding Co.

REFERENCES

1. Yang, G., Pastore, C. Tsai, Y, Soebroto, H., Ko, F. "CAD/CAM of Braided Preforms for Advanced Composites". Proceedings of The Third Annual Conference on Advanced Composites, Detroit, MI, 15-17 Spetember 1987, p 103-107.

2. Ko, F.K. and Soebroto, H.B. "Braided Thermoplastic Composites for Bone Implants". ASTM Composites Workshop, Atlanta, GA November 8, 1988

CONCLUSION

With the capability of laying 80 lb/hr, braiding is a high speed fabrication method for composite preform. The spool of yarns in the braiding machines can be depleted in under two hours at full speed when using heavy yarns. Composite with complex contour can be braided to shape but accessories may be required depending on complexity of of the shape. Fibers of low to medium elastic modulie can be braided by using different type of yarn carriers or tension spring. The structure of braid is similar to a weave. The process is highly adaptable for high volume production of simple shapes requiring both excellent longitudinal and torsional properties.

CHARACTERIZATION OF INTRALAMINAR HYBRID LAMINATES

Carl H. Luther
General Dynamics
Troy, MI USA

ABSTRACT

Mechanical and impact properties of quasi-isotropic intralaminar carbon/glass/epoxy hybrids have been characterized. Tension, compression and shear properties show relatively linear behavior with hybridization. Impact behavior of the hybrid materials exhibited significant non-linearity.

THE COST OF COMPOSITE MATERIALS is currently a key element prohibiting a more widespread use of composite structures in industry. Hybrid materials, those composed of more than one fiber type, offer unique cost and performance advantages. This study focused on determining the mechanical and impact performance of quasi-isotropic carbon/glass/epoxy intralaminar hybrids and comparing those properties to those of the carbon/epoxy and glass/epoxy laminates.

The two fibers chosen for use in the study were Amoco T-500 carbon fiber and Owens Corning S-2 Fiberglas. Dow Tactix 123/H41 epoxy resin was used as the matrix. The properties of these materials are given in Tables 1 and 2.

The following laminate properties were measured:

Tensile Modulus
Tensile Strength
Tensile Poisson's Ratio
Tensile Strain at Failure
Compression Modulus
Compression Strength
Compressive Poisson's Ratio
Compressive Strain at Failure
Shear Modulus
Shear Strength
Shear Strain at Failure
Penetration Energy
Incipient Damage Energy
Specific Gravity

All tests were performed according to ASTM standards. Of particular interest was the possibility of obtaining synergistic effects from combining the carbon and glass materials in a intralaminar fashion.

FABRICATION PROCEDURE

Panels were fabricated by filament winding circumferential layers onto a flat mandrel, removing the material from the mandrel, orienting the layers into a quasi-isotropic laminate and then curing in a compression press.

The carbon, S-2 glass and hybrid laminates were wet filament wound with alternating tows for the hybrids. Amoco T-500 was used in 12K form and the S-2 glass in 20-end rovings. The resin content of the windings were closely controlled by the wet winding bath.

The laminates were wound over a polyethelene film during fabrication. The film and material were cut and removed from the mandrel. The laminates were constructed by hand laying plies of material in the required directions and removing the backing film. Laminates for the mechanical tests were 0.100 in thick and were composed of eight plies in a (0, +/-45, 90)s configuration. Laminates for the impact tests had the same configuration but were .060 in. thick.

The laminates were cured in a oil heated mold in a compression press. The cure cycle used, as recommended by Dow, was 1 hour at 130°C and 2 hours at 160°C.

TEST RESULTS

The test results have been divided into four categories: tension, compression, shear and impact. The averaged test results are expressed in Table 1 and Figures 1-4.

TENSION - Tabbed tensile specimens were fabricated and tested according to ASTM D-3039. Strain gages, 0-90 rosettes, were applied to the specimens and attached to the Instron testing machine through Wheatstone bridges. A total of six specimens were tested for each laminate.

COMPRESSION - Compression property data was generated in accordance with ASTM D-3410. Specimens were instrumented with 0-90 rosettes and evaluated at room temperature for strength and stiffness.

The IITRI compression fixture was used for this series of tests. The fixture incorporates trapezoidal wedge grips which eliminates the problem of line contact and permits individual specimen tab thickness variations. Prestressing of the specimen tabs transverse to the specimen is accomplished by bolting across the wedges, which prevents slippage of the tabs early in the load cycle. Lateral alignment of the fixture top and bottom halves is assured with two parallel roller bushings in the upper half of the fixture and two corresponding bushing shafts in the lower half of the fixture. This insures against misalignment in loading.

SHEAR - Shear tests were conducted according to ASTM D-4255 and utilized a two rail fixture in the compression mode. The ASTM standard specifies a torque of 70 ft-lb in the fixture bolts. With this torque slippage of the specimen occurred. This problem was resolved by roughing the rail area of the fixture with a sandblaster and using high strength bolts at higher torques. A 0-90 strain gage rosette was applied and oriented at +/-45 degrees with respect to the longitudinal axis of the specimen.

IMPACT - Impact testing was performed on a Dynatup Drop Weight Impactor, Model 8200, which was interfaced with an IBM PC. One specimen at a time was placed on a 6" X 6" steel holder and securely clamped. This provided an effective unsupported area of 5" X 5" area for impact. To insure penetration, the panels were made 0.060 inch thick for impact testing, rather than the 0.100 inch thick panels used in mechanical testing. A 1/2 inch diameter impact tup was used during the testing.

The object was to use an iteration procedure to narrow the gap between penetration and non-penetration to less than 5 percent. The test procedure began by impacting the panels of a given material with increasing energy levels until penetration was achieved. The minimum energy required to penetrate the panels therefore lies between the highest non-penetration impact and the lowest penetration impact. The mid-point between these values is reported as the penetration energy in Table 3.

Data from each impact event was taken and stored on the IBM computer for analysis and data reduction. For each impact event load vs. time vs. energy and load vs. deflection vs. velocity curves were generated. The load vs. deflection were used to determine the point at which incipient damage occurred. This point was determined as the first sharp discontinuity on the load trace. The energy associated with each impact event was calculated from the integration of the load vs. deflection curve.

CONCLUSIONS

The material properties of S-2 fiberglass, carbon and hybrid laminates have been characterized. As was expected, the strength and moduli of these laminates show basically a linear behavior versus hybridization. Impact properties exhibited nonlinear behavior. The laminates characterized offer a wide range of stiffness, strength and impact properties and the selection of a given laminate is dependent on specific application requirements.

Table 1 - Properties of Composite Fibers

Properties	S-2 Glass	T-500 Carbon
Strand Tensile Strength, ksi	665	500
Tensile Modulus, msi	12.6	35
Elongation, %	4.8	1.35
Density, lb/in^3	0.090	0.063

Table 2 - Properties of Dow Tactix H123/41

Property	Value
T_g, °C	166.0
Flex. Strength, ksi	15.1
Flex. Modulus, ksi	380
Tensile Strength, ksi	11.1
Tensile Modulus, ksi	430
Ult. Tensile Elongation, %	5.5

Table 3 - Summary of Test Results

PROPERTY	_0_%	25%	Carbon Content, % 50%	75%	100%
Tensile D-3039:					
Tensile Strength, ksi	55.8	53.5	66.6	67	78.1
Tensile Modulus, msi	2.69	3.63	5.12	5.85	6.61
Poisson's Ratio	0.304	0.316	0.355	0.36	0.36
Strain at Failure, %	1.87	1.53	1.51	1.28	1.28
Cal. Shear Modulus	1.03	1.89	1.89	2.15	2.43
Compressive D-3410:					
Comp. Strength, ksi	49.0	50.6	55.7	50.3	55.8
Comp. Modulus, msi	4.13	4.85	5.40	6.41	7.29
Poisson's Ratio	0.311	0.339	0.331	0.333	0.391
Strain at Failure, %	1.45	1.23	1.25	0.83	0.86
Cal. Shear Modulus	1.57	1.81	2.03	2.40	2.62
Shear D-4255:					
Shear Strength, ksi	26.4	30.8	31.1	31.4	32.1
Shear Modulus, msi	1.43	1.99	2.10	2.84	2.74
Strain at Failure, %	1.79	1.57	1.48	1.14	1.31
Impact Prop. D-3029:					
Penetration, avg.	28.48	22.51	19.52	12.40	8.66
Penetration, act.	31.24	22.70	19.63	12.58	8.81
Incipient Damage	19.75	12.53	14.77	4.79	4.30
Specific Gravity	1.837	1.552	1.561	1.512	1.444

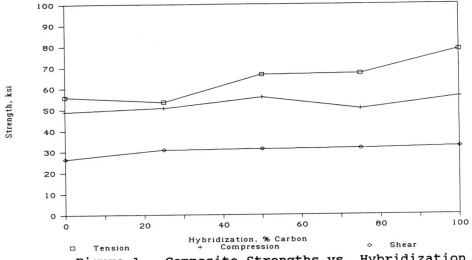

Figure 1 - Composite Strengths vs. Hybridization

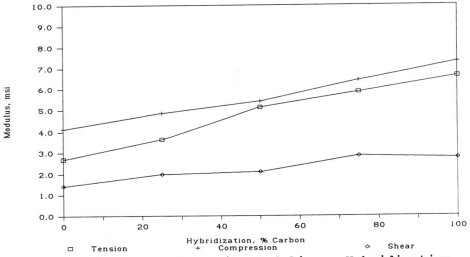

Figure 2 - Composite Moduli vs. Hybridization

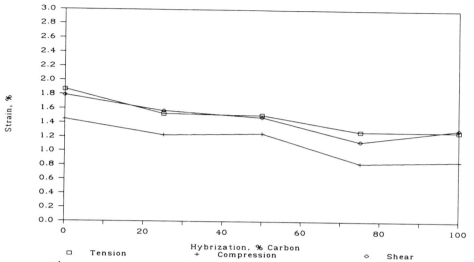

Figure 3 - Composite Failure Strains vs. Hybridization

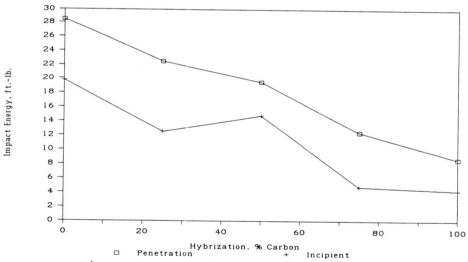

Figure 4 - Impact Properties vs. Hybridization

NEW TECHNIQUES IN ULTRASONIC IMAGING FOR EVALUATION OF COMPOSITE MATERIALS

Bong Ho, Roland Zapp
Department of Electrical Engineering
Michigan State University
East Lansing, MI USA

ABSTRACT

The conventional ultrasonic B and C scans have limits such as the range resolution, the type of information retrieved and the mechanism of scanning. In the past few years, we have developed techniques to overcome these drawbacks. The on-going research efforts are as follows: 1. Techniques for high range resolution: High sampling rate digitizing of echo return for signal detection rather than the conventional threshold detection is used. Together with a method to synchronize the sending and receiving pulses, a range resolution approaching the operating wavelength is achieved; 2. Acoustical attenuation imaging: The acoustic imaging on material attenuation property rather than acoustical boundary has been developed and tested; 3. Video pulse techniques for material characterization: With a narrow pulse excitation, the echo contains a broad band of frequencies. Each frequency component has different velocity and attenuation characteristics. The material properties can be determined by displaying these characteristics in both time and frequency domains; 4. A real-time C scan of materials by pulsed laser-acoustic interaction: Instead of using transducer array or mechanical scanning, the cross-sectional image of material can be obtained by diffracting a laser beam with an ultrasound beam. Both theoretical and experimental work have been preliminarily developed; 5. Non-invasive probing of temperature profile inside materials: Using thermistor and thermal couple probes are of invasive nature. A truly noninvasive technique of mapping temperature variation inside materials by detecting the velocity change of ultrasonic wave has been developed and tested.

ULTRASONIC IMAGING FOR EVALUATION OF COMPOSITE MATERIALS has experienced tremendous growth in recent years. Of special interest is the ability to identify voids and defects inside materials. Both B and C scans are used for these purposes from the commercially available systems. However, the range resolution is inherently limited by the operating frequency, pulse width and detection technique. Furthermore, we observed that the range resolution is also affected by the noncoherence of trigger time from the microprocessor control unit. In the past few years, we have developed several techniques to overcome some of these shortcomings. Specifically, to use a high sampling rate to digitize the echo return, the the phase information of the signal can be retrieved. A small physical reference is placed in front of the transducer such that the echo from it can be used as a time reference for the remaining echo train. The recorded data points are then precisely in time sequence with respect to the pulse trigger. Our results show that the range resolution is approaching the operating wavelength, which is the theoretical limit.

Composites, or fiber reinforced materials, are inherently inhomogeneous due to the fabrication process, environmental exposure and handling damage. In the auto and aerospace industries, they are subjected to intense structural demands. In order to evaluate reliability, ultrasonic nondestructive testing can provide valuable information about the mechanical properties of the sample. Typically, the pulse-echo mode is used for static testing, while acoustic emission is employed for dynamic evaluation.

Most of the commercially available ultrasonic imaging systems employ amplitude detection of signal echo returns. The location of the boundaries are identified by the appearance of signals above a preset amplitude threshold. As a result, the boundary location has a range of uncertainty depending on the threshold setting. This uncertainty could be as wide as the pulse width which is in the order of several operating wave lengths. In addition, we have recently observed that there exists a range accuracy problem related to the limit coherence of trigger time from the microprocessor controlled ultrasonic pulse transmission. This timing jitter could greatly deteriorate the range accuracy.

Imaging systems for material characterization are mostly based on pulse echo data. The quality of the image is limited by the fact that the information of attenuation and backscatter can not be uniquely determined separately. Consequently, it is difficult, if not impossible, to detect the nonhomogeneity of material structure as well as nonlocalized damages.

To eliminate these drawbacks, we have recently developed several techniques to improve range resolution and enhance material characterization.

TECHNIQUES

To minimize the range resolution uncertainty due to amplitude detection by threshold setting, phase processing of the return echo is performed. To retrieve the phase information, a high sampling rate of digitizing the signal is required. In order to achieve such a fast wave recording, two methods have been employed: the charge-coupled devices CCD and a fast analog-to-digital converter with buffer storage. For short signal length, the charge-coupled device works well. However, when the sample is thick, the data stored in CCD start to distort due to the diffusion of charge carriers. For thick sample, we use a monolithic analog-to-digital conversion chip available commercially. The digital output is stored in a temporary buffer memory of 1 K bytes. Since the sampling frequency (20 mega samples per second) is much higher than the operating frequency (2.25 MHz) and the clock frequency (4 MHZ) of the microprocessor control unit, a circuit was designed to transfer the digital output data into the microcomputer memory at a much lower rate (100 KHz). A signal processing subroutine is used to analyze the data to determine the boundary location. Finally, both B and C scans are displayed simultaneously for rapid identification of voids and damages.

To resolve the range inaccuracy due to jitter in the transmitted pulses, a small physical reference is placed in front of the transducer. The echo return from this reference is then used as a time reference for the remaining echo train. The recorded data points are then precisely in time sequence with respect to the reference marker, so that the uncertainty of echo time shifting is completely eliminated.

A technique has been developed and tested in our laboratory to obtain acoustic images based on the attenuation properties of materials. The method uses two transducers on opposite sides of the sample. The impulse response of each is decoupled to resolve the attenuation coefficients of a layered structure. The image obtained is not based solely on acoustic impedance variation across the boundaries. Thus, nonuniform material as well as voids can be detected by its attenuation characteristics.

Almost all the existing ultrasonic imaging systems are using echo return from the boundaries with changing impedances. The image shows the outline of the boundaries rather than the characteristics of materials. The material characterization can be extracted from the shape of the echo return when the transducer is excited by a narrow video pulse. The pulse width used is shorter than the operating wavelength of the transducer. In such a way, the transmitting pulse contains a broad band of frequencies. Due to the dispersive nature of materials, each frequency component propagates through the material with different velocity and attenuation. The echo return is the summation of all these frequency components being reflected. From the shape as well as the frequency spectrum of the return pulse, one should be able to retrieve information about the characteristics of the target materials. Various materials with known properties such as density, compressibility, and velocity of propagation are used in the simulation to verify the theory. The accuracy of this technique is heavily dependent on the echo wave recording. High sampling rate is essential. A novel way of increasing sampling rate has been developed for this application. Temperature sensing has many important applications such as in material processing. At the present time invasive methods using thermistor and thermalcouple are commonly employed. Temperature measurements by ultrasound have been investigated. The theoretical base is that the speed of acoustic wave in material is a function of temperature. Therefore, a variation of temperature can be determined from the change of velocity of propagation. Typically, the change in temperature is minute and the corresponding change in acoustic velocity is extremely small. If envelope detection scheme is used, it is almost impossible to detect such small velocity change with reasonable accuracy.

The work reported here is a special technique we have developed recently to monitor the shift in echo due to temperature change in material. The technique involves the use of a reference signal to minimize the range uncertainty and a high sampling rate circuit to give fine resolution in velocity change. An 0.2 mm range accuracy has been achieved.

Instead of using transducer array or mechanical transducer scanning, cross-sectional images of a material's internal structure can be obtained by diffracting a laser beam with an ultrasound beam. It requires the synchronized pulsing of both systems. A theoretical approach is being developed. The attractive features of such an imaging system are that the scanning ultrasound beam is not needed, which gives the possibility of a truly real-time imaging; and that the C scans of various depths can easily be obtained by setting the time delay between the pulsings of the laser beam and the ultrasound beam.

RESULTS

Both B and C scans of a layered composite material are shown in Figure 1. The 5-layered graphite composite sample is 4.5 mm thick. The sample was damaged by high speed projectile impact. Damages caused by impact are mainly layer delamination and fiber fracture. Theoretically, the damage patterns have bipolar shape. For the sample examined, the fiber directions are displaced by $90°$ from consecutive layers. The C scan indicated that variation.

To evaluate the range accuracy, a target model mode of plexiglass is scanned by A-mode ultrasonic pulses. An 2.25 MHz transducer and a 15.4 MSPS sampling are used. The time interval between sampling points is approximately 65 nanoseconds which corresponds to 0.1 mm in water. Figure 2 shows the echo data from the reference, and the front and back surfaces of the flat plexiglass plate. It is observed that the front face echoes have a variation of 1.386 microseconds in time. However, the time duration between the front and back surface echoes remains the same for all classes. It is therefore, for high range resolution imaging, all echo returns should be taken with respect to a fixed spatial reference rather than from the electrical triggered pulse from the computer controller. To demonstrate this technique, a plexiglass staircase model was used. Figure 3 shows the improvement of surface boundary identification. Figure 4 shows the temperature profile of a layered model with temperature gradient in each region.

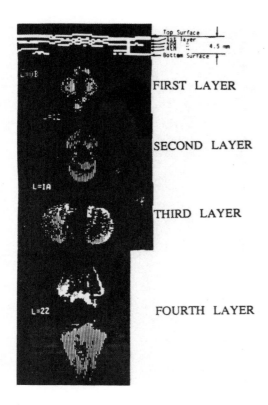

Fig. 1 C scan layered display of a 5 layered graphite composite with impact damage

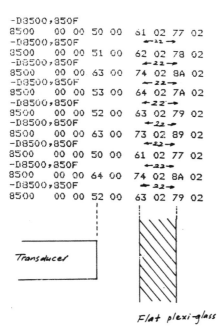

Fig. 2 Noncoherency of trigger time from microprocessor controlled system

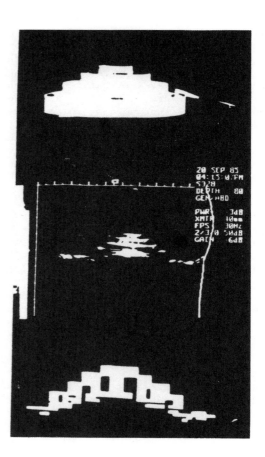

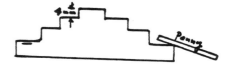

Fig. 3 Improvement of surface boundary identification

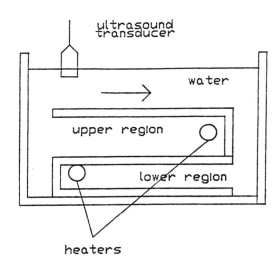

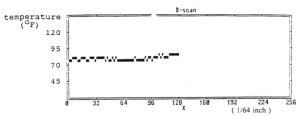
(a) Temperature Profile of Upper Region without Heating

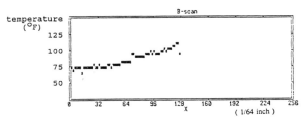
(b) Temperature Profile of Upper Region with Heating

Fig. 4 Temperature profiles of a multilayer model

HIGH STRENGTH SHEET MOLDING COMPOUND—PROPERTY/PROCESSING INTERACTION

Jon Collister
Premix, Incorporated
North Kingsville, OH USA

Abstract

An investigation of the rheological characteristics and mold fill behavior of a 50% glass vinyl ester SMC has been conducted. A squeeze flow rheological tester and a servo controlled hydraulic compression press have been used to determine the flow behavior of this SMC. The results indicate that 50% glass vinyl ester SMC shows substantial differences in flow characteristics than SMC with lower glass content. The pressures required to initiate flow behavior are much higher than previously reported and a convex flow front was observed which has only been observed in low glass content SMC at extremely low pressing velocities. A comparison of the squeeze flow rheological behavior of 30% and 50% glass SMC formulations and resultant mold fill behavior are presented to demonstrate the differences in process and equipment to mold these materials.

THERE HAS BEEN A VARIETY OF NEW APPLICATIONS for high glass content vinyl ester SMC reported in literature.[1][2] These formulations have become a commercially viable commodity and are becoming the materials of choice for semi-structural and mild-structural components on automotive applications. Because the molding of SMC involves flow forming prior to the thermosetting reaction, the understanding of the rheological behavior of these materials will ensure proper usage and assist in the development of the optimum processing conditions for these applications.

A considerable amount of research in recent years has been reported on the flow behavior of SMC. This work has been in several categories: 1) rheological behavior of SMC resins and pastes;[3] 2) attempts to describe the rheology of fully compounded SMC;[4][5] and, 3) techniques have been developed which will predict the mold fill behavior of sheet molding compound.[6] Although this research has given users of SMC a much better understanding of the material's response in molds, the bulk of the work has been reported on lower glass content SMC formulations which are primarily intended for automotive body panel applications. Although the SMC formulations and process are similar in basic characteristics, the dominating rheological feature of SMC is the long glass reinforcement and the increase of glass content from 30% to 50-65% glass has a very large effect on the behavior of the SMC. The commercially important effect is a notable increase in the physical properties of the material which is the reason for its use in the semi- to mild-structural applications; but the requirements of the impregnation process of SMC demand that the filler content be reduced while the glass content is increased, therefore the resultant rheological behavior of these materials is quite different. Furthermore, the formulations of the higher glass content SMCs usually employ different resins which may result in completely different rheological behavior. The predominant resin used in high glass content SMC is vinyl ester which can vary considerably depending on the source of the resins used. In this study, we have used an isocyanurate based vinyl ester resin whose chemistry and properties have been discussed by Phipps.[7]

One of the primary considerations in molding a structural grade sheet molding compound would be the required press capacity. It is generally believed in the industry that higher glass content SMC requires higher pressures to flow and fill molds, although the quantification of this has not been reported or has been left in the domain of individual molder's proprietary information. Recently, techniques have become available which will perform actual squeeze flow testing on SMC formulations to give information regarding the pressure requirements and viscosities of actual SMC formulations during squeeze flow tests. In a recent SPI meeting, Allen[8] reported on the development of a prototype squeeze flow tester which was the result of work done by the Edison Polymer Innovation Corporation (EPIC) and is intended to provide rapid processability information concerning SMC in its fully-formulated, ready-to-mold state. The details and the equations used to derive the rheological behavior from squeeze flow testing are presented by Meinecke.[9]

The final technique used in this characterization of high glass content SMC was the actual molding of SMC with a servo hydraulic controlled compression press. This compression press was similar to presses described in the literature[10] and was developed as a cooperative effort between personnel at the University of Loughborough in Loughborough, England (Dr. Barry Fisher, Dr. Graham Chapman, and Nigel Henson), and in-house engineering and computer science personnel at Premix. The press is instrumented for monitoring hydraulic pressures and position. The position information is fed back to control the closure profile of the press during mold flow. This information can then be used to calculate critical pressing pressures during mold fill.

THE ORIGINAL INTENTION OF THIS INVESTIGATION was to employ different pressing conditions on 50% glass SMC and measure the resultant physical properties. But, upon molding this material with our squeeze flow tester and our servo controlled hydraulic press, we found that the flow behavior of this 50% glass SMC was considerably different than previously tested SMCs and therefore establishing the conditions of the process with which we wanted to measure physical properties was impossible. We studied the flow behavior of this SMC more completely to allow us to characterize the fill behavior of this material and compared these results to 28% glass SMC. The physical properties of the basic material without process consideration can be found in the literature[7] and future work will be devoted toward doing this process/property interaction.

The initial testing with the squeeze flow rheological device was done at the University of Akron which has a prototype unit. We quickly found that although this equipment was satisfactory for lower glass content SMC, we had extreme difficulty with the 50% glass material. Initially using 75 mm. diameter plates, we exceeded the load capacity of the squeeze flow tester and therefore could not control the rate of closure. This instrument was then modified to have 50 mm. plates and still found that the 50% glass SMC exceeded the limits of the device. Typical information from the squeeze flow tester is presented in Figure #1, which shows results for both 50% glass and a 28% glass automotive body panel formulation. In the case of the 28% glass material, the stress rises and then develops into a stable flow region. When the displacement is stopped, a typical stress relaxation of the material is observed. In contrast to this, the 50% glass material shows a non-linear closure, a stress build to the limit of the operating pressure of the machine, and a linear flow-out of the material at the maximum pressure. The normal treatment of the data for SMC flow is presented in Figure #2 which shows that an apparent viscosity can be calculated from the linear flow region of the 28% glass and demonstrates a typical relationship of viscosity to shear rate as the press closes. However, with the 50% glass no calculations of viscosity are possible since it never entered a stable flow region. Although disappointing, this information was used in the design of subsequent squeeze flow testers which will have increased capacity.

Since we had an inability to measure the force required for consistent flow-out of 50% glass SMC, we then proceeded to make measurements with the servo controlled hydraulic press. The experience with this press was similar to the squeeze flow tester in that the initial closure with 28% glass SMC produced a linear closure down to full close, while the 50% glass material deviated very quickly from the linear closure behavior after contact of the charge was made. Figure #3 shows the closure behavior and applied force for a 50% glass SMC using a force of 78.5 tons. As can be seen in Figure #3, the actual closure data deviates considerably from the intended closure profile. In Figure #4 the results are shown at 78.5 tons for the 28% glass material which indicates a good agreement of the predicted and the actual closure profiles. This indicates that the press could supply sufficient force to move the SMC according to the program control. To gain further insight into the required force to move this 50% glass at the intended closure rate, the pressure on the system was increased to allow for 118 tons, which is shown in Figure #5. Again, deviation from the intended closure profile is observed at approximately the same position as in the 78.5 ton example. The force was increased on the press to 157 tons and a closure profile shown in Figure #6 was observed which closely agrees with the intended closure profile. In a rough approximation, twice as much force is required to move 50% glass SMC at the same rate as a 28% glass automotive body panel formulation.

Although the material does not flow at the intended rate according to the servo control program, the material does flow out and fill the mold at the lower pressures, but the time taken to fill the mold is not according to the input to the press. To further investigate the mechanism of flow, we molded non-filled panels using stops on the mold to allow the observation of the developed flow front during the partial fill situation. This technique of observing the flow front can be a useful mechanism to assess the flow behavior of SMC. It has long been known that the outer plies of an SMC charge flow preferentially because of the heating effect of the charges that have been in contact with the mold.[11] There have also been demonstrations that the rheological behavior of the plies will also determine the influence of the relative velocity of the plies during flow.[12] Fisher indicated that based on the velocity of closure, the material can develop completely different flow front profiles which should determine the morphology of the glass fibers in the resultant part and therefore may influence the physical properties[6]. Upon examination of the flow front profiles of this 50% glass SMC, we have observed the concave flow front profile as discussed by Fisher which was observed in low glass content SMC only at extremely low velocity flow situations. This may cause some problems due to the tumbling effect of the glass fibers during this concave flow front development and should be carefully investigated in critical structural applications. A representation of the flow front profile is given in Figure #7.

WE HAVE SEEN THAT 50% GLASS SMC requires much more force to flow at similar rates than low glass content SMC. A rough approximation is that it takes approximately twice the pressing force to achieve controlled velocity flow in a servo controlled hydraulic press for this 50% glass vinyl ester SMC. Furthermore, we have seen the development of a concave flow front at normal pressing forces which is a result of the heated plies at the mold surfaces exhibiting preferential flow resulting in a tumbling behavior of the glass fibers in advance of the flow front.

Further information regarding the ultimate force requirements for testing 50% glass SMC in a squeeze flow test apparatus has been obtained and appropriate action is being taken on the build of a prototype instrument. After sufficient force has been designed into a machine, we will be able to calculate apparent viscosities of 50% glass SMC during actual squeeze flow experiments.

ACKNOWLEDGEMENTS

I would like to acknowledge Paula Allen for her contributions on the squeeze flow tester, and I would like to acknowledge Dennis Boyle for his contributions to the controlled velocity compression molding experiments.

Notes

1. Phipps, J.J., and Miskech, P., "44th Annual Technical Conference," Reinforced Plastics/Composite Institute, SPI 1989, Section 18-C, p. 1-6.

2. Vernyi, B., Plastics News, "Plastics Wheel May Get a Whirl," June 19, 1989, p. 1.

3. Gruskiewicz, M., and Collister, J., "35th Annual Technical Conference," Reinforced Plastics/Composites Institute, SPI 1980, Section 7-E, p. 1-7.

4. Collister, J., and Gruskiewicz, M., Short Fiber Reinforced Composite Materials, ASTM STP 772, "Dynamic Mechanical Characterization of Fiber Filled Unsaturated Polyester Composites," 1982, p. 183-207.

5. Kau, H.T., and Hagerman, E.M., Polymer Composites "Experimental and Analytical Procedures for Flow Dynamic of Sheet Molding Compound (SMC) in Compression Molding," Vol. 8, No. 3, June 1987, p. 176-187.

6. Silva-Nieto, R.J., Fisher, B.C., and Birley, A.W., 34th Annual Technical Conference, Reinforced Plastics/Composites Institute, 1979, Section 7-B, p. 1-12.

7. Phipps, J.J., and Collister, J., "35th Annual Technical Conference," Reinforced Plastics/Composites Institute, SPI 1989, Section 16-B, p. 1-11.

8. Allen, P., et. al., "Development of a Processability Tester for Sheet Molding Compound," SPI Press Molders Meeting, Spring 1989.

9. Meinecke, E.A., "EPIC Research Progress Report," Dec. 1986, UAP 103.

10. Chapman, G.M., Fisher, B.C., and Kanagendra, "Servo Hydraulic Control Applies to SMC Compression Moulding," Proc. RP Congress, Brittish Plastics Federation, Brighton, Nov. 1982.

11. Barone, M.R., Caulk, D.A., Polymer Composites, 1985, Vol. 6, 105.

12. Collister, J., "Ply Morphology of Molded SMC/Relationship to Rheological Properties and Process Conditions," Proceedings of the European Physical Society Symposium, June 9, 1986, Naples.

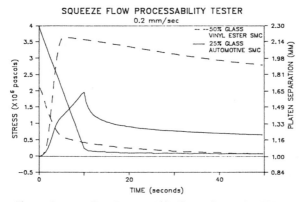

Fig. 1. Squeeze Flow Processability Tester Result for 28% and 50% glass content SMC showing overloading of instrument with 50% glass SMC.

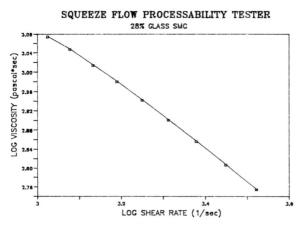

Fig. 2. Apparent Viscosity of 28% glass SMC using stable flow region of the Squeeze Flow test experiment.

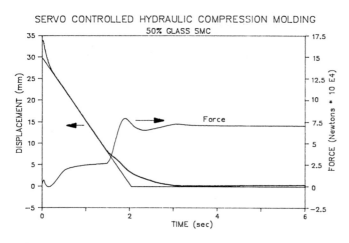

Fig. 3. Closure Results from Compression Molding @ 78.5 tons showing deviation of closure profile from set program

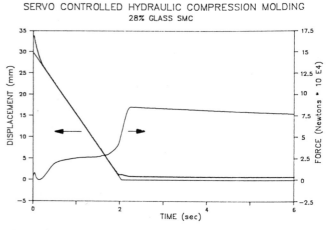

Fig. 4. Closure Results from Compression Molding @ 78.5 tons showing deviation of closure profile from set program

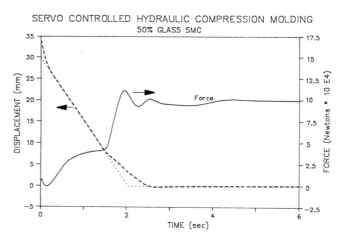

Fig. 5. Closure Results from Compression Molding @ 118 tons showing deviation of closure profile from set program

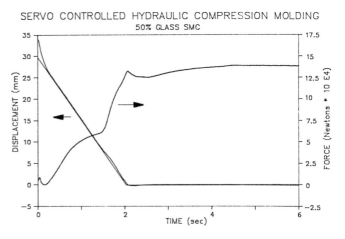

Fig. 6. Closure Results from Compression Molding @ 157 tons showing actual closure approximating programmed closure profile

Fig. 7. Flow Front Structure Produced From Short Shot Experiment Epicting Concave Flow Front Development

USING LASER DOPPLER VELOCIMETRY FOR THE DYNAMIC EVALUATION OF DAMAGE IN COMPOSITE MATERIALS

James P. Nokes, Gary Cloud
Michigan State University
East Lansing, MI 48824 USA

Abstract

This paper describes a method to improve the sensitivity of vibrational NDE techniques on composite structures. Using Laser Doppler Velocimetry (LDV) to measure the system response allows higher modes to be utilized in the evaluation of composite damage.

Initial work will be presented comparing the damage response of a composite beam when measured using an eddy current probe, a low mass accelerometer, and the LDV technique.

Composite materials can be found in a wide range of structural applications. One reason is that the structural properties of a composite can be modified locally in response to specific design parameters. With this flexibility there is an increase in the complexity of the design process for composite parts. Some design variables, such as the physical layup of the laminate, are relatively simple to analytically determine and to control during processing. With other design factors, such as manufacturing flaws and service damage, experimental methods must be used to verify the design performance. Methods for the Non-destructive evaluation (NDE) of composite materials have been the focus of a large research effort, particularly in the areas of Acoustic emission, Ultrasonics and Radiography (1,2,3). These methods are powerful, but they do not meet all the needs for the NDE of composite materials. One technique that may be able to address some of these gaps is Dynamic Material Evaluation, which uses modal testing techniques to provide simple data acquisition and enhanced sensitivity to material properties. In general, the modal technique utilizes the response of a system to an known excitation in order to characterize system behavior.

There are different ways to utilize the dynamic information obtained from a modal test. The simplest is by measuring changes in the frequency response of a structure. The shifting of frequency peaks is an artifact of damage that has occurred in the component. Tracy et al.(4) compared the change in the frequency response of a damaged plate with that of an undamaged one in order to detect impact damage (4). A waveform generator was used to excite a composite panel while its response was measured using accelerometers at selected locations. Figure 1 shows a comparison of the transfer function between a damaged panel and an undamaged one. Results obtained by Tracy as well as other researchers suggest that the higher order modes should be examined to obtain information on the damage as well as to locate the damage in the plate.

An extension of this basic modal testing technique is to use changes in the dynamic stiffness and damping capacity as the measure of damage. In general, damage in a fiber reinforced composite will result in an increased damping capacity and a decrease in the stiffness. Mantena et.al. (5) used the damping capacity to examine matrix cracking in a composite beam. The information gathered from this type of study is potentially more valuable than the simple frequency shift because it provides information about the material condition. The relative sensitivity of damping to various types of damage has been explored by other researchers such as Lee (6) and Suarez (7). Lee et al. used the modal technique to examine delaminations and notches in cantilever beams and documented a correlation between the damage and the damping. The results from the studies by Lee and Suarez are limited by the eddy current probe used to measure the frequency response of the beam. An eddy current motion transducer can measure accurately only the lower vibration modes. This limitation in the transducer is critical since there is a consensus among researchers that the higher modes contain the information on small scale damage. The measurement of these higher modes requires a different experimental approach. An ideal method would be non-contacting like the eddy current probe but with the higher frequency response that is found using accelerometers. One such method is the Laser Doppler technique.

Laser Doppler Velocimetry (LDV) is based on the frequency shift of coherent light when it is scattered from a moving target. It incorporates many of the experimental advantages found in other interferometric techniques (8). It is a true non-contacting method, so the response of the specimen is not affected by the addition of mass or the response limitations of a transducer. In addition to the high frequency response, LDV is sensitive to small amplitude vibrations (9). The ability to accurately measure small amplitude vibrations should allow smaller scale damage to be detected. LDV can also be used at high temperatures and for remote velocity measurements. The major limitations for the technique are the coherence length of the laser and controlling the displacements of the target to minimize the amplitude modulation of the signal. To create the doppler signal, a laser beam is scattered from a moving target. The frequency of the scattered light is shifted by an amount $d\nu$ according to the doppler equation.

$$d\nu = \frac{2(v)}{\lambda}$$

where v is the velocity of the target and λ is the wavelength of the laser beam

By heterodyning a reference beam of frequency ν with the shifted object beam, a signal is generated that has a beat frequency equal to the $d\nu$. This heterodyned signal is easier to detect than the direct shift, and is directly proportional to the velocity of the target. The experimental system used in this study is a variation of the Michelson Interferometer and is shown in figure 2. In this configuration the vibrometer does not provide information on the direction of the target velocity, as a result the doppler signal output is frequency modulated with a beat frequency that is exactly twice the true vibration frequency. Directional information can be provided by shifting the frequency of the reference arm of the LDV system to an appropriate value. before mixing with the object beam. The shift can be generated by a number of means such as a Bragg cell or Diffraction grating. This equipment was not incorporated into this feasibility study but will be utilized in future work.

In this initial study LDV was used to measure frequency shifts in cantilever beams due to induced damage. These results were then compared to data collected using an eddy current probe and a low mass accelerometer. The test specimens for this experiment were made of a fiber glass-epoxy laminate with woven fibers(R1500/1581). Each cantilever beam was 25.4 mm wide, 160 mm long and 3.5 mm thick (13 plies). For this test two beams were used to examine the effect of damage location on the frequency shift. Damage was generated by cutting a groove in the specimen which reduced the effective cross-sectional area of the beam. In the first beam a cut was made 10 mm from the base of the beam and in the second it was located 45 mm from the base. In each case data was recorded with cross-section reductions of 0%,20%,40%,and 60% respectively.

PROCEDURE

The specimen was excited using a modal hammer. In order to insure the consistency of the impulse the hammer was mounted to a pendulum. By controlling the hammer drop an appropriate excitation signal for each type of transducer could be generated. The frequency measurements were recorded on an HP 3582 spectrum analyzer. For each frequency range an average of 16 impulses were used. After recording the initial data from all three techniques, the cross section was reduced by 20 % using a saw blade. The damaged frequency response was then measured and recorded. This was repeated for each case. A percentage change in the first three vibration modes was then calculated for each of the damage amounts.

RESULTS

The percent frequency shift in the first mode with the cut at x = 10 mm is shown in figure 3. At the first fundamental of 72 Hz the eddy probe and the LDV provide almost identical results. Since both are non-contacting good low frequency agreement in the data was expected. The accelerometer data showed the strong influence that the additional mass had on the response as the damage became more severe. This trend was seen through out the results. Figure 4 is a plot of the 2nd mode (445 Hz) with damage at 10 mm. and again shows the same trends as the first mode data only the magnitudes of the shift have changed. This magnitude is determined in part by the location of the damage so it can provide a way to locate damage. Figure 5 which is the 3rd mode (1250 Hz) shows the sensitivity of the eddy probe dropping off when compared to the LDV. Again the accelerometer mass exaggerates the shift at the higher damage states. These plots match very well with results reported by Lee (10) for cantilever beams with a similar damage type.

CONCLUSIONS

This work shows that the LDV technique provides information comparable with the eddy current probe as well as accelerometers, but with the added potential to provide a greatly enhanced sensitivity to low amplitude vibrations and higher response frequencies.

REFERENCES

1. Hamstad Marvin A, "A Review: Acoustic Emission, A Tool for Composite Materials Studies," Experimental Mechanics, March 1986, 7-13.
2. Nestleroth J.B., J.L. Rose, M. Bashyam, and K. Subramanian, "Physically Based Ultrasonic Feature Mapping for Anomaly Classification in Composite Materials," MaterialsEvaluation,43, April 1985,541-546.
3. Bar-Cohen, Y. "NDT of Fiber-reinforced Composite Materials - A Review," Materials Evaluation, 44, March 1986,446-454.
4. Tracy,J. J. ,D. J. Dimas and G.C. Pardoen, " Using Modal Analysis to Determine Damage in Advanced Composite Materials,"McDonnell Douglas Astronautics Company, Huntington Beach, California. Report MDC G9032, October 1983.
5. Mantena, R. T.A. Place, R.F. Gibson, "Characterization of MatrixCracking in Composite Laminates by the Use of Damping Capacity Measurements," ASM Metals/Materials Technology Series, October 1985.
6. Lee B.T.,C.T. Sun, Liu ,D., "An Assessment of Damping Measurement in the Evaluation of Integrity of Composite Beams," Journal of Reinforced Plastics and Composites, Vol.6, April,1987, pp. 114-125.
7. Suarez S.A. and Gibson R.F.,"Improved Impulse-Frequency Response Techniques for the Measurement of Dynamic Mechanical Properties of Composite Materials," Journal of Testing and Evaluation, Vol. 15, No. 2, March 1987, pp. 114-121.
8. Drain L.E.,The Laser Doppler Technique, Wiley and Sons New York 1980.
9. Wlezien R.W., D.K. Miu and V. Kibens, "Characterization of Rotating Flexible Disks Using a Laser Doppler Vibrometer," Optical Engineering, July/August 1984, Vol.23, Mo. 4, 436-442.
10. Lee,Bai-Tang Measurements of Damping for Nondestructively Assessing the Integrity of Fiber Reinforced Composites, Ph.D Dissertation, Dept. of Engineering Sciences, University of Florida (May 1985).
11. Cawley,P., R.D. Adams, " The Location of Defects in Structures From Measurements of Natural Frequencies," Journal of Strain Analysis, Vol. 14, No 2, 1979.
12. Lin,D.X., R.G. Ni, R.D. Adams, " Prediction and Measurement of the Vibrational Damping Parameters of Carbon and Glass Fibre-Reinforced Plastic Plates," Journal of Composite Materials, Vol. 18, March 1984.
13. Kobayashi, Albert S.,Handbook of Experimental Mechanics , Prentice-Hall New York (1987).
14. Lang,G.F. "Understanding Vibration Measurements," Sound and Vibration, 10, (3), 26-37 (March,1976).
15. Whaley, P.W and Chen P.S.,"Experimental Measurement of Material Damping using Digital Test Equipment," Shock and Vibration Bulletin No.53 part 4, Damping and Machinery Dynamics, May, 1983.

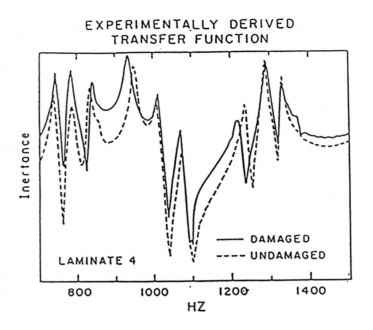

Figure 1. Comparison of Transfer Function Between a Damaged and Undamaged Plate. (Ref. 4)

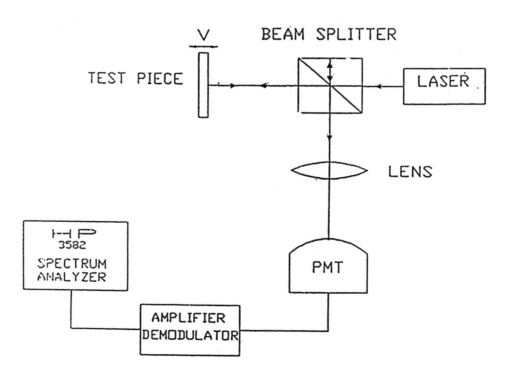

Figure 2. Schematic Diagram of LDV System.

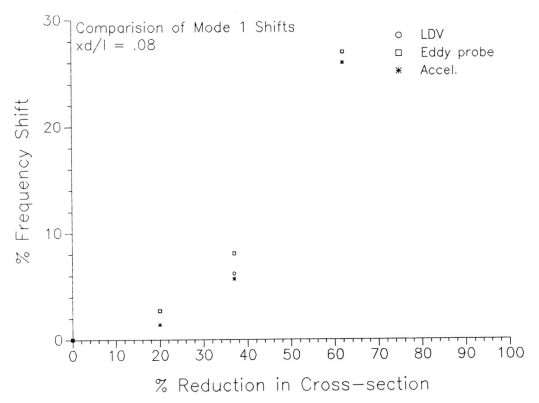

Figure 3.   Comparison of First Mode Frequency Shifts x=10mm

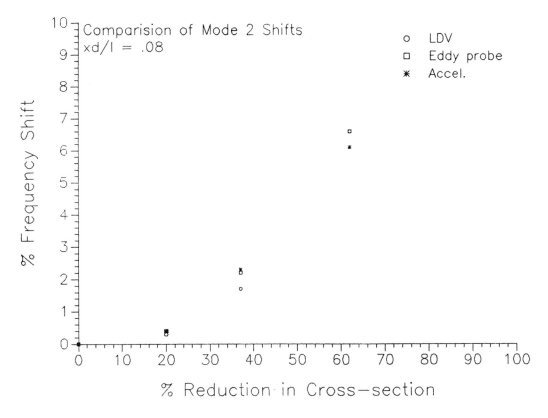

Figure 4.   Comparison of Second Mode Frequency Shifts x=10mm

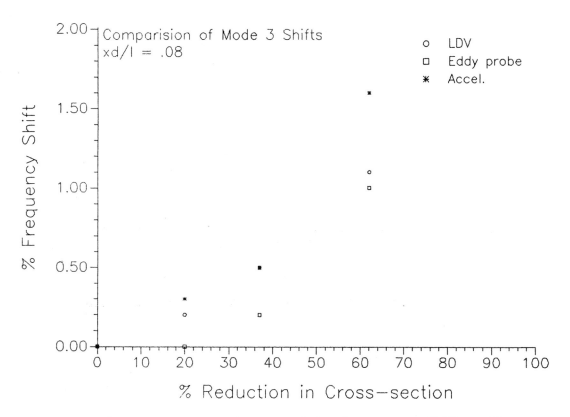

Figure 5.   Comparison of Third Mode Frequency Shifts x=10mm